CHANCE AND CHANGE

2nd Edition

G. Viglino and E. Viglino

Ramapo College of New Jersey

September 2013

CONTENTS

PART 1
PROBABILITY

PART 2

CALCULUS

PREFACE

This text is designed to introduce the reader to two distinct topics in mathematics:

PROBABILITY (Part 1) and CALCULUS (Part 2)

(Each part is totally independent of the other)

PROBABILITY PART

The family tree of probability can be traced to the gambling tables of the sixteenth century. In spite of its shady beginnings, however, the discipline has evolved into an indispensable branch of mathematics—one with numerous applications in many diverse fields.

The innocent nature of probability motivated the development of this part of the text. You will find that counting and plain old reasoning are the cornerstones of our approach. No one can reason for you, but we can and do illustrate how to count large quantities:

The number of possible lottery tickets, of poker hands, and so on.

With these counting techniques at hand, you will be able to determine the probability of winning a lottery, of being dealt a full house, and many other interesting and challenging problems.

CALCULUS PART

In ancient Roman days, a "calculus" was a pebble used for counting. Centuries later, the verb "calculate" surfaced.

The current Calculus, formalized in the sixteenth century, has become an essential cornerstone of numerous and diverse disciplines. In this part of the text we offer a mere glimpse of its power and versatility.

Our first section lays an algebraic foundation for what follows. You will enter calculus in Section 2, through the doorway of the limit concept.

Calculus is composed of two related parts: The differentiable calculus and the integral calculus. Sections 3 through 7 scratch the surface of the differentiable component of the discipline, while the last three sections touch upon the integral part.

We have made every effort to provide a leg-up for the journey you are about to take. Along the way, you will encounter numerous **Check Your Understanding** boxes designed to challenge your understanding of each newly introduced concept. Detailed solutions to each of the **Check Your Understanding** problems appear in the back of the text, but you should only turn to our solutions after making a valiant effort to solve the given problem on your own, or with others. In the words of Descartes:

> ***We never understand a thing so well, and make it our own, when we learn it from another, as when we have discovered it for ourselves.***

Graphing calculator glimpses primarily designed to illuminate concepts appear throughout the text. In the final analysis, however, one cannot escape the fact that

MATHEMATICS DOES NOT RUN ON BATTERIES

A case in point:

> While graphing calculators can certainly graph most functions better and faster than any of us, learning to sketch them by hand requires an understanding of important concepts, and serves to reinforce those concepts.

Finally, we wish to thank our colleagues James Gillespie, Peter Kaufmann, Harold Katz, Sara Kuplinsky, Mark Martino, Donovan McFeron, Maria-Cristina Niederstrasser, and Bernie Rothman for their invaluable input. We are especially grateful to Professor Marion Berger, who has contributed greatly throughout the development of the text.

Probability Part

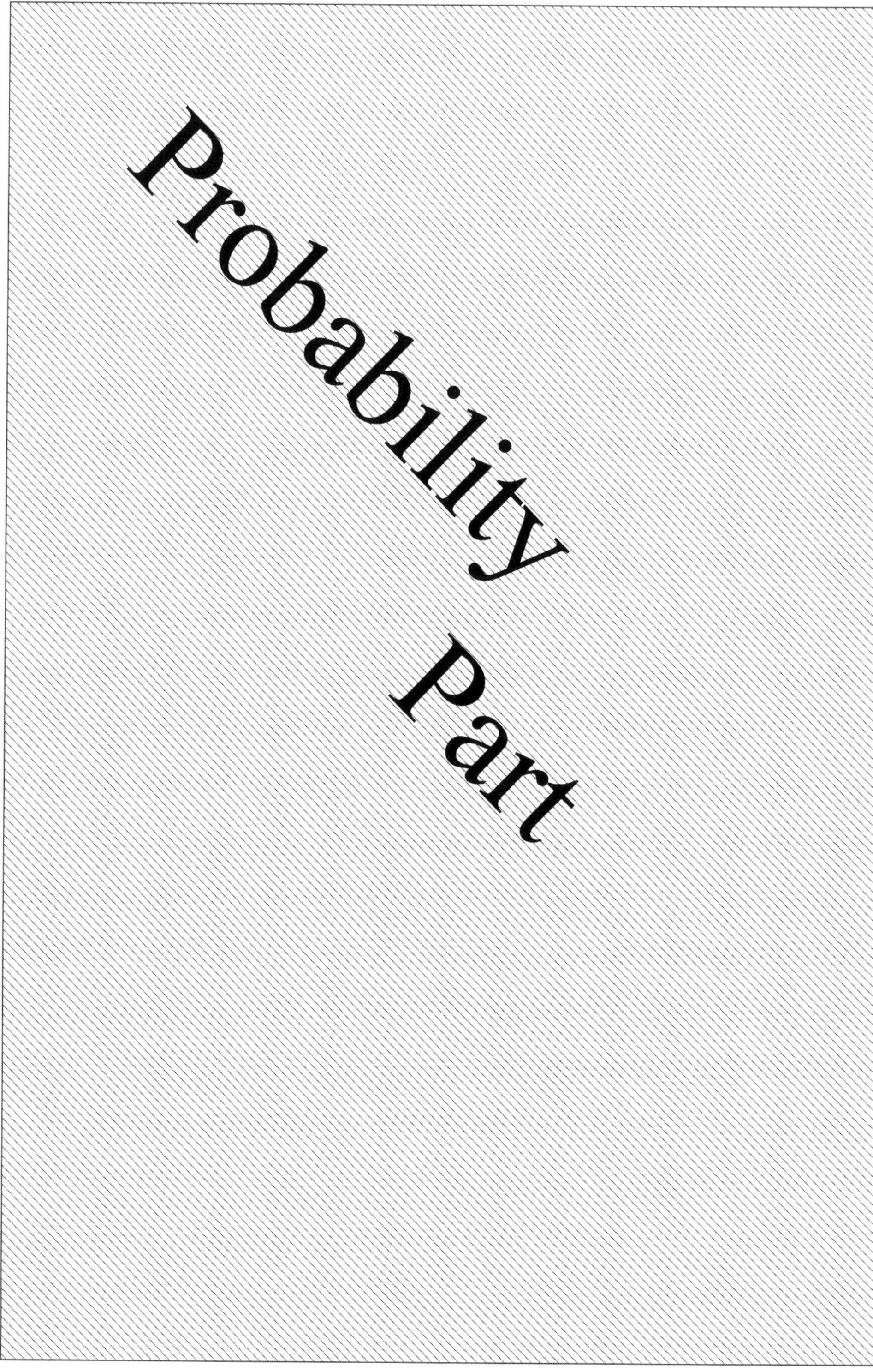
Probability
Part

§1. Definitions and Examples

Roll a die. What is the probability that you roll a 3?

Chances are that you responded with "1 out of 6," or "$1/6$," basing your answer on the fact that there are 6 possibilities: $\{1, 2, 3, 4, 5, 6\}$, with one of them being a 3.

Draw a card from a standard deck.
What is the probability that you draw an ace?

If you are familiar with a deck of cards (Figure 1.1), then you probably answered "4 out of 52," or "$4/52$," for there are 52 cards in a deck, and exactly four of them are aces.

In probability, the form of a number often "tells a story." The form $\frac{4}{52}$ for example, nicely displays the probability of drawing an ace; more so than its reduced form $\frac{1}{13}$. If you want to reduce, go ahead, but you should first display the "story-form:"

$$\frac{4}{52} = \frac{1}{13} \approx 0.077$$

↑ "story form"

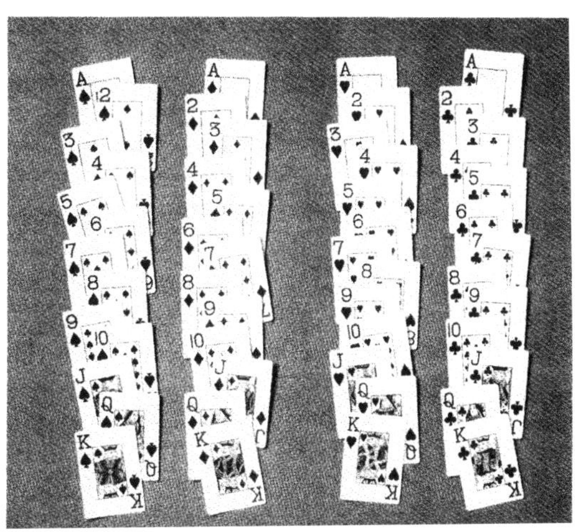

A deck of cards consists of four suits:
Spades ♠, Diamonds ♦, Hearts ♥, and Clubs ♣.
Spades and Clubs are black, Diamonds and Hearts are red.
Each suit consists of 13 cards: Ace, 2, 3,...,10, J, Q, K.

Jacks, Queens and Kings are called face cards
(aces are **not** face cards).

Figure 1.1

In probability, the word **experiment** is used for the activity under consideration (the rolling of a die, or the drawing of a card). The word **success**, or **event**, will be used to denote a specified outcome of the experiment: (rolling a 3, or drawing an ace). In both of the above illustrations, the probability of an event (or success) turned out to be the number of successes divided by the number of possibilities:

$$Pr(\text{rolling a 3}) = \frac{1}{6} \begin{array}{l} \leftarrow \text{number of successes} \\ \leftarrow \text{number of possibilities} \end{array}$$

$$Pr(\text{drawing an ace}) = \frac{4}{52} \begin{array}{l} \leftarrow \text{number of successes} \\ \leftarrow \text{number of possibilities} \end{array}$$

So far so good. Now roll two dice. What is the probability that you end up with a sum of 7?

Proceeding as above, we consider the 11 possible sums:

$$S = \{2, 3, 4, 5, 6, 7, 8, 9, 10, 11, 12\}$$

Noting that 7 appears but once in S, we might be tempted to conclude that:

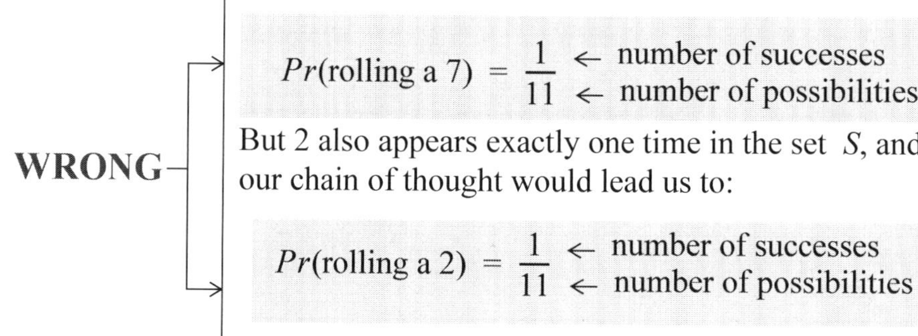

At this point, you should be feeling a bit uncomfortable with our "findings," since it is certainly more likely to roll a sum of 7 with a pair of dice (6 and 1; 5 and 2; and so on) than it is to roll a sum of two, which can only be done in one way: 1 and 1. The problem stems from the fact that when you roll one die, any of the six numbers is as likely to come up as any other. But when you roll two dice, a sum of 7 is more likely to occur than is a sum of 2. The time has come to be a bit more precise:

DEFINITION 1.1
EQUIPROBABLE SAMPLE SPACE

An **equiprobable sample space** for an experiment is a set that represents all of the possible outcomes of the experiment, with each outcome **as likely** to occur as any other.

Note that while $\{1, 2, 3, 4, 5, 6\}$ is an equiprobable sample space for the rolling of a single die (any number is as likely to come up as any other), $\{2, 3, 4, 5, 6, 7, 8, 9, 10, 11, 12\}$ is **not** an equiprobable sample space representing the sum of the roll of two dice (a sum of 2, for example, is less likely to occur than a sum of 7).

Our informal approach to probability:

$$\frac{\textbf{Number of Successes}}{\textbf{Number of Possibilities}}$$

is fine, providing counting takes place in a setting where every outcome of the experiment is as likely to occur as every other:

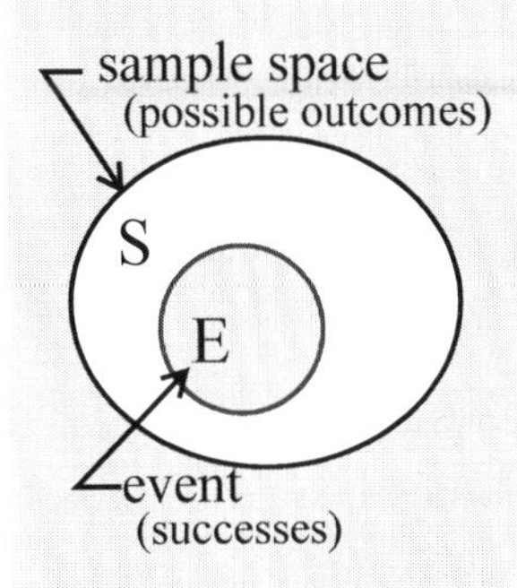

DEFINITION 1.2

PROBABILITY

(When the Sample Space contains finitely many elements)

Let S be an <u>equiprobable</u> sample space for a given experiment, and let E denote the subset of S (called an event) representing the successes of the experiment. The **probability of a success occurring**, or the **probability of the event *E***, is given by:

$$Pr(E) = \frac{\#(E)}{\#(S)}$$

where $\#(E)$ and $\#(S)$ denote the number of elements in the sets E and S, respectively.

CHECK YOUR UNDERSTANDING 1.1

(a) Verify that for any event E: $0 \le Pr(E) \le 1$.
(b) Give an example for which $Pr(E) = 1$.
(c) Give an example for which $Pr(E) = 0$.

Answers: Page A-1.

EXAMPLE 1.1

ROLLING A PAIR OF DICE

Write down an equiprobable sample space for the rolling of a pair of dice. With that sample space at hand, determine the probability of:

(a) Rolling a sum of 7 (b) Rolling a sum of 2

SOLUTION: They may look alike, but two white dice are thrown. Let's paint one of them red. On rolling the dice, the red die will show a face of 1 through 6, with any of the six faces as likely to show as any other; as will the white die. Denoting, for example, the outcome "1 on the red die and 4 on the white die" in the compact form "$(1, 4)$," we find ourselves face to face with the following 36-element equiprobable sample space:

1 on the red and 4 on the white ↓

$$S = \left\{ \begin{array}{cccccc} (1,1) & (1,2) & (1,3) & \mathbf{(1,4)} & (1,5) & (1,6) \\ (2,1) & (2,2) & (2,3) & (2,4) & (2,5) & (2,6) \\ (3,1) & (3,2) & (3,3) & (3,4) & (3,5) & (3,6) \\ (4,1) & (4,2) & (4,3) & (4,4) & (4,5) & (4,6) \\ (5,1) & (5,2) & (5,3) & (5,4) & (5,5) & (5,6) \\ \mathbf{(6,1)} & (6,2) & (6,3) & (6,4) & (6,5) & (6,6) \end{array} \right\}$$

↑ 6 on the red and 1 on the white

Figure 1.2

(a) Defining a success to be that of rolling a sum of 7, and noting that there are six such successes in the sample space of Figure 1.2: (1,6), (2,5), (3,4), (4,3), (5,2), and (6,1), we have:

$$Pr(\text{rolling a sum of 7}) = \frac{6}{36} \begin{array}{l} \leftarrow \text{successes} \\ \leftarrow \text{possibilities} \end{array}$$

(b) Since there is but one sum of 2 in the equiprobable sample space (the "(1, 1)"), we have:

$$Pr(\text{rolling a sum of 2}) = \frac{1}{36} \begin{array}{l} \leftarrow \text{successes} \\ \leftarrow \text{possibilities} \end{array}$$

EXAMPLE 1.2
FLIPPING A COIN

Flip a coin three times, what is the probability that exactly 2 heads are flipped?

SOLUTION: Once you have an equiprobable sample space in front of you:

H H H

H H T **H T H** **T H H** ← successes

T T H T H T H T T

T T T

you need but count to arrive at the answer:

$$Pr(\text{exactly 2 H}) = \frac{3}{8} \begin{array}{l} \leftarrow \text{successes} \\ \leftarrow \text{possibilities} \end{array}$$

EXAMPLE 1.3
MARBLE FROM HAT

A marble is chosen at random from a hat containing 5 red marbles, 4 white marbles, and 3 blue marbles. What is the probability that the marble chosen is white?

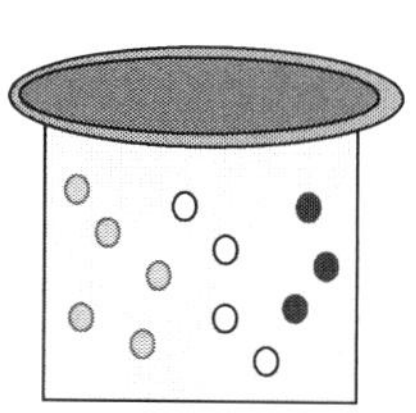

SOLUTION: Since the hat contains a total of 12 marbles (the sample space, of which 4 are white (the successes), we have:

$$Pr(W) = \frac{4}{12} \begin{array}{l} \leftarrow \text{successes (4 white marbles)} \\ \leftarrow \text{possibilities} \end{array}$$

CHECK YOUR UNDERSTANDING 1.2

Five slips of paper, marked 1 through 5, are placed in a hat. Two slips are drawn without replacement (so you can't draw the same number twice). Write down an equiprobable sample space for the experiment (similar to the two-die sample space of Example1.1), and then use it to determine the probability that:

(a) The first number drawn is odd.

(b) The second number drawn is greater than the first.

Answers: (a) $\frac{12}{20}$ (b) $\frac{10}{20}$

Empirical Probability

We began this section by agreeing that the probability of rolling a 3 with a fair die is $Pr(3) = \frac{1}{6}$. We never rolled any die, but took the **theoretical** approach: the die has six faces, and only one of them shows a 3. An **empirical** (or experimental, or statistical approach) would be to make no assumption about the die whatsoever, but to simply roll it a "large" number of times, counting the number of successes along the way. If, for example, a 3 comes up exactly 103 times out of 600 rolls, we would then say that the (empirical) probability of rolling a 3 is $\frac{103}{600} \approx 0.17$.

One should certainly question the fairness of the die, if a 3 came up, say, 200 out of 600 times.

It stands to reason that the more times an experiment is conducted, the more confidence one can place on the resulting empirical probability. Suppose, for example, that you open a carton of twelve eggs and start dropping them, one at a time, from a height of two inches, and that exactly 5 of the eggs break. From this you may conclude that the probability of an egg breaking under such an activity is approximately $\frac{5}{12} \approx 0.42$. A "stronger conclusion" would ensue were you to drop 144 eggs instead of 12, but that would be a gross waste of eggs.

Since empirical probabilities are, by their very nature, approximations, they will be expressed in decimal form.

EXAMPLE 1.4

GRADE IN COURSE

There are 26 students currently enrolled in Professor Chalk's Math 101 course. Professor Chalk has taught that course four times before and here is the grade distribution of those four sections:

Section	A's	B's	C's	D's	F's
1	5	3	16	3	1
2	4	2	18	5	0
3	2	5	12	1	2
4	1	4	13	2	2
Sum:	12	14	59	11	5

Based on the above data, determine the (empirical) probability that a student chosen at random, from the current class, receives an A in the course.

SOLUTION: The number of students in the current class plays no role whatsoever in determining the probability of a randomly chosen student receiving an A in the class, as that probability is based on the data of the instructor's previous four sections.

We calculate the total number of students in those previous sections (the sample space), by summing the total number of grades:

$$12 + 14 + 59 + 11 + 5 = 101$$

In that sample space of 101 students, there were a total of 12 A's (the successes); bringing us to:

$$Pr(\text{A}) = \frac{12}{101} \approx 0.12$$

EXAMPLE 1.5
EXIT POLL

In an exit poll, 620 people indicated that they voted for Proposition 9, and 329 said that they voted against the proposition. Based only on that survey, determine, to two decimal places, the empirical probability that the next person to vote will vote in favor of Proposition 9.

SOLUTION: A total of $620 + 329 = 949$ individuals voted (the sample space), of which 620 voted "Yes" (the success). Consequently:

$$Pr(\text{Yes}) = \frac{620}{949} \approx 0.65$$

↑ empirical probability

CHECK YOUR UNDERSTANDING 1.3

A firecracker manufacturer tested 900 lady-fingers and found that all but 32 of them exploded. Determine, to two decimal places, the empirical probability that a randomly chosen lady-finger from the production line will function properly.

Answer: $\frac{868}{900} \approx 0.96$

ODDS

This section began with:

Roll a die. What is the probability that you roll a 3?

Chances are that you answered the above question correctly; for you probably already "knew" that:

$$\textbf{Probability}: \frac{\#(\text{successes})}{\#(\text{possibilities})}$$

The odds that you will answer the following questions correctly, however, are not so good:

Roll a die. What are the odds in favor of rolling a 3?

What are the odds against rolling a 3?

Let's remedy the situation:

$$\textbf{Odds (in favor)}: \frac{\#(\text{successes})}{\#(\text{failures})} \qquad \textbf{Odds (against)}: \frac{\#(\text{failures})}{\#(\text{successes})}$$

(providing denominators are not zero)

On a more formal level: Let $P(E)$ be the probability of an event E. Then:

$$\text{Odds in favor E} = \frac{Pr(E)}{1 - Pr(E)}$$

$$\text{Odds against E} = \frac{1 - Pr(E)}{Pr(E)}$$

(See Exercise 70)

A statement such as *the odds are 30 to 1* is generally understood to mean "odds against." In other words: "against" is the default setting in an odds statement.

n particular:

$$\text{Odds in favor of rolling a 3} = \frac{1 \leftarrow \text{successes}}{5 \leftarrow \text{failures}}$$

$$\text{Odds against rolling a 3} = \frac{5 \leftarrow \text{failures}}{1 \leftarrow \text{successes}}$$

Actually, one typically says:

The odds in favor of rolling a 3 are ***one to five***; written "**1:5**"

and

The odds against rolling a 3 are ***five to one***; written "**5:1**"

EXAMPLE 1.6

DEFECTIVE COMPUTER

A shipment of 45 computers contains 2 units that are defective. You purchase one of the units.

(a) What is the probability that you will get a defective unit?

(b) What are the odds in favor of you getting a defective unit?

(c) What are the odds against you getting a defective unit?

SOLUTION:

(a) $Pr(\text{defective}) = \frac{2}{45}$

(b) Odds (in favor) $= \frac{2}{43}$, or 2:43 (2 to 43)

(c) Odds (against) $= \frac{43}{2}$, or 43:2 (43 to 2)

CHECK YOUR UNDERSTANDING 1.4

Draw a card from a standard deck.

(a) What is the probability of drawing a club?

(b) What are the odds in favor of drawing a club?

(c) What are the odds against drawing a club?

Answers: (a) $\frac{13}{52}$ (b) 13 to 39. (c) 39 to 13.

EXAMPLE 1.7

FROM PROBABILITY TO ODDS AND THE OTHER WAY AROUND.

(a) What are the odds in favor of drawing a red marble from a certain urn, if the probability of drawing a red marble is $\frac{23}{79}$?

(b) What is the probability of drawing a blue marble, if the odds against drawing a blue marble are 12 to 5?

SOLUTION:

(a) $\begin{array}{r} \text{successes} \rightarrow \\ \text{possibilites} \rightarrow \end{array} \frac{23}{79} \Rightarrow$ 23 successes and $79 - 23 = 56$ failures

$\Rightarrow$ Odds in favor $= \frac{23}{56}$ (or 23:56)

(b) $\begin{array}{r} \text{failures} \rightarrow \\ \text{successes} \rightarrow \end{array} \frac{12}{5} \Rightarrow$ 5 successes and $12 + 5 = 17$ possibilities

$\Rightarrow$ Probability $= \frac{5}{17}$

CHECK YOUR UNDERSTANDING 1.5

Mathematically Deprived is running in a derby. What are the odds against her winning the race, if the probability of her winning is 0.35?

Answer: 13:7

EXERCISES

Exercises 1-10. Determine an equiprobable sample space for the given experiment.

1. Selecting a month of the year.
2. Flipping 2 coins.
3. Choosing one of three ice cream flavors: chocolate, vanilla, and strawberry, along with one of two toppings: fudge, whipped cream.
4. Answering a three question, True or False test.
5. Flipping a coin and rolling a die.
6. Stacking three blocks, (block A, block B, and block C), one on top of the other.
7. Answering a two-question multiple choice test, with three choices for each question.
8. Mixing the contents of two of four test tubes together.
9. Lining up three ducks, one behind the other.
10. Taking three of four courses: Math, Art, English, Biology.

Exercises 11-14. (Cards) Draw a card from a standard deck (Figure 1.1). What is the probability that the card is:

11. Red?
12. A face card?
13. A red face card?
14. Not a face card?

Exercises 15-18. (Dice) You roll two dice. What is the probability that:

15. You roll doubles (same number on both dice)?
16. The first die is a five?
17. The second die is a five?
18. At least one die is a five?

Exercises 19-22. (Selection) You draw two slips of paper from a hat containing five slips, marked 1 through 5, **without replacement**. What is the probability that:

19. The number five is chosen?
20. The number five is not chosen?
21. Exactly one of the numbers is a five?
22. Both numbers are fives?

Exercises 23-26. (Selection) You draw two slips of paper from a hat containing five slips, marked 1 through 5, **with replacement** (the first number drawn is put back into the hat before drawing the second). What is the probability that:

23. The number five is chosen?
24. The number five is not chosen?
25. Exactly one of the numbers is a five?
26. Both numbers are fives?

Exercises 27-34. (Coins) You have a penny, a nickel, a dime, and a quarter in your pocket, and grab two of the coins at random (without replacement). What is the probability that you are holding:

27. Thirty five cents?
28. More than 35 cents?
29. At least 15 cents?
30. More than 15 cents?
31. Less than 10 cents?
32. Less than 25 cents?
33. More than 10 cents?
34. Exactly 10 cents?

Exercises 35-46. (Coins) You have 2 pennies, a nickel, a dime, and a quarter in your pocket, and grab two of the coins at random, without replacement. (In developing your sample space, make sure you distinguish between the two different pennies). What is the probability that you are holding:

35. More than 25 cents?
36. More than 26 cents?
37. More than 27 cents?
38. Less than 25 cents?
39. Two cents?
40. 6 cents?
41. Exactly one of the two pennies?
42. At least one of the two pennies?
43. Neither of the pennies?
44. Less than 10 cents?
45. More than 10 cents?
46. Exactly 10 cents?

Exercises 47-50. (Blocks) Little Louie has 2 identical blocks, with faces colored red, white, blue, green, yellow, and black. He randomly stacks one block upon the other. What is the probability that:

47. The top of the two-block stack is red?
48. The bottom of the two-block stack is not red?
49. The bottom of the two-block stack is of the same color as the top of the stack?
50. The bottom color of the two-block stack differs from the color at the top of the stack?

51. **(Matching)** You pick a number from 1 to 5, and your friend does the same. What is the probability that the two of you pick the same number?

Exercises 52-56. (Lining Up) Johnny, Mary, and Billy are randomly lined up, one behind the other. What is the probability that:

52. Johnny and Mary are next to each other?
53. Johnny is immediately behind Mary?
54. Johnny and Mary are not next to each other?
55. Billy is first?
56. Billy is in the middle?

Exercises 57-59. (Baseball) In the first five games of the season, Bobbie Baseball hit safely 6 of 20 times at bat.

57. What is the (empirical) probability that Bobbie will hit safely the next time at bat?

58. Suppose Bobbie gets a hit in Exercise 57. Calculate the new empirical probability that he will hit safely the next time at bat.

59. Suppose Bobbie makes an out in Exercise 57. Calculate the new empirical probability that he will hit safely the next time at bat.

Exercises 60-62. (Baseball) Late in the season we find that Bobbie Baseball hit safely 100 of 300 times at bat.

60. What is the (empirical) probability that Bobbie will hit safely the next time at bat?

61. Suppose Bobbie gets a hit in Exercise 60; calculate the new empirical probability that he will hit safely the next time at bat.

62. Suppose Bobbie strikes out in Exercise 60; calculate the new empirical probability that he will hit safely the next time at bat.

63. **(Work Survey)** A survey taken of the work schedules of a number of students at Ivy College is tabulated below.

	Work-Hours per Week (WH)			
	$WH < 5$	$5 \le WH < 10$	$10 \le WH < 20$	$20 \le WH$
Freshman	175	210	115	55
Sophomore	200	178	95	83
Junior	115	205	101	75
Senior	250	225	55	50

Based on the above data, determine the empirical probability that a student chosen at random:

(a) Is a junior.

(b) Works at least 20 hours per week.

(c) Works at least 10 hours per week.

(d) Is a senior who works at least 20 hours per week.

64. **(Political Survey)** Political affiliations of a number of registered voters in a Florida district are tabulated below.

	Democrat	Republican	Independent
Male	520	630	110
Female	620	450	90

Based on the above data, determine the empirical probability that a voter chosen at random from that Florida district is:

(a) A female. (b) A male. (c) A Republican. (d) A Democrat. (e) An independent.

65. **(Odds)** Draw a card from a standard deck (Figure 1.1).
    (a) What are the odds in favor of drawing an ace?
    (b) What are the odds against drawing a face card?

66. **(Odds)** Flip a coin.
    (a) What are the odds in favor of flipping a head?
    (b) What are the odds against flipping a head?

67. **(Odds to Probability)** What is the probability that your ship will come in, if the odds in favor of your ship coming in are 3 to 5?

68. **(Odds to Probability)** What is the probability that you will receive an A in this course, if the odds in favor of your receiving an A are two to one?

69. **(Probability to Odds)** What are the odds against you securing an interesting position upon graduation, if the probability of securing an interesting position is 0.75?

70. **(Theory)** Show that for an event E:

$$\text{Odds in favor of } E = \frac{Pr(E)}{1-Pr(E)} \quad \text{and} \quad \text{Odds against } E = \frac{1-Pr(E)}{Pr(E)}$$

§2 Unions and Complements of Events

We begin by introducing a bit of set notation:

Roughly speaking, a **set** is a collection of objects, or elements. The order in which elements of a set are represented is of no consequence. For example, the sets $\{1, 2, 3\}$ and $\{2, 3, 1\}$ are one and the same set.

A set can be described by listing its elements (inside brackets), as is done with the set A below (this is called the **roster method**).

$$A = \{-2, 5, 11, 99\}$$

The **descriptive method** can also be used to describe a set. In this method, a statement or condition is used to specify the elements of the set, as is done with the set O below:

$$O = \{x | x \text{ is an odd positive integer}\}$$

Read: O is the set of all x **such that** x is an odd positive integer

For a given set A, $x \in A$, is read: ***x* is an element of *A*** (or x is contained in A), and $x \notin A$ is read: x is **not** an element of A. For example:

$$5 \in \{-2, 5, 11, 99\} \quad \text{while} \quad 9 \notin \{-2, 5, 11, 99\}$$

Just as you can add or multiply numbers to obtain other numbers, you can also combine sets to arrive at other sets:

If someone asks you if you want tea or coffee, you are being offered one or the other, but not both (the **exclusive-or** is used). In mathematics, however, the **inclusive-or** is generally used. In particular, "x is in A or B;" is true even if x is in both A and B.

DEFINITION 1.3

INTERSECTION AND UNION OF SETS

For sets A and B, the **intersection** of A and B, written $A \cap B$, is the set consisting of the elements common to both A and B. That is:

$$A \cap B = \{x | x \in A \text{ and } x \in B\}$$

[see Figure 1.3(a)]

The **union** of A and B, written $A \cup B$, is the set consisting of the elements that are in A or in B. That is:

$$A \cup B = \{x | x \in A \text{ or } x \in B\}$$

[see Figure 1.3(b)]

The adjacent visual representations of sets are called **Venn diagrams**. [John Venn, English logician (1834-1923)].

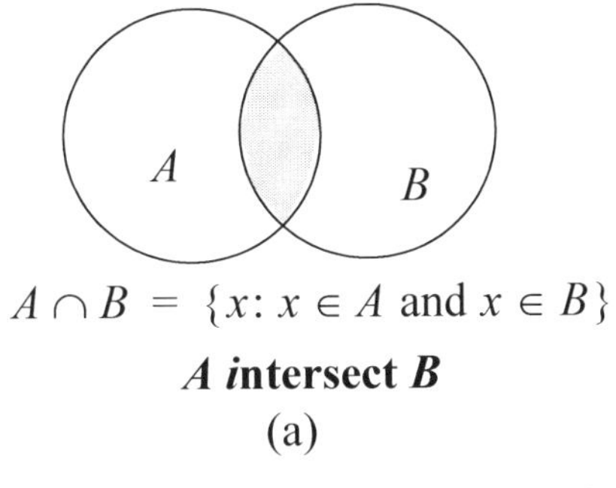

$A \cap B = \{x : x \in A \text{ and } x \in B\}$

A intersect B

(a)

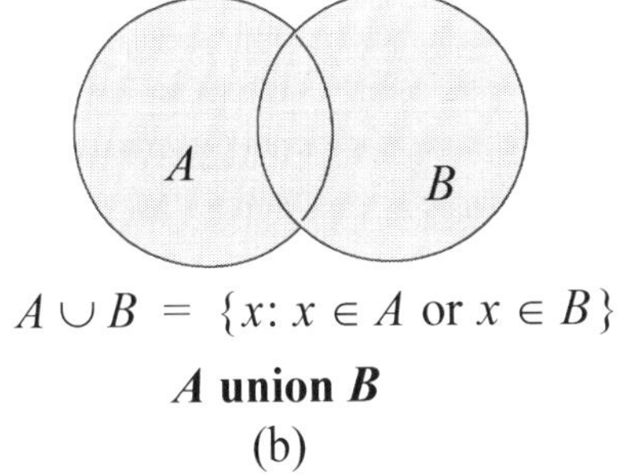

$A \cup B = \{x : x \in A \text{ or } x \in B\}$

A* union *B

(b)

Figure 1.3

For example, if $A = \{3, 5, 9, 11\}$ and $B = \{1, 2, 5, 6, 10, 11\}$, then:

$$A \cap B = \{3, \mathbf{5}, 9, \mathbf{11}\} \cap \{1, 2, \mathbf{5}, 6, 10, \mathbf{11}\} = \{\mathbf{5}, \mathbf{11}\}$$

and:

$$A \cup B = \{3, 5, 9, 11\} \cup \{\mathbf{1}, \mathbf{2}, 5, \mathbf{6}, \mathbf{10}, 11\} = \{3, 5, 9, 11, \mathbf{1}, \mathbf{2}, \mathbf{6}, \mathbf{10}\}$$

EMPTY SET

What is the intersection of the set $\{1, 2, 3\}$ with the set $\{4, 5\}$? We need a symbol to denote a set which contains nothing, and here it is: $\varnothing$. It is appropriately called the **empty** or **null set**. In particular:

$$\{1, 2, 3\} \cap \{4, 5\} = \varnothing$$

COMPLEMENT

When dealing with sets, one typically has a **universal set** U in mind: a set consisting of all elements under consideration. As is indicated in Figure 1.4, if A is a **subset** of U (if every element of A is contained in U), then A^c is used to denote the **complement** of A: those elements in U that are **not** in A.

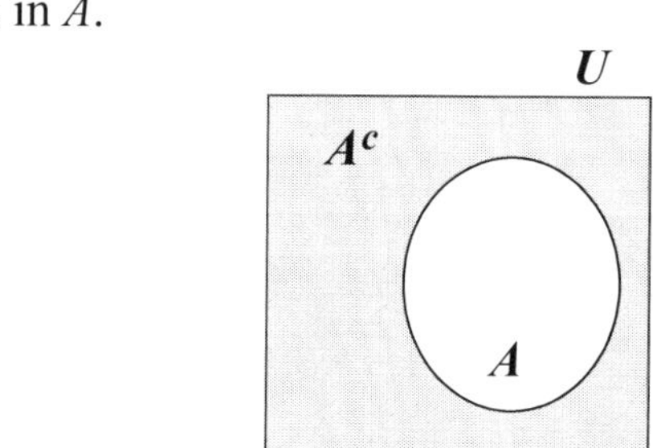

$$A^c = \{x | x \in U \text{ and } x \notin A\}$$

(or simply $\{x | x \notin A\}$ when U is understood)

The complement of A

Figure 1.4

For example, if the universal set is $U = \{\mathbf{1}, \mathbf{2}, \mathbf{3}, \mathbf{4}, \mathbf{5}, \mathbf{6}, \mathbf{7}, \mathbf{8}, \mathbf{9}\}$, then $\{\mathbf{1}, \mathbf{3}, \mathbf{5}, \mathbf{7}, \mathbf{8}\}^c = \{\mathbf{2}, \mathbf{4}, \mathbf{6}, \mathbf{9}\}$.

THE COMPLEMENT THEOREM

Draw a card from a standard deck. What is the probability that it is not a King? That's right:

$$Pr(\text{not a King}) = \frac{48}{52}$$

In solving the above problem, you did not arrive at the number 48 by adding up all of the cards that are not Kings. Rather, you simply subtracted the 4 Kings from the 52 cards:

$$Pr(\text{not a King}) = \frac{52 - 4}{52}$$

Breaking the above quotient into two parts, we have:

$$\boldsymbol{Pr}(\textbf{not K}) = \frac{52-4}{52} = \frac{52}{52} - \frac{4}{52} = 1 - \frac{4}{52} = \mathbf{1} - \boldsymbol{Pr}(\mathbf{K})$$

Generalizing:

THEOREM 1.1

PROBABILITY OF THE COMPLEMENT OF AN EVENT

E^c

E

For any event E,

$$Pr(\text{not E}) = 1 - Pr(\text{E})$$

More formally:

$$Pr(\text{E}^c) = 1 - Pr(\text{E})$$

Or:

$$Pr(\text{E}) + Pr(\text{E}^c) = 1$$

IN WORDS: The probability that something does not occur is one minus the probability that it does occur.

AND: The probability that something does occur is one minus the probability that it does not occur.

EXAMPLE 1.8
NEGATIVE REACTION

The probability that an individual will have a negative reaction to a medication is 0.07. What is the probability that an individual will not have a negative reaction?

SOLUTION: We simply apply Theorem 1.1:

$$Pr(\text{no neg. reaction}) = 1 - Pr(\text{neg. reaction}) = 1 - 0.07 = 0.93$$

CHECK YOUR UNDERSTANDING 1.6

Draw a card from a standard deck. What is the probability that it is not a face card?

Answer: $\frac{40}{52}$

THE UNION THEOREM

The following example will serve to usher in our next theorem.

EXAMPLE 1.9
DRAW A CARD

Draw a card from a standard deck. What is the probability that it is a Club or a King?

SOLUTION: There are 13 Clubs and there are 4 Kings, but the King of Clubs is **not to be counted twice**:

$$Pr(\text{Club or King}) = \frac{16}{52} \quad \longleftarrow \text{(13 Clubs + 4 Kings – the King of Clubs)}$$

Breaking down the above quotient we find that:

$$Pr(\text{C or K}) = \frac{16}{52} = \frac{13+4-1}{52} = \frac{13}{52} + \frac{4}{52} - \frac{1}{52} = Pr(\text{C}) + P(\text{K}) - Pr(\text{C and K})$$

Generalizing, we have:

PROBABILITY OF THE UNION OF TWO EVENTS

THEOREM 1.2 For events E and F,

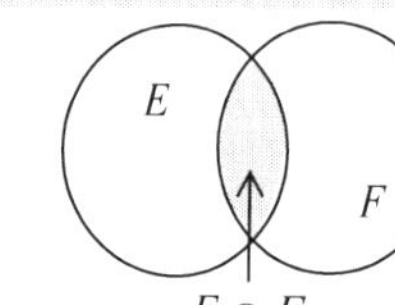

$$Pr(\text{E or F}) = Pr(\text{E}) + Pr(\text{F}) - Pr(\text{E and F})$$

More formally:

$$Pr(\text{E} \cup \text{F}) = Pr(\text{E}) + Pr(\text{F}) - Pr(\text{E} \cap \text{F})$$

IN WORDS: The probability that one or another thing happens (possibly both), is the probability that one happens, plus the probability that the other happens, minus the probability that they both happen.

EXAMPLE 1.10
DEFECTIVE LIGHTS AND BRAKES

The probability that a car will have at least one defective light is 0.031, and the probability that a car will have defective brakes is 0.020. Assuming that there is a 0.005 probability that a car will have both a defective light and defective brakes, determine the probability that a car will have either a defective light or defective brakes.

SOLUTION: Let L and B denote the events that a car has a defective light or brakes, respectively. Applying Theorem 1.2, we have:

$$\underset{\text{or}}{} \quad Pr(L \cup B) = Pr(L) + Pr(B) - Pr(L \cap B) \quad \underset{\text{and}}{}$$

$$= 0.031 + 0.020 - 0.005 = 0.046$$

CHECK YOUR UNDERSTANDING 1.7

Draw a card from a standard deck. What is the probability that it is red or a face card?

Answer: $\frac{32}{52}$

DEFINITION 1.4
MUTUALLY EXCLUSIVE EVENTS

Two events E and F are **mutually exclusive** if they have no element in common:

$$\text{E} \cap \text{F} = \varnothing$$

the empty set

The following result is a special case of Theorem 1.2.

PROBABILITY OF THE UNION OF TWO MUTUALLY EXCLUSIVE EVENTS

THEOREM 1.3 For **mutually exclusive events** E and F,

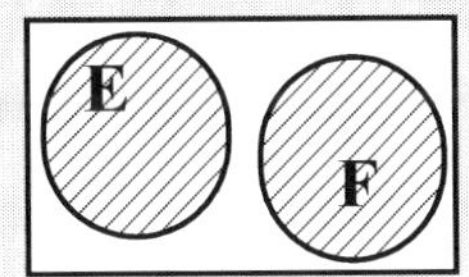

$$Pr(\text{E or F}) = Pr(\text{E}) + Pr(\text{F})$$

More formally:

$$Pr(\text{E} \cup \text{F}) = Pr(\text{E}) + Pr(\text{F})$$

PROOF: Since $\text{E} \cap \text{F} = \varnothing$, $Pr(E \cap F) = 0$. Applying Theorem 1.2, we have:

$$\begin{aligned} Pr(\text{E} \cup \text{F}) &= Pr(\text{E}) + Pr(\text{F}) - Pr(E \cap F) \\ &= Pr(E) + Pr(F) - 0 = Pr(E) + Pr(F) \end{aligned}$$

EXAMPLE 1.11
DRAW A CARD

Draw a card from a standard deck. What is the probability that it is a King or a Queen?

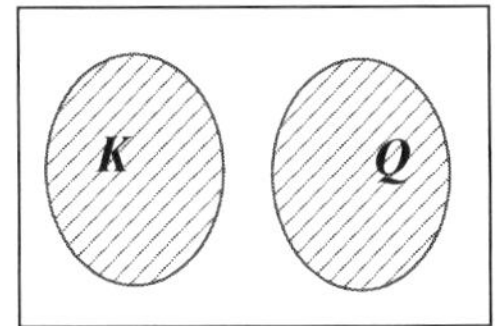

SOLUTION: There are 4 Kings and 4 Queens, and if you add them up, nothing is counted twice (so nothing has to be subtracted from the sum):

$$Pr(K \cup Q) = Pr(K) + Pr(Q) = \frac{4}{52} + \frac{4}{52} = \frac{8}{52}$$

CHECK YOUR UNDERSTANDING 1.8

Draw a card from a standard deck. What is the probability that it is a black face card or a red ace?

Answer: $\frac{8}{52}$

EXAMPLE 1.12
VOTERS IN DISTRICT

The following table gives a summary of Male (M) and Female (F) registered voters in a Florida district:

Sex and Age	Democrat (D)	Republican (R)	Independent (I)
$18 \le M < 38$	593	655	290
$18 \le F < 38$	642	583	139
$38 \le M < 58$	515	712	110
$38 \le F < 58$	708	525	125
$M \ge 58$	249	830	27

What is the probability that a randomly chosen registered voter in the district is:

(a) Not a Democrat of age 58 or older?

(b) A male or an Independent?

(c) A Democrat or a Republican.

SOLUTION: (a) Summing up all columns and rows we come to:

SEX AND AGE	Democrat	Republican	Independent	SUM
$18 \le M < 38$	593	655	290	1538
$18 \le F < 38$	642	583	139	1364
$38 \le M < 58$	515	712	110	1337
$38 \le F < 58$	708	525	125	1358
$M \ge 58$	249	830	27	1106
$F \ge 58$	430	515	72	1017
SUM	**3137**	**3820**	**763**	**7720**

Note that summing down the first 6 rows of the SUM column (last column), or across the first three columns of the SUM row (last row) must yield the same result. Why?

(a) Using Theorem 1.1:

$$Pr(\text{not: } D \ge 58) = 1 - Pr(D \ge 58) = 1 - \frac{249 + 430}{7720} \approx 0.91$$

(b) Using Theorem 1.2:

There is more than one way to pet a cat:

$$Pr(M \cup I) = \frac{\overbrace{1538 + 1337 + 1106}^{\text{males}} + \overbrace{139 + 125 + 72}^{\text{female independents}}}{7720}$$

Male or Independent
↓

$$Pr(M \cup I) = Pr(M) + Pr(I) - Pr(M \cap I)$$
$$= \frac{1538 + 1337 + 1106}{7720} + \frac{763}{7720} - \frac{290 + 110 + 27}{7720} \approx 0.56$$

(c) Using Theorem 1.3:

Democrat or Republican
↓

$$Pr(D \cup R) = Pr(D) + Pr(R) = \frac{3137}{7720} + \frac{3820}{7720} \approx 0.90$$

CHECK YOUR UNDERSTANDING 1.9

Referring to the data of Example 1.12, determine the probability that an individual selected at random is a male who is not a Democrat.

Answer: $\frac{2624}{7720} \approx 0.34$

COUNTING WITH VENN DIAGRAMS

Sometimes, for counting purposes, one has to first partition a sample space into disjoint pieces. Consider the following example.

EXAMPLE 1.13

STUDENTS AND COURSES

In a group of 100 students, 55 are taking Math (M), 63 are taking English (E), and 25 are taking both Math and English. A student is chosen at random from the class. Determine the probability that the student is:

(a) Taking Math but not English.

(b) Taking neither Math nor English.

(c) Taking exactly one of the two courses.

SOLUTION:

The given information is compactly represented in Figure 1.5, where "M E" is used to denote the more formal $M \cap E$. We also positioned the number 25 in the "intersection piece" of the Venn diagram, and the number 100 for the universe space, as given.

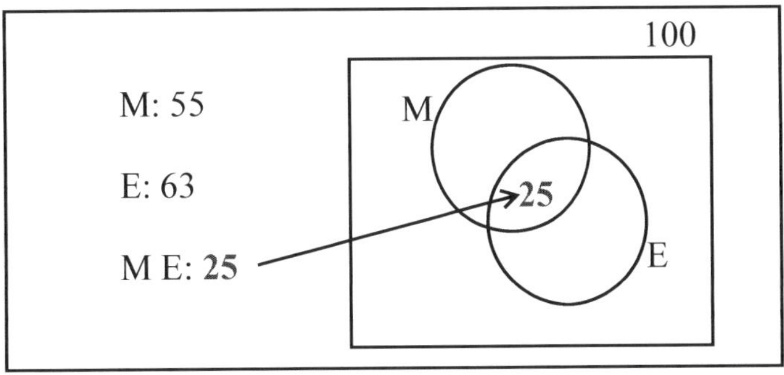

Figure 1.5

We now fill in the remaining pieces of the above puzzle:

There are 55 in all of M, and 25 have already been accounted for; so there must be $55 - 25 = \mathbf{30}$ in the piece "M but not E" [see (*) of Figure 1.6].

By the same token, there must be $63 - 25 = \mathbf{38}$ in the piece "E but not M" [see (**) of Figure 1.6.

One more piece of the puzzle remains: students that are taking neither M nor E, and there are $100 - (30 + 25 + 38) = 7$ in that piece [see (***) of Figure 1.6].

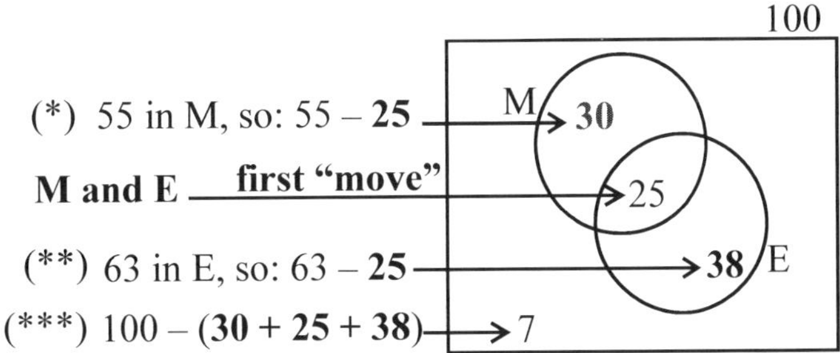

Figure 1.6

Using the Venn diagram of Figure 1.6, we conclude that:

More formally:

$Pr(M \cap E^c) = \frac{30}{100}$

(a) $Pr(\text{Math and not English}) = \frac{30 \leftarrow \text{successes}}{100 \leftarrow \text{possibilities}}$

$Pr(M^c \cap E^c) = \frac{7}{100}$

(b) $Pr(\text{neither Math nor English}) = \frac{7 \leftarrow \text{successes}}{100 \leftarrow \text{possibilities}}$

$Pr[(M \cap E^c) \cup (E \cap M^c)] = \frac{68}{100}$

(c) $Pr(\text{just Math or just English}) = \frac{30 + 38}{100} = \frac{68 \leftarrow \text{successes}}{100 \leftarrow \text{possibilities}}$

CHECK YOUR UNDERSTANDING 1.10

Of 1500 students, 135 are business majors, 78 are math majors, and 3 have a double major in business and math.

(a) Construct a Venn diagram, as in the previous example.

Determine the probability that a randomly chosen student:

(b) Is neither a math nor a business major?

(c) Is a business major but not a math major?

Answers: (a) See A-3. (b) $\frac{1290}{1500}$ (c) $\frac{132}{1500}$

The following example is a bit more challenging than the previous one, in that its associated Venn diagram will have 8 rather than 4 disjoint pieces.

In general, n categories ("circles") in a Venn diagram will serve to decompose the Venn diagram into 2^n (disjoint) pieces:

2 circles $\rightarrow 2^2 = 4$ pieces

3 circles $\rightarrow 2^3 = 8$ pieces

4 circles $\rightarrow 2^4 = 16$ pieces

EXAMPLE 1.14

STUDENTS AND COURSES

A survey of students found that 19 of them were taking a Math course and 17 were taking an Art course. Twelve were taking a Math and an English course, while 11 were taking an Art and an English course. Five were taking all three courses. Nine were taking only English, and 3 were taking only Math. Fourteen were not taking any of the courses.

(a) Construct a complete Venn diagram representing the above situation.

(b) How many students were surveyed?

(c) What is the probability that a student chosen at random is taking exactly one of the three courses?

(d) What is the probability that a student chosen at random is not taking either a Math course or an Art course?

SOLUTION:

On the right, we compactly represented the given information in an informal fashion. We also filled in two of the "obvious" pieces, and ask you to fill in as many of the remaining boxes as you can before moving on.

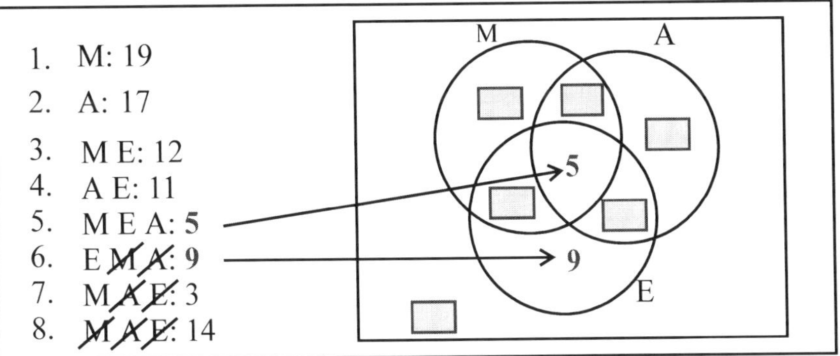

Figure 1.7

(a) Let M, A, and E represent the sets of students taking Math, Art, and English courses, respectively. Imagine a solitaire game involving the 8 "cards" at the left of Figure 1.7. As you work down the deck, you are unable to play the first 4 cards, but can play the 5th, and 6th cards (Figure 1.7). The 7th and 8th cards can also be played as shown in Figure 1.8(a):

Card 7 (M, ~~A~~, ~~E~~: 3): 3 elements inside *M* but outside of both *A* and *E*.

Card 8 (~~M~~, ~~A~~, ~~E~~: 14): 14 elements outside of all three circles.

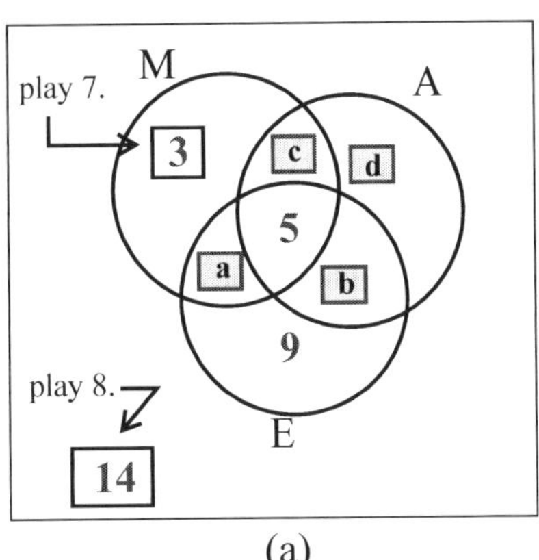

(a)

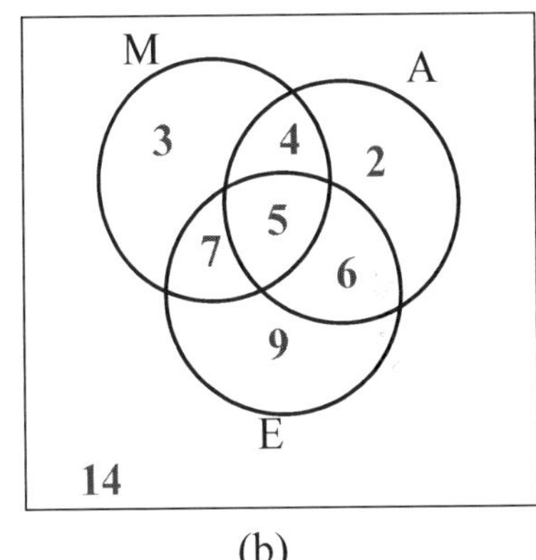

(b)

Figure 1.8

We can now fill in the remaining 4 shaded boxes [labeled a through d in Figure 1.8(a)]:

Card 3 (M, E: 12): 12 = [a] + 5 → [a] = 7

Card 4 (A, E: 11): 11 = [b] + 5 → [b] = 6

Card 1 (M: 19): 19 = [c] + 5 + 7 + 3 → [c] = 4 (where 7 ↑ [a])

Card 2 (A: 17): 17 = [d] + 6 + 5 + 4 → [d] = 2 (where 6 ↑ [b], 4 ↑ [c])

With the completed Venn diagram of Figure 1.8(b), we can address any and all questions pertaining to the situation at hand. In particular:

(b) Students surveyed: $3+4+2+6+5+7+9+14 = 50$

(c) $Pr(\text{Exactly one of the 3 courses}) = \frac{3+2+9}{50} = \frac{14}{50}$

(d) $Pr(\text{Neither Math nor Art}) = \frac{14+9}{50} = \frac{23}{50}$

CHECK YOUR UNDERSTANDING 1.11

Of 320 physical education majors, it was found that 155 play tennis, 120 play baseball, and 60 play handball. Fifty play tennis and baseball, and only 3 of those 50 play handball. Seven play handball and baseball, and 5 play tennis and handball.

(a) Construct a complete Venn diagram (fill in all of the pieces) representing the above situation.

(b) How many play exactly two of the three sports?

(c) What is the probability that a randomly chosen physical education major plays at least two of the three sports?

Answers: (a) See A-3. (b) 53 (c) $\frac{56}{320}$

EXERCISES

Exercises 1-6. (Card) Draw a card from a standard deck. What is the probability that you draw:

1. A face card?
2. Not a face card?
3. Neither a face card nor an ace?
4. Either a face card or an ace?
5. A face card or a club?
6. Neither a face card nor a club?

Exercises 7-12. (Dice) A pair of dice is tossed. What is the probability that their sum is:

7. Odd or greater than 5?
8. Odd and greater than 5?
9. Divisible by 2 or 3?
10. Divisible by 2 and 3?
11. Not divisible by 2 and not divisible by 3?
12. Not divisible by 2 or not divisible by 3?

Exercises 13-18. (Blocks) Becky-Boo has three letter-blocks: block-A, block-B, and block-C. She stacks the three blocks, one on top of the other. What is the probability that:

13. The A-block is in the middle?
14. The A-block is not in the middle?
15. The A- or B-block is in the middle?
16. Neither the A- nor the B-block is in the middle?
17. The A-block is either at the bottom or on the top?
18. The letters read ABC in either direction (reading from the bottom up, or from the top down)?

Exercises 19-27. (Urn) A marble is drawn at random from an urn containing 4 red marbles numbered 1 through 4, five white marbles numbered 1 through 5, and two blue marbles numbered 5 and 10. Determine the probability that the marble drawn:

19. Is red or displays an even number.
20. Is red and displays an even number.
21. Is not red and displays an even number.
22. Is not red or does not display an even number.
23. Displays the number 10 and is blue.
24. Is not blue and displays the number 10.
25. Is blue or displays the number 10.
26. Is blue and does not display the number 10.
27. Is not blue and does not display the number 10.

28. **(Grade Distribution)** The following table summarizes the grade distribution of students enrolled in Math 101 at State University for the last five years. The survey involved Matriculated Male student Commuters (MMC), Matriculated Female student Commuters (MFC), Matriculated Male students Living on campus (MML), Matriculated Female students Living on campus (MFL), Non-matriculated Male students (NM), and Non-matriculated Female students (NF). Assume all non-matriculated students commute.

	A	B	C	D	F
MMC	22	32	221	19	22
MFC	19	30	197	14	10
MML	43	41	335	21	25
MFL	29	32	257	35	19
NM	25	19	115	8	3
NF	32	24	124	4	2

Determine the probability that a student chosen at random from the above surveyed group:

(a) Was matriculated.
(b) Lived on Campus.
(c) Was a non-matriculated male student who received an A in the course.
(d) Was a non-matriculated female student or received an A in the course.
(e) Passed the course (A through D).
(f) Did not fail the course.
(g) Was a matriculated female student who received an A or a B in the course.
(h) Lived on campus and failed the course.
(i) Did not live on campus and failed the course.
(j) Did not receive an A or a B in the course.
(k) Was a non-matriculated student or received an A or a B in the course.
(l) Was a matriculated student who did not receive an A or a B in the course.

29. **(Workers)** The following table gives the number (in thousands) of 20 years and over U.S. employed and unemployed workers from 1995 through 1999.

	1995	1996	1997	1998	1999
Men					
Employed	60,085	64,897	66,524	67,134	67,761
Unemployed	3,239	3,147	2,826	2,580	2,433
Women					
Employed	54,396	53,310	57,647	57,278	58,655
Unemployed	2,819	2,783	2,187	2,424	2,285

Source: U.S. Department of Labor Statistics (2000) [Current Population Survey: Web: stats.bis.gov.]

[How would you account for the fact that, in any year, if you sum up the employed and unemployed men, you end up with a number considerably greater than the sum of the employed and unemployed women?]

Take a very large hat and put in it slips of paper containing all of the names and associated years and employment status of the individuals represented in the above table. Draw one of the slips of paper. What is the probability that the slip of paper:

(a) Contains the name of a man.
(b) Contains the year 1996.
(c) Contains the name of a woman and the year 1997.

(d) Contains the name of an unemployed individual.

(e) Does not contain the name of an employed woman.

(f) Contains the name of an unemployed person and the year 1999.

(g) Does not contains the name of a man but does contain the years 1999.

30. **(Dogs and Cats)** Dogs reside in 12% of households, and cats reside in 15% of households. Both dogs and cats are found in 3% of households. A house is chosen at random. What is the probability that:

(a) A dog does not reside there?
(b) A dog or a cat resides there?
(c) Neither a dog nor a cat resides there?
(d) Only a cat resides there?

31. **(Cars and Motorcycles)** A survey found that 575 individuals owned a car and that 55 owned a motorcycle. All but 5 of the individuals owning a motorcycle also owned a car. Five of the surveyed individuals owned neither a car nor a motorcycle. An individual is chosen at random from the group surveyed. What is the probability that he or she:

(a) Owned either a car or a motorcycle?
(b) Did not own a motorcycle?
(c) Owned a car but not a motorcycle?
(d) Owned a motorcycle but not a car?

32. **(Media)** A merchant surveyed 400 people to determine the way they learned about an upcoming sale. The survey showed that 180 learned about the sale from the radio, 190 from television, 190 from the newspaper, 80 from radio and television, 90 from radio and newspapers, 50 from television and newspapers, and 30 from all three sources.

(a) How many people learned of the sale from radio or television but not both?

(b) How many people learned of the sale from only one of the three media?

(c) What is the probability that a randomly chosen individual learned of the sale from at least two of the three media?

33. **(Tutors)** There are 21 math tutors at the tutoring center. Five of them tutor Calculus, 14 tutor Precalculus, and 9 tutor Transitional Math. No one that tutors Calculus tutors Transitional. Four tutor Transitional and Precalculus, and 3 tutor Calculus and Precalculus.

(a) How many tutor only one course?

(b) How many tutor all three courses?

(c) What is the probability that a randomly chosen individual tutors exactly two courses?

34. **(Dogs, Cats, and Birds)** In a survey of 110 individuals it is found that 50 own dogs; 40 own cats, and 41 own birds. 5 own dogs, cats, and birds; 15 own dogs and cats; 25 own dogs and birds but not cats; 10 own birds and cats.

(a) How many of the individuals own only cats?

(b) How many of the individuals do not own a cat?

(c) What is the probability that a randomly chosen individual owns a bird but not a cat?

35. **(Dogs, Cats, and Birds)** A survey of 750 individuals found that 350 owned dogs, 90 owned birds, 275 owned cats, 175 owned cats and dogs, 75 owned dogs and birds, 30 owned cats and birds, and 25 owned all three. How many of the individuals surveyed:

(a) Do not own a dog?
(b) Own a dog or a bird.
(c) Own at least one of the three animals?
(d) Own none of the three animals?

36. **(Newspapers)** Out of 500 individuals, 130 read the New York Times, 180 read the Wall Street Journal. Twenty-five read those two papers, as well as the San Francisco Chronicle. Seventy-five read only the Times, 80 read only the Journal. Twenty-five read the Times and the Journal, but not the Chronicle. Two hundred read none of the three papers. One of the individuals is chosen at random. What is the probability that he or she:

    (a) Reads only the Chronicle?
    (b) Reads exactly one of the three papers?
    (c) Reads at least one of the three papers?
    (d) Reads at most two of the three papers?

37. **(Courses)** In a survey of 110 college students, it was found that 23 were taking Math, 10 were taking Math and Art, 18 were taking Art, 9 were taking Math and History, 14 were taking History, 8 were taking Art and History, and 5 were taking all three courses. What is the probability that a student selected at random from those surveyed:

    (a) Is taking only Math?
    (b) Is taking exactly two of the three courses?
    (c) Is taking none of the three courses?
    (d) Is taking exactly one of the three courses?
    (e) Is taking at most one of the three courses?
    (f) Is taking at least one of the three courses?

38. **(Courses)** In a group of 60 students, 20 are taking Math, 13 are taking English, and 20 are taking Biology. Half of the students are not taking any of the three courses, and 10% of them are taking English but not Math. Four students are taking Math and English but not Biology. Every student taking English is also taking Math or Biology. What is the probability that a student selected at random from those surveyed:

    (a) Is taking only Math?
    (b) Is taking exactly two of the three courses?
    (c) Is taking exactly one of the three courses?
    (d) Is taking at most one of the three courses?
    (e) Is taking at least one of the three courses?

39. **(Vices)** In a group of 1000 individuals, 165 smoke. There are 520 females in the group, 26 of whom have an alcoholic addiction. Ten of the women have an alcoholic addiction and also smoke. Thirty of the men have an alcoholic addition. Seventy-five of the men do not have an

alcoholic addiction but do smoke. Five of the men have an alcoholic addiction and smoke. What is the probability that a individual selected at random from the group:

(a) Either smokes or has an alcoholic addiction?

(b) Does not smoke and does not have an alcoholic addiction?

(c) Does not smoke but has an alcoholic addiction?

(d) Has an alcoholic addiction and smokes?

(e) Is a male smoker?

(f) Either smokes or has an alcoholic addiction, but not both?

(g) Is a male who neither smokes nor has an alcoholic addiction?

(h) Is a female who neither smokes nor has an alcoholic addiction?

40. **(Cars, Motorcycles, and Bicycles)** A survey of 1000 individuals found that 750 owned a car, 50 owned a motorcycle, and 550 owned a bicycle. Forty owned a car, a motorcycle, and a bicycle. Everyone that owned a motorcycle also owned a bicycle. Three hundred owned a car and a bicycle. An individual is chosen at random from the group surveyed. What is the probability that he or she:

(a) Owned either a car or a motorcycle?

(b) Did not own any of the three?

(c) Owned a car but not a motorcycle?

(d) Owned a motorcycle but not a car?

(e) Owned a bicycle but not a motorcycle?

(f) Owned a motorcycle but not a bicycle?

(g) Did not own a bicycle?

(h) Owned only a bicycle?

(i) Did not own a car or a bicycle?

(j) Owned only a car or only a bicycle?

§3. Conditional Probability and Independent Events

You get a glimpse of a card as it is dealt to you from a standard deck, and see that it is a Face card. What is the probability that the card is a King? Not $\frac{4}{52}$, since the added information of it being a Face card restricts the 52 card sample space to the 12 Face cards. In that restricted sample space there remain four successes (the four Kings). Hence:

$$Pr(\text{King given that it is a Face Card}) = \frac{4}{12}$$

Notation: $Pr(\text{King}|\text{Face Card}) = \frac{4}{12}$

(the vertical bar "|" is read: **given**)

The above is an example of a **conditional probability** problem, in which a stated condition serves to restrict the possible outcomes of an experiment. The first step in solving such a problem is to determine the **restricted sample space**, and then look for the successes within that restricted sample space.

EXAMPLE 1.15
ROLLING A PAIR OF DICE

What is the probability of rolling a pair of fives with two dice, given that at least one die shows a 5.

SOLUTION: First, find the restricted sample space:

$$\left\{\begin{matrix} (1,1) & (1,2) & (1,3) & (1,4) & \mathbf{(1,5)} & (1,6) \\ (2,1) & (2,2) & (2,3) & (2,4) & \mathbf{(2,5)} & (2,6) \\ (3,1) & (3,2) & (3,3) & (3,4) & \mathbf{(3,5)} & (3,6) \\ (4,1) & (4,2) & (4,3) & (4,4) & \mathbf{(4,5)} & (4,6) \\ \mathbf{(5,1)} & \mathbf{(5,2)} & \mathbf{(5,3)} & \mathbf{(5,4)} & \mathbf{(5,5)} & \mathbf{(5,6)} \\ (6,1) & (6,2) & (6,3) & (6,4) & \mathbf{(6,5)} & (6,6) \end{matrix}\right\} \Rightarrow \left\{\begin{matrix} & & & & \mathbf{(1,5)} & \\ & & & & \mathbf{(2,5)} & \\ & & & & \mathbf{(3,5)} & \\ & & & & \mathbf{(4,5)} & \\ \mathbf{(5,1)} & \mathbf{(5,2)} & \mathbf{(5,3)} & \mathbf{(5,4)} & (5,5) & \mathbf{(5,6)} \\ & & & & \mathbf{(6,5)} & \end{matrix}\right\}$$

Unrestricted Sample Space
(see Example 1.1, page 3)

Restricted Sample Space

Looking at the restricted space of 11 possibilities at the right, we see that there is but one success [the **(5,5)**]. Thus:

$$Pr(\text{Two fives} \mid \text{at least 1 five}) = \frac{1}{11}$$

EXAMPLE 1.16
VOTERS IN DISTRICT

The following table gives a decomposition of Male (M) and Female (F) registered voters in a Florida district:

SEX AND AGE	Democrat (D)	Republican (R)	Independent (I)
$18 \le M < 38$	593	655	290
$18 \le F < 38$	642	583	139
$38 \le M < 58$	515	712	110
$38 \le F < 58$	708	525	125
$M \ge 58$	249	830	27
$F \ge 58$	430	515	72
Sum	3137	3820	763

What is the probability that a randomly chosen registered voter in the district is:

(a) A female, given that the individual is a Democrat?

(b) A Democrat, given that the individual is a female?

SOLUTION: (a) Focusing first on the given conditions: ***Democrat***, we see the restricted sample space:

SEX AND AGE	Democrat (D)	Republican (R)	Independent (I)
$18 \le M < 38$	593	655	290
$18 \le F < 38$	**642**	583	139
$38 \le M < 58$	515	712	110
$38 \le F < 58$	**708**	525	125
$M \ge 58$	249	830	27
$F \ge 58$	**430**	515	72
Sum	3137	3820	763

Looking at the surviving Democrat column (restricted sample space), we then identify the successes (Females). This brings us to our answer:

$$Pr(F|D) = \frac{642 + 708 + 430}{3137} = \frac{1780}{3137} \approx 0.57$$

(b) Restricting the sample space by the given condition: ***Female***, we come to:

SEX AND AGE	Democrat (D)	Republican (R)	Independent (I)	SUM
$18 \leq M < 38$	593	655	290	1538
$18 \leq F < 38$	**642**	583	139	1364
$38 \leq M < 58$	515	712	110	1337
$38 \leq F < 58$	**708**	525	125	1358
$M \geq 58$	249	830	27	1106
$F \geq 58$	**430**	515	72	1017

Bringing us to:

$$Pr(D|F) = \frac{642 + 708 + 430}{1364 + 1358 + 1017} = \frac{1780}{3739} \approx 0.48$$

EXAMPLE 1.17
A CON GAME

Mr. Shill, standing on a corner, displays three cards. One of the cards is red on both sides, one is blue on both sides, and the remaining card is red on one side and blue on the other. Handling the three cards briskly, Mr. Shill slaps one of them on the table, and the side facing you is red.

"Well," says Mr. Shill, "one thing is certain, the card on the table is not the blue-blue card. It is therefore either the red-red card, or the red-blue card, and it is therefore just as likely that the other side of the card is blue as it is that it is red. Tell you what I'm going to do," says our friend, "I'll give you $5 if the card on the other side is blue, if you give me $4 if it is red." Should you accept the wager?

SOLUTION: No. There are three cards, each with two sides. There is no problem distinguishing the two sides of the blue-red card, for they are of different color. To distinguish the two sides of the blue-blue card, we will label one side B_1, and the other side B_2. Similarly the two sides of the red card will be labeled R_1 and R_2. The six possible ways that the three cards can be slapped down on the table is depicted in Figure 1.9(a).

up face	down face		red face up	down face
B	R		R	B
R	B		R_1	R_2
B_1	B_2		R_2	R_1
B_2	B_1			
R_1	R_2			
R_2	R_1			

Unrestricted Sample Space (a) $\Rightarrow$ **Restricted Sample Space** (b)

Figure 1.9

You observe that the face-up card is red, and this restricts the sample space to that of Figure 1.9(b). That restricted sample space has 3 elements, only one of which ($\mathbf{R}_{\text{up}}$- B_{down}) is in your favor. We conclude that the probability that you will win the bet is:

$$Pr(\text{blue down} \mid \text{red up}) = \frac{1}{3}$$

In other words: If you accept the wager, then your odds of winning the bet are 1 to 2: you can expect to lose two times for every one time you win.

CHECK YOUR UNDERSTANDING 1.12

Roll a pair of dice. (See Figure 1.2, page 3.)

(a) What is the probability that their sum is 7, given that at least one die is a 5?

(b) What is the probability that at least one die is a 5, given that their sum is 7?

Answers: (a) $\frac{2}{11}$ (b) $\frac{2}{6}$

PROBABILITY OF TWO EVENTS OCCURRING

By now, you should have little difficulty in answering the following question:

Draw a card from a standard deck. What is the probability that the card is a King, given that the card is Red?

Answer: $Pr(\text{K} \mid \text{R}) = \frac{2}{26}$ ← there are 2 red kings ← there are 26 red cards

Dividing numerator and denominator by 52, we arrive at a representation for the conditional probability as a quotient of "regular" probabilities:

$$Pr(K|R) = \frac{\frac{2}{52}}{\frac{26}{52}} \begin{array}{l} \leftarrow \text{probability of drawing a red king} \\ \leftarrow \text{probability of drawing a red card} \end{array}$$

Which is to say:

$$Pr(\text{K} \mid \text{R}) = \frac{Pr(\text{K and R})}{Pr(\text{R})} = \frac{Pr(\text{K} \cap \text{R})}{Pr(R)}$$

Generalizing, we have:

THEOREM 1.4 Let E and F be events, with $Pr(F) \neq 0$. Then:

$$Pr(E|F) = \frac{Pr(E \cap F)}{Pr(F)}$$

A more useful form of Theorem 1.4 is obtained by multiplying both sides of the equation in Theorem 1.4 by $Pr(F)$, thereby expressing $Pr(E \cap F)$ in terms of $Pr(E|F)$ and $Pr(F)$:

PROBABILITY OF TWO EVENTS OCCURRING

Note that Theorem 1.5 holds even if $Pr(F) = 0$.

THEOREM 1.5 Let E and F be two events. Then:

$$Pr(E \cap F) = Pr(F) \cdot Pr(E|F)$$

IN WORDS: The probability of two events occurring is the probability that one occurs, times the probability that the other occurs **given** that the one has taken place.

As is illustrated in the following example, the above theorem can sometimes be used to break down an experiment into two steps:

EXAMPLE 1.18
DRAWING TWO CARDS

Two cards are drawn from a standard deck without replacement. What is the probability that:

(a) You end up with two Kings?

(b) You end up with a King and a Queen?

To solve this problem using Definition 1.2 (page 3) we would need to determine the number of all possible two-card combinations (a big number), along with the number of two-King combinations.
Note how Theorem 1.5 enables us to focus on much smaller sample spaces: the first consisting of a full deck (52 cards) and the second consisting of 51 cards.

SOLUTION:

(a) To end up with two Kings, you must be dealt a King on the first card (K_{1st}) followed by another King on the second card (K_{2nd}). Turning to Theorem 1.5, we have:

$$Pr(\text{K}_{1st} \text{ and } \text{K}_{2nd}) = Pr(\text{K}_{1st}) \cdot Pr(\text{K}_{2nd}|K_{1st})$$

$$= \frac{4}{52} \cdot \frac{3}{51} \approx 0.00452$$

at this stage, there are 51 cards left, with 3 of them Kings.

(b) It is important to note that you can end up with a King and a Queen by being dealt a King first and then a Queen, **or** by being dealt a Queen first and then a King. Both of these possibilities must be taken into account:

$$Pr(\text{K and Q}) = Pr(\text{K}_{1st} \text{ and } \text{Q}_{2nd}) \text{ or } Pr(\text{Q}_{1st} \text{ and } \text{K}_{2nd})$$

$$= \frac{4}{52} \cdot \frac{4}{51} + \frac{4}{52} \cdot \frac{4}{51} \approx 0.012066$$

Taking the ratio of the probability of drawing a King and a Queen with the probability of drawing two Kings:

$$\frac{0.012066}{0.004525} \approx 2.67$$

we find that it is more than twice as likely to be dealt a King and a Queen than it is to be dealt a pair of Kings (in a 2 card hand).

CHECK YOUR UNDERSTANDING 1.13

Two marbles are drawn, without replacement, from a box containing 5 red, 4 blue, and 7 white marbles. Determine the probability of ending up with

(a) Two red marbles. (b) A red and a blue marble.

Answers: (a) $\frac{1}{12}$ (b) $\frac{1}{6}$

INDEPENDENT EVENTS

Roll a die 7 times. Clearly, what happens on, say, the first four rolls of the die will not influence what happens on the next roll. Each roll is independent of the others, in that the outcome of any one of the rolls has no effect on the outcome of any of the others.

In general, to say that two activities are independent, is to say that the occurrence of either of the activities is not dependent on the occurrence of the other. To put it more precisely:

The probability of E given F, is simply the probability of E, since E "doesn't care" about F; and ditto for the probability of F given E.

DEFINITION 1.5 INDEPENDENT EVENTS Two events E and F are **independent** if:

$$Pr(E|F) = Pr(E) \text{ and } Pr(F|E) = Pr(F)$$

We recall Theorem 1.5 which gave the probability of two events occurring:

$$Pr(E \cap F) = Pr(E)P(F|E)$$

When E and F are independent events, $Pr(F|E) = Pr(F)$, and the theorem takes the following simpler form:

PROBABILITY OF TWO INDEPENDENT EVENTS OCCURRING

THEOREM 1.6 For **independent** events E and F:

$$Pr(E \cap F) = Pr(E) \cdot Pr(F)$$

IN WORDS: The probability of two (or more) **independent** events occurring is simply the product of their probabilities.

EXAMPLE 1.19
DIE AND COIN

Roll a die and flip a coin. What is the probability that you roll a 3 and flip a Head?

SOLUTION: One approach is to consider a sample space for the experiment of rolling a die and flipping a coin:

success ↓

Sample Space: 1H 2H **3H** 4H 5H 6H
1T 2T 3T 4T 5T 6T

Conclusion: $Pr(3 \text{ and H}) = \frac{1}{12}$ ← successes / ← possibilities

We are still considering sample spaces, but now they are smaller: 6 possible outcomes for the roll of a die, and 2 possible outcomes for the flip of a coin.

Another approach is to think in terms of two independent events, and apply Theorem 1.6:

$$Pr(3 \text{ and H}) = Pr(3) \cdot Pr(\text{H}) = \frac{1}{6} \cdot \frac{1}{2} = \frac{1}{12}$$

EXAMPLE 1.20
SECURITY CHECK

The probability that a metal object will not be detected at an airport scanning station is 0.01. To improve security, each passenger must pass through three such stations. What is the probability that a metal object will not be detected at any of the three stations?

SOLUTION: All three (independent) stations must "fail" (each with probability 0.01). Applying Theorem 1.6, we conclude that:

$$Pr(\text{All fail}) = (\text{1st fails})(\text{2nd fails})(\text{3rd fails}) = (0.01)^3 = 0.000001$$

We see that the probability of a security breach drops from 1 out of 100 when only one machine is used, to 1 out of a million when three scanning tests are made.

CHECK YOUR UNDERSTANDING 1.14

Roll a die five times. What is the probability that you roll a 4 each time?

Answer: $\left(\frac{1}{6}\right)^5$

EXAMPLE 1.21
HITTING STREAK

Determine the probability that a baseball player with a batting average of .333 (hits safely once out of every three times at bat) will get at least one hit in each of his next 56 consecutive games. Assume that the player comes to bat four times per game.

In 1941, Joe Di Maggio of the New York Yankees did hit safely in 56 consecutive games.

Solution: Before we can calculate the probability that the player will hit safely in 56 consecutive games, we first have to determine the probability that he will hit safely during any given game. Using the complement theorem (Theorem 1.1, page 15) we have:

$$Pr(\text{hits safely in } \mathbf{1} \text{ game}) = 1 - \underbrace{Pr(\text{No hit in } \mathbf{4})}_{*}$$

To find (*), we use the complement theorem again, and first determine the probability that the player will **not** get a hit when he comes to the plate:

$$Pr(\text{does Not get a hit in } \mathbf{1} \text{ time at bat}) = 1 - Pr(\text{hit}) = 1 - \frac{1}{3} = \frac{2}{3}$$

Assuming that whether or not he gets a hit in one time at bat will not influence his next time at bat (independent events), we employ Theorem 1.6, and find the value of (*):

$$\underbrace{Pr(\text{No hit in } \mathbf{4})}_{*} = Pr(\text{Not}) \cdot Pr(\text{Not}) \cdot Pr(\text{Not}) \cdot Pr(\text{Not}) = \left(\frac{2}{3}\right)^{4}$$

Thus:

$$Pr(\text{hits safely in } \mathbf{1} \text{ game}) = 1 - \left(\frac{2}{3}\right)^{4} \approx 0.8025$$

Alright then, there is a good chance that the player will get a hit in any given game; but to do it 56 games in a row, that's pretty tough. Indeed, assuming independence once more, we find that:

$$Pr(\text{hits safely in } \mathbf{56} \text{ games}) = \overbrace{Pr(\text{safe in } \mathbf{1}) \cdot Pr(\text{safe in } \mathbf{1}) \cdots Pr(\text{safe in } \mathbf{1})}^{\text{56 times}}$$

$$= Pr(\text{safe in } \mathbf{1})^{56} = \left[1 - \left(\frac{2}{3}\right)^{4}\right]^{56} \approx 0.00000445$$

CHECK YOUR UNDERSTANDING 1.15

At a carnival game you get two shots at a basket. If you play the game five times and if you get at least one basket each of the five times, then you win a large cuddly bear. Assuming that your probability of making a basket is 0.4, determine the probability that you will win the large cuddly bear.

Answer: $[1 - (0.6)^2]^5 \approx 0.11$

EXERCISES

1. **(Card)** Draw a card from a standard deck. Determine the probability that:
   (a) It is a spade. (b) It is a spade, given that it is black.
   (c) It is a spade, given that it is not a diamond.
2. **(Die)** Roll a die. Determine the probability that:
   (a) You roll a 5. (b) You roll a 5, given that the number rolled is odd.
   (c) You roll a 5, given that the number rolled is even.
3. **(Urn)** A marble is drawn at random from an urn containing 4 red marbles, five white marbles, and two blue marbles. Determine the probability that the marble drawn:
   (a) Is red. (b) Is red, given that the marble is not white.
4. **(Urn)** A marble is drawn at random from an urn containing 4 red marbles numbered 1 through 4; five white marbles numbered 1 through 5; and two blue marbles numbered 5 and 10. Determine the probability that the marble drawn:
   (a) Is red. (b) Is red, given that the marble is not white.
   (c) Is red, given that the marble displays the number 4.
   (d) Displays the number 4, given that the marble is red.
   (e) Is not red, given that the marble does not display the number 4.
5. **(Dice)** Roll a pair of dice. Determine the probability that the sum is:
   (a) Odd. (b) Odd, given that at least one die is odd.
   (c) Odd, given that exactly one die is odd.
6. **(Dice)** Roll a pair of dice. Determine the probability that the sum is:
   (a) Divisible by 3. (b) Divisible by 3, given that it is divisible by 4.
   (c) Divisible by 5, given that it is divisible by 3.
7. **(Tax Audit)** Agent Eager will audit one of 500 tax forms. Fifty of those forms list a charitable deduction of up to \$500; 250 list a deduction of more than \$500 but not more than \$1000; and the rest, including that of George Generous, list a deduction of more than \$1000. What is the probability that Mr. Generous will be audited if agent Eager chooses the form:
   (a) At random?
   (b) From the forms that list charitable contributions in excess of \$500?
   (c) From the forms that list charitable contributions in excess of \$1000?
8. **(Grade Distribution)** The following table summarizes the grade distribution of students enrolled in Math 101 at State University for the last five years. The survey involved Matriculated Male student Commuters (MMC), Matriculated Female student Commuters (MFC), Matriculated Male students Living on campus (MML), Matriculated Female students Living on campus (MFL), Non-matriculated Male students (NM), and Non-matriculated Female students (NF). Assume all Non-matriculated students commute.

	A	B	C	D	F
MMC	22	32	221	19	22
MFC	19	30	197	14	10
MML	43	41	335	21	25
MFL	29	32	287	35	19
NM	25	19	115	8	3
NF	32	24	124	4	2

Determine the probability that a student chosen at random from Math 101:

(a) Is matriculated, given that the student is a male.

(b) Is a male, given that the student is matriculated.

(c) Is matriculated, given that the student is a male commuter.

(d) Received an A, given that the student is a female.

(e) Is a female, given that the student received an A.

(f) Did not receive a D or an F, given that the student is male or is a non-matriculated student.

9. (**Political Distribution**) The following table summarizes the political distribution, in four districts.

	District A	District B	District C	District D
MALE				
Democrat	2500	1863	1952	2430
Republican	1400	2400	2100	1300
Independent	750	920	1011	951
FEMALE				
Democrat	2620	2001	1973	2456
Republican	1100	2122	2351	1342
Independent	624	830	1121	1012

Determine the probability that an individual chosen at random from one of the four districts:

(a) Is a Democrat, given that he or she is registered in District A.

(b) Is registered in District A, given that the individual is a Democrat.

(c) Is a male, given that the individual is a Democrat.

(d) Is not a Democrat, given that the individual is not in District C.

(e) Is a female, given that the individual is not an Independent and is in either District A or C.

(f) Is not an Independent, given that the individual is not a male in District A and is not a female in District C.

10. **(Urn)** Two marbles are drawn, without replacement, from an urn containing 4 red marbles, five white marbles, and two blue marbles. Determine the probability that:
    (a) Both are red.
    (b) The first marble drawn is red, and the second is blue.
    (c) The first marble drawn is blue, and the second is red.
    (d) One of the marbles is red and the other is blue.

11. **(Urn)** Three marbles are drawn, without replacement, from an urn containing 4 red marbles, five white marbles, and two blue marbles. Determine the probability that:
    (a) All are red.
    (b) None is red.
    (c) The first marble drawn is red, and the other two are not.
    (d) Exactly one of the marbles drawn is red.

12. **(Two Cards)** You draw two cards from a standard deck without replacement. Determine the probability that:
    (a) Both are Diamonds.
    (b) Neither is a Diamond.
    (c) Both are of the same suit.
    (d) Both are of the same suit, given that the first card drawn is a Club.
    (e) They are not of the same suit.
    (f) They are not of the same suit, given that the first card drawn is a Club.

Exercises 13-26. (Three Cards) You draw three cards from a standard deck without replacement. Determine the probability that:

13. All are Diamonds.
14. None is a Diamond.
15. All are of the same suit.
16. The first card is an Ace, the second is a King, and the third is a Queen.
17. The first two cards are Aces, and the third is not.
18. The first two cards are not Aces, and the third is.
19. All are Clubs, given that all are face cards.
20. All are face cards, given that all are Clubs.
21. All are face cards, given that all are of the same suit.
22. All are face cards, given that none is a Club.
23. All are face cards, given that none is a Jack.
24. None is a face card, given that none is a Jack.
25. You draw an Ace, a King, and a Queen, in that order, given that no Jack is drawn.
26. You draw three Aces, given that no Jack is drawn.

27. **(Flags)** There are six different colored signal flags which can be hoisted onto a mast, including a red and a green flag. Two of the six flags are randomly selected and hoisted. Determine the probability that:
    (a) The red flag is hoisted first.
    (b) The red flag is hoisted first, given that the green flag is not hoisted.
    (c) The red flag is hoisted last, given that the green flag is not hoisted.
    (d) The red flag is hoisted last, given that the green flag is hoisted first.

28. **(Die, Coin, and Card)** You roll a die, flip a coin, and draw a card. What is the probability that you roll a 5, flip a Head, and draw an Ace.

29. **(Cards)** Five people draw a card from five different decks. What is the probability that all draw:
    (a) An Ace? (b) An Ace or a King?

30. **(Die)** You roll a die 5 times. What is the probability that:
    (a) The number 1 is rolled each time? (b) An odd number is rolled each time?

31. **(Picking a Number)** Five individuals pick a number from 1 to 10, inclusive. What is the probability that:
    (a) All choose the number 7? (b) All choose the same number?

32. **(Marksman)** Based on past performance, we know that Dead Eye Dick will hit the bull's-eye with probability 0.94. He fires 15 shots. Determine, to two decimal places, the probability that:
    (a) He will hit the bull's-eye each time?
    (b) He will hit the bull's-eye each time, given that he has hit the bull's-eye on his first 12 shots?
    (c) He will hit the bull's-eye exactly 14 times, given that he has hit the bull's-eye with 11 of his first 12 shots?

33. **(Baseball)** Of the 355 batters he faced, pitcher Every Whichway retired all but 75. Determine, to four decimal places, the probability that Every will:
    (a) Retire the next 2 batters he faces.
    (b) Not retire either of the next two players he faces.
    (c) Rethink the above answers, in light of the fact that Every's retirement average will change slightly after he faces the next batter.

34. **(Bowling)** A perfect game in bowling is throwing 12 consecutive strikes. Assume that the probability that Penelope Pinsdown will throw a strike is 0.92. Determine, to two decimal places, the probability that:
    (a) Her next game will be a perfect game.
    (b) Her next game will not be a perfect game.

(c) Her current game will be a perfect game, given that she began the game with 5 consecutive strikes.

(d) Her current game will not be a perfect game, given that she began the game with 10 consecutive strikes.

35. **(Roll a Die)** Roll a die until you roll a 5, at which time you stop. Determine the probability that you roll the die:

(a) Exactly three times. (b) At most three times. (c) At least three times.

36. **(Draw a Card)** You draw a card from a deck. If it is an Ace, you stop. If not, you then replace the card, shuffle the deck, and try again. What is the probability that the game stops on the:

(a) First draw. (b) Second draw. (c) Third draw. (d) Fourth draw.

(e) Can the game go on "forever?" **If not**, what is the maximum number of draws before the game will end?

37. **(Draw a Card)** Repeat Exercise 36, but now without replacing any card.

38. **(Draw a Card)** You draw a card from a standard deck. If it is a face card, then you stop. If not, then you replace the card, shuffle the deck, and draw a second card. If the second card is a face card, then you stop. If not, then you proceed as before and draw a third card, and continue the process until a face card is drawn. Determine the probability that you draw:

(a) Exactly three cards. (b) At most three cards. (c) At least three cards.

39. **(Draw a Card)** Repeat Exercise 38, but now without replacing any card.

40. **(Keys and Doors)** To get into her office, Catherina has to unlock the door to her building, and then the door to her office. She has 7 keys in her pocket, all of which look alike. One of the seven keys is for the building, and another is for her office. Determine, to three decimal places, the probability that she will enter her office:

(a) With the second key (the first key chosen opens the building, and the second key, chosen from the remaining 6, opens her office).

(b) With the third key.

41. **(Tournament)** In a single-elimination handball tournament, 16 players are arbitrarily assigned a number from 1 through 16. The winner of the tournament will be awarded the prestigious golden gidget, and the second-place winner will receive the silver gidget. In the first round, player 1 plays player 2, player 3 plays 4, and so on. In the second round the winner of 1-versus-2 plays the winner of 3-versus-4, and so on. Assuming that in each game the better player will win, determine the probability that:

(a) The golden gidget will be awarded to the best player in the group.

(b) The silver gidget will be awarded to the second best player in the group.

42. **(Defective Components)** Three electrical components are to be linked in series, as is depicted. If any of the three components fail, the unit will malfunction. Assume that component A is chosen from a collection of 10, and that (exactly) one of the 10 is faulty; that component B is chosen from 15, and that two of them are faulty; and that 2 of 9 of component C are faulty. What is the probability that the unit will:

    (a) Not malfunction? (b) Malfunction?

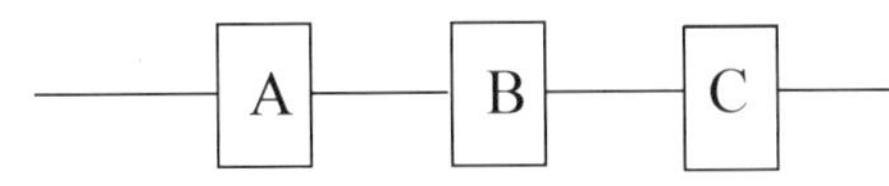

43. **(Defective Components)** Three electrical components are to be linked as is depicted. If A fails, or if both B and C fail, the unit will malfunction. Assume that component A is chosen from a collection of 10, and that (exactly) one of the 10 is faulty; that component B is chosen from 15, and that two of them are faulty; and that 2 of 9 component C are faulty. What is the probability that the unit will:

    (a) Not malfunction? (b) Malfunction?

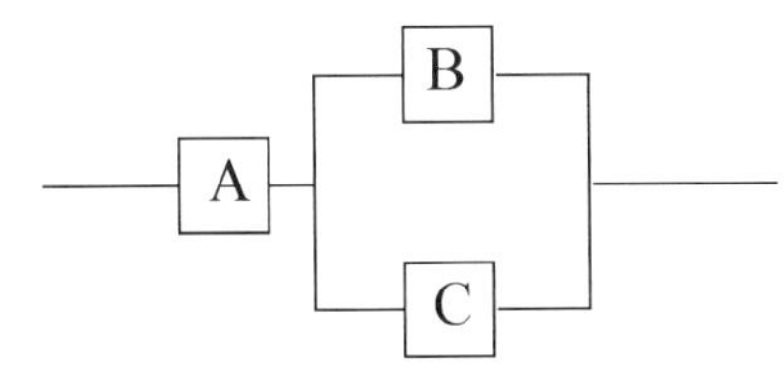

44. **(Playing without a Full Deck)** You draw a card from a deck that has an unknown missing card. By considering two cases [(1) the missing card is a King, and, (2) the missing card is not a King] prove that the probability of drawing a King remains at $\frac{4}{52}$. Suppose 13 cards are missing? Suppose 51 cards are missing?

§4. Trees and Bayes' Formula

In CYU 1.13, page 33, you encountered the following problem:

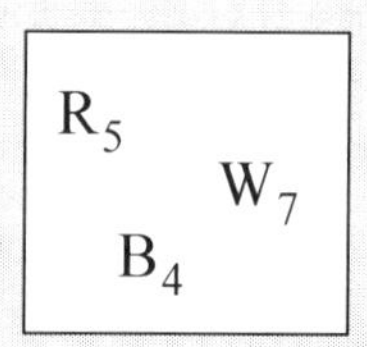

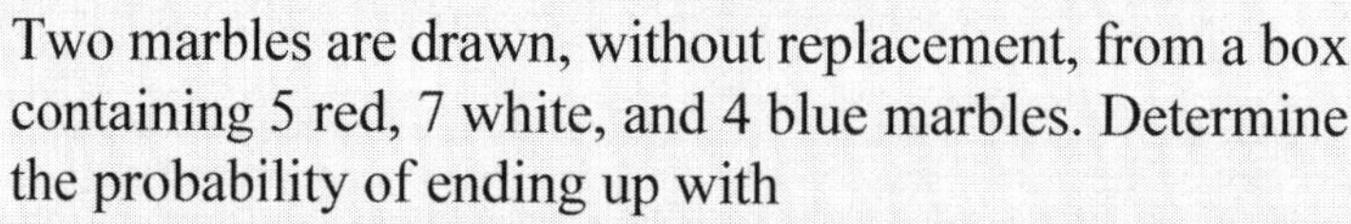
Two marbles are drawn, without replacement, from a box containing 5 red, 7 white, and 4 blue marbles. Determine the probability of ending up with
(a) Two red marbles. (b) A red and a blue marble.

The following visual representation, called a **tree diagram**, nicely displays the activity of drawing two marbles from that urn:

Note that the probabilities at each node of the tree must sum to 1. For example:
At the root:
$$\frac{5}{16}+\frac{7}{16}+\frac{4}{16}=1$$
At the "Blue node":
$$\frac{5}{15}+\frac{7}{15}+\frac{3}{15}=1$$

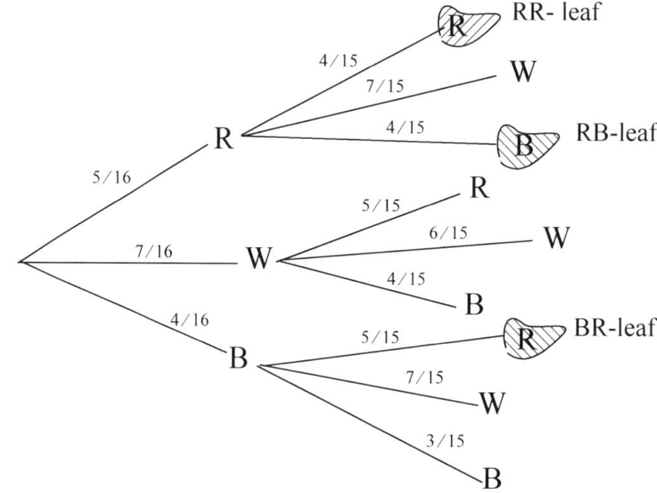

As you can see, three branches stem from the root of the above tree, and each is augmented with the probability of taking that (color) branch (the probability of initially drawing a red marble, for example, is $5/16$). From each initial branch sprout three other branches which also display the probability of taking that branch.

There is but one **RR**-leaf (the top leaf), and using Theorem 1.5 (page 32) we find that the probability of getting to that leaf is:

$$Pr(\text{R and R})=\frac{5}{16}\cdot\frac{4}{15}$$

Note that there is a RB- and a BR-leaf in the tree, and between them they represent the event of ending up with a red and blue marble (in either order). Using Theorem 1.5 together with Theorem 1.3 of page 17, we have:

$$Pr(\text{R and B})=\underbrace{\frac{5}{16}\cdot\frac{4}{15}}_{\text{RB-leaf}}+\underbrace{\frac{4}{16}\cdot\frac{5}{15}}_{\text{BR-leaf}}$$

EXAMPLE 1.22

TAX AUDIT

A survey showed that 16% of tax returns with adjusted gross income of \$250,000 or more were subjected to an audit, as opposed to 5% of the tax returns with adjusted gross income of less than \$250,000. What is the probability that a return chosen at random will be audited, if 4% of returns are in the \$250,000 or higher bracket?

SOLUTION:

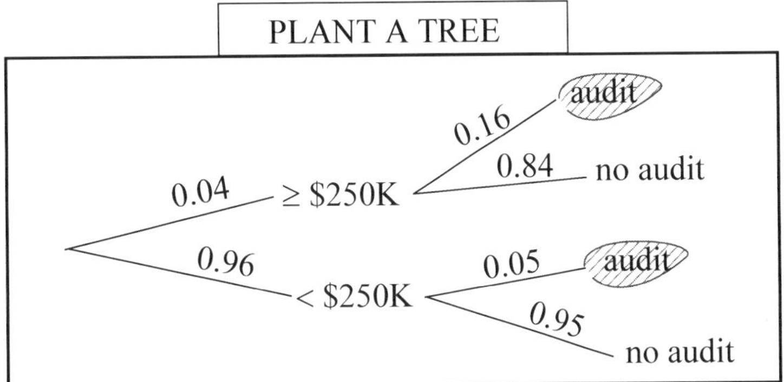

There are two audit-leaves. Using Theorem 1.5, page 32, we find that the probability of ending up at the top audit leaf is $(0.04)(0.16)$, and that the probability of ending up at the lower audit-leaf is $(0.96)(0.05)$. Using Theorem 1.3, page 17, we can easily determine the probability of ending up at one or the other of the audit-leaves:

$$Pr(\text{audited}) = (0.04)(0.16) + (0.96)(0.05) = 0.0544$$

CHECK YOUR UNDERSTANDING 1.16

Fifty-five percent of the student body at Ramapo College are females, and 21% of them are in the Teacher-Education program. Seventeen percent of the male students are in that program. What is the probability that a randomly chosen student is not in the Teacher-Education program?

Answer:
$(0.55)(0.79) + (0.45)(0.83) = 0.81$

Bayes' Formula

Theorem 1.4, page 32, expresses a conditional probability as a quotient of probabilities; specifically, for given events E and F:

$$Pr(E|F) = \frac{Pr(E \cap F)}{Pr(F)}$$

The above formula is used in the next example where we appear to be "predicting the past:"

EXAMPLE 1.23

Two urns

Urn U_1 contains 2 white marbles and 3 black marbles, while urn U_2 contains 4 white marbles and 2 black marbles. A card is drawn from a standard deck. If it is a face card, then a marble is drawn from U_1, otherwise a marble is drawn from U_2. If the marble drawn is white, what is the probability it came from U_1?

Solution:

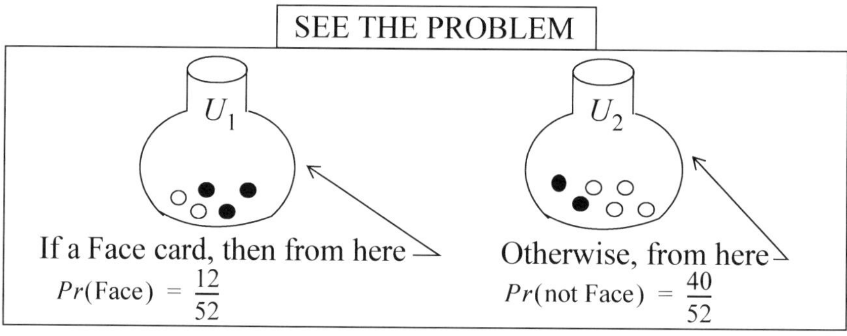

We begin by constructing a probability tree:

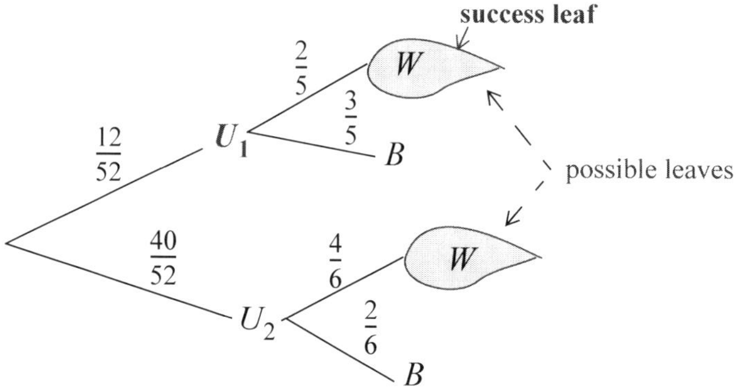

We now determine the probability that the marble drawn came from U_1, **given** that the marble drawn is white:

$$Pr(U_1|W) = \frac{Pr(U_1 \cap W)}{Pr(W)} = \frac{\frac{12}{52}\cdot\frac{2}{5}}{\frac{12}{52}\cdot\frac{2}{5}+\frac{40}{52}\cdot\frac{4}{6}} \approx 0.15$$

(Annotations: "Theorem 1.4" under the first equals sign; "success leaf" over the numerator; "possible leaves" under the denominator.)

IN WORDS: Divide the probability of ending up with a white marble via U_1 (the "success leaf"), by the sum of the probabilities of ending up with a white marble by either branch (the "possible leaves").

To make sure you understand the above procedure, let's throw another urn into the mix:

EXAMPLE 1.24

THREE URNS

There are three urns: U_1, U_2, and U_3. U_1 contains 2 white marbles and 3 black marbles, U_2 contains 4 white marbles and 2 black marbles, and U_3 contains 1 white marble and 2 black marbles. A card is drawn from a standard deck. If it is an ace, then a marble is drawn from U_1, if it is a face card, then a marble is drawn from urn U_2, otherwise, a marble is drawn from U_3. If the marble drawn is white, what is the probability that it came from U_2?

SOLUTION:

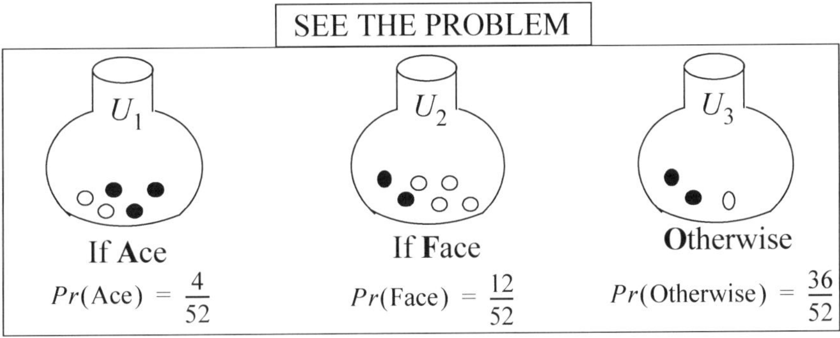

We again construct a suitable probability tree:

Note, once more, that the sum of the probabilities at every node in the tree equals 1.

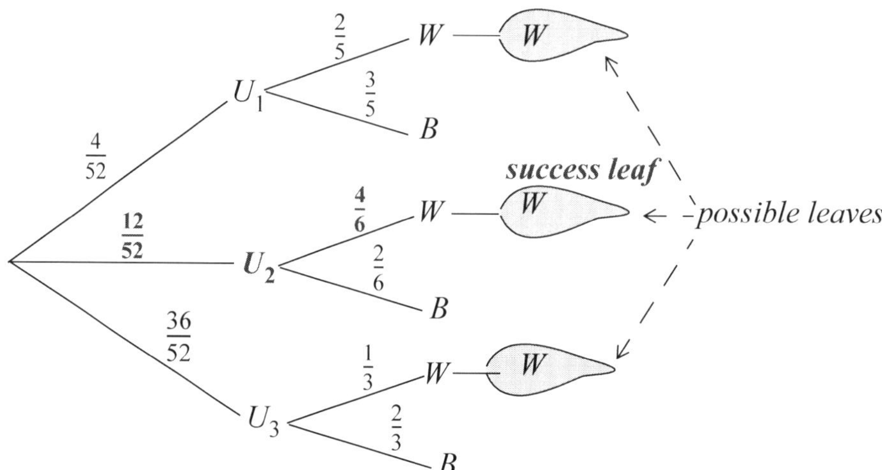

Employing Theorem 1.4 and Theorem 1.5 of page 32, we have:

$$Pr(\boldsymbol{U_2}|W) = \frac{Pr(U_2 \cap W)}{Pr(W)}$$

$$= \frac{Pr(U_2)Pr(W|U_2)}{Pr(U_1)Pr(W|U_1) + Pr(U_2)Pr(W|U_2) + Pr(U_3)Pr(W|U_3)}$$

$$= \frac{\frac{12}{52}\cdot\frac{4}{6}}{\frac{4}{52}\cdot\frac{2}{5}+\frac{12}{52}\cdot\frac{4}{6}+\frac{36}{52}\cdot\frac{1}{3}} \approx 0.37$$

← "success leaf" (numerator)

← "possible leaves" (denominator)

CHECK YOUR UNDERSTANDING 1.17

Referring to the situation of Example 1.24, determine, to two decimal places, the probability that U_3 was chosen, given that the marble drawn is black.

Answer: 0.79

Were you able to follow our solution to Example 1.24, and to solve the problem in the above CYU box? If so, then you can take comfort in knowing that you already know how to use the following somewhat intimidating looking Theorem:

Thomas Bayes, English theologician and mathematician, (1702-1761)

THEOREM 1.7
BAYES' FORMULA

Let $U_1, U_2, \ldots, U_n$ be n mutually exclusive events, each with positive probability, whose union is the sample space S. Let E be an event in S such that $Pr(E) \neq 0$. Then for any $1 \leq i \leq n$:

$$Pr(U_i|E) = \frac{Pr(U_i \cap E)}{Pr(U_1 \cap E) + Pr(U_2 \cap E) + \ldots + Pr(U_n \cap E)}$$

You need not worry too much about the above intimidating formula. To apply the theorem, just construct an appropriate probability tree, and then:

> Divide the probability of ending up at the success leaf (or leaves), by the sum of the probabilities of all leafs satisfying the given condition.

Consider the following examples.

EXAMPLE 1.25
DIAGNOSTIC TEST

A diagnostic test for disease A is found to be 94% accurate when administered to a patient afflicted with that disease, and 97% accurate when administered to a patient that does not have the disease. What is the probability that an individual testing positive does not have the disease (a false-positive result), given that 2% of the population is afflicted with that disease?

Do not confuse "accurate" with "positive." If the patient does not have disease A, then since in this situation the test is 97% accurate, there is a 3% probability that the test will come back positive. Think about it.

SOLUTION: Let $\boldsymbol{A}$ denote the event that the patient has disease A, and $\boldsymbol{NA}$ denote the event that the patient is not afflicted. Let $\boldsymbol{P}$ denote the event that the test comes back positive, and **N** denote the event that the test comes back negative. We then have:

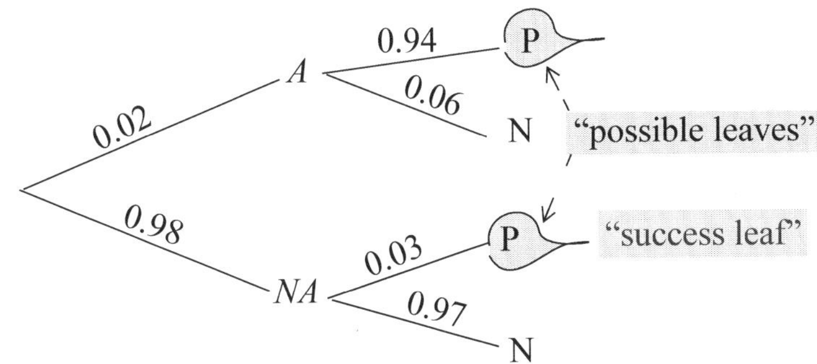

Dividing the probability of getting to a P-leaf via NA, by the probability of getting to either of the P-leaves, we find that:

$$Pr(A|\boldsymbol{P}) = \frac{(\mathbf{0.98})(\mathbf{0.03})}{(0.02)(0.94)+(0.98)(0.03)} \approx 0.61$$

We see that roughly 60% of the time a positive result will actually be false-positive. The large percentage is due, in part, to the fact that only a small number of the population is infected.

EXAMPLE 1.26

DEFECTIVE CAMERAS

A company produces digital cameras at three plants. Records indicated that 3% of the cameras produced at plant A are defective, that 2% of those produced at plant B are defective, and that 4% of those produced at plant C are defective. On a given week, plant A produced 300 cameras, plant B produced 215 cameras, and plant C produced 250 cameras. A camera is selected at random from the total production. Determine the probability that:

(a) The camera is defective.

(b) The camera was produced at plant C, **given** that the camera is defective.

SOLUTION: Let $\boldsymbol{A}$, $\boldsymbol{B}$, and $\boldsymbol{C}$ denote the events that the camera comes from plants A, B, and C, respectively. Let D denote the event that a camera is defective, and N denote the event that a camera is not defective. Noting that a total of $300 + 215 + 250 = 765$ were manufactured, we have:

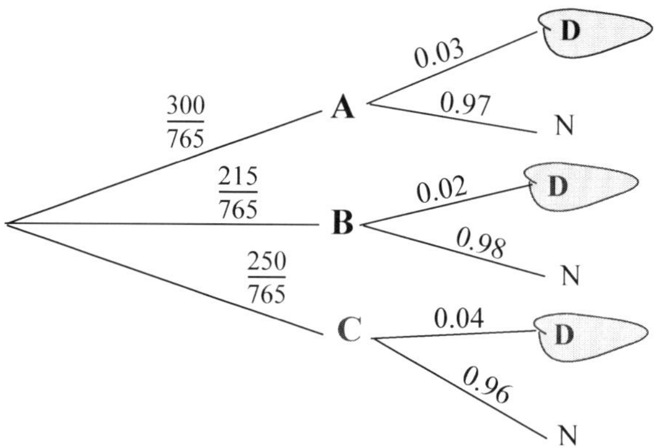

(a) Any leaf is a possible outcome of the experiment. To find the probability that the chosen camera is defective, we simply have to sum the probabilities of ending up at a **D**-leaf:

$$Pr(\boldsymbol{D}) = (\tfrac{300}{765})(0.03) + (\tfrac{215}{765})(0.02) + (\tfrac{250}{765})(0.04) \approx 0.03$$

(b) (Bayes' formula) The given condition restricts the possible outcomes of the experiment to the D-leaves, with the bottom D-leaf being the "success-leaf." Thus:

"success leaf"

$$Pr(\boldsymbol{C}|\boldsymbol{D}) = \frac{(\tfrac{\mathbf{250}}{\mathbf{765}})(\mathbf{0.04}) \leftarrow - - \lrcorner}{(\tfrac{300}{765})(0.03) + (\tfrac{215}{765})(0.02) + (\tfrac{250}{765})(0.04) \leftarrow - \urcorner} \approx 0.43$$

sum of "possible leaves"

CHECK YOUR UNDERSTANDING 1.18

With reference to the situation of Example 1.26, determine, to two decimal places, the probability that the unit was produced at plant C, **given** that the unit is not defective.

Answer: 0.32

In the following example, the numbers are nice enough to enable us to solve the problem in two ways.

EXAMPLE 1.27
BUSINESS MAJORS

60% of a student body are males. 25% of the males are business majors and 20% of the females are business majors. A student is chosen at random. What is the probability that:

(a) The student is a business major.

(b) The student is a female, given that the student is a business major.

SOLUTION: (without words):

	Business	Non-Bus.
Male (60)	15 25% of males	45 75% of males
Female (40)	8 20% of females	32 80% of females

(a) $Pr(\text{Bus.}) = \dfrac{15+8}{100} = 0.23$

(b) $Pr(\text{F}|\text{Bus.}) = \dfrac{8}{23} \approx 0.35$

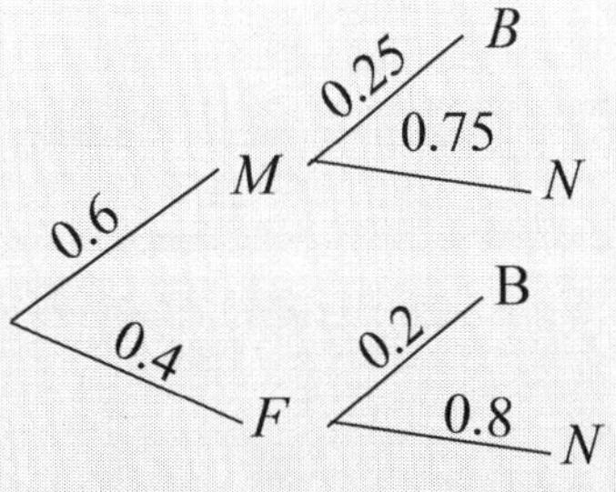

(a) $Pr(B) = (0.6)(0.25)+(0.4)(0.2) = 0.23$

(b) $Pr(F|B) = \dfrac{(0.4)(0.2)}{(0.6)(0.25)+(0.4)(0.2)} \approx 0.35$

Exercises

1. **(Urn)** An urn contains 5 Red, 4 White, and 6 Blue marbles. Two marbles are drawn at random, without replacement. Use both Theorem 1.5, page 32, and the probability tree process of this section to determine, to three decimal places, the probability that:
   (a) Neither marble is Red. (b) Both marbles are Red.
   (c) A Blue marble is drawn and then a White marble is drawn.
   (d) A Blue and a White marble are drawn, in either order.

2. **(Quality Control)** A store receives a shipment of 100 cell phones, five of which are defective. Two of the phones are tested. Determine, to three decimal places, the probability that:
   (a) The first phone tested is defective. (b) Both phones are defective.
   (c) Exactly one of the two phones is defective.
   (d) At least one of the two phones is defective.

3. **(Quality Control)** A store receives a shipment of 100 cell phones, five of which are defective. Three of the units are tested. Determine, the probability that:
   (a) All three phones are defective?
   (b) The first two phones tested are defective, and the third is not?
   (c) The first two phones tested are not defective but the third is?
   (d) None of the three phones are defective?
   (e) Exactly one of the three phones is defective.

4. **(Flipping a Coin)** What is the probability that no consecutive heads occur on three tosses of a coin?

5. **(Flipping a Coin)** What is the probability that two or more consecutive heads occur on three tosses of a coin?

6. **(Urn)** An urn contains 5 Red, 4 White, and 6 Blue marbles. Two marbles are drawn at random, without replacement. Determine the probability that:
   (a) A Red marble is drawn, and then a White marble.
   (b) A White marble is drawn, and then a Red marble.
   (c) A Red and a White marble are drawn (in either order).
   (d) The second marble drawn is white, given that the first marble drawn is white.
   (e) The first marble drawn is white, given that the second marble drawn is white.

7. **(Urn)** Repeat exercise 6, assuming that the two marbles are drawn with replacement.

8. **(Urns)** An urn contains 5 red marbles, 4 white marbles, and 3 blue marble. A marble is drawn at random. Whatever marble is drawn, is returned to the urn, along with two additional marbles of the same color. Another marble is then drawn from the resulting urn. What is the probability that the second marble drawn is blue?

9. **(Flipping a Coin)** You are holding two genuine quarters, and a fake quarter which has heads on both sides. You flip one of the coins at random. Determine the probability that:
   (a) You flip a Head. (b) You flipped the fake coin, given that you flipped a Tail.
   (c) You flipped the fake coin, given that you flipped a Head.

10. **(Quality Control)** A store receives a shipment of 100 cell phones, five of which are defective. Three of the phones are tested. Determine, to three decimal places, the probability that:
    (a) The third phone tested is defective, given that neither of the first two is defective.
    (b) The first phone tested is defective, if the second and third phones are defective.
    (c) The first phone tested is not defective, if the second and third are defective.
    (d) The first phone tested is defective, if only one of the next two phones is defective.

11. **(Urns)** Urn 1 contains 5 Red, 4 White, and 6 Blue marbles. Urn 2 contains 12 Red, 2 White, and 3 Blue marbles. One of the urns is chosen at random, and then one marble is drawn at random from that urn. Determine the probability that:
    (a) A Red marble is drawn. (b) A Red marble is drawn, given that Urn 2 was chosen.
    (c) Urn 2 was chosen, given that a Red marble was drawn?
    (d) Urn 1 was chosen, if a Red marble was not drawn?

12. **(Urns)** Urn 1 contains 5 Red, 4 White, and 6 Blue marbles; Urn 2 contains 12 Red, 2 White, and 3 Blue marbles; and Urn 3 contains 20 Red, 1 White, and 1 Blue marbles. One of the three urns is chosen at random, and then one marble is drawn at random from that urn.
    (a) Determine the probability that a Red marble is drawn.
    (b) Determine the probability that a Red marble is drawn, given that Urn 2 was chosen.
    (c) What is the probability that Urn 2 was chosen, given that a Red marble was drawn?
    (d) What is the probability that Urn 1 was chosen, given that a Red marble was not drawn?
    (e) What is the probability that Urn 3 was not chosen, given that a Red marble was not chosen?

13. **(Urns)** Urn 1 contains 5 Red, 4 White, and 6 Blue marbles. Urn 2 contains 12 Red, 2 White, and 3 Blue marbles. One of the two urns is chosen at random, and then two marbles are drawn without replacement from that urn.
    (a) Determine the probability that two Red marble are drawn, given that Urn 1 was chosen.
    (b) Determine the probability that two Red marbles are drawn.
    (c) What is the probability that Urn 1 was chosen, given that two Red marbles are drawn?
    (d) What is the probability that Urn 2 was chosen, given two Red marble were not drawn?

14. **(Voters)** The registered voters in a certain district break down as follows: 45% Democrats, 42% Republicans, and 13% Independents. In a certain election, 22% of the Democrats, 25% of the Republicans, and 83% of the Independents voted in favor of a proposition. Determine the probability that a randomly chosen voter:
    (a) Is an Independent. (b) Voted in favor of the proposition.
    (c) Is an Independent, if the individual voted in favor of the proposition.
    (d) Not a Democrat, given that the individual did not vote in favor of the proposition.

15. **(Students)** In a class of 100 students, 60 are females. Of the females, 10 are physics majors, 20 are business majors, 15 are art majors, and the rest are undecided. Of the males, 12 are physics majors, 7 are communication majors, 4 are business majors, 5 are education majors and the rest are undecided. A student is chosen at random from the class. Determine the probability that the individual is:

    (a) A business major. (b) A female business major. (c) An education major.

    (d) A female, given that the student is a business major.

    (e) A business major, given that the student is a female.

    (f) A female, given that the student is an art major.

16. **(DVD Drives)** A computer company obtains its DVD drives from three different companies: 25% from company A, 35% from company B, and 40% from company C. Experience has shown that 1.9% of company A's, 1.5% of company B's, and 0.07% of company C's drives are defective. A computer is shipped to a customer. Determine the probability that:

    (a) The unit has a faulty drive.

    (b) The unit has a faulty drive, given that the drive was supplied by company A.

    (c) The drive came from company A, given that the drive turns out to be faulty.

    (d) The unit has a faulty drive, given that the drive was not supplied by company A.

    (e) The drive did not come from company A, given that the drive turns out to be faulty.

    (f) The drive did not come from company A, given that the drive did not turn out to be faulty.

17. **(Pregnancy Test)** A home-pregnancy test is 95% accurate when the woman is pregnant, and 85% accurate when the woman is not pregnant. Assume that 90% of the women taking the test are pregnant. A woman takes the test. Determine the probability that:

    (a) The test is accurate. (b) The test comes back positive.

    (c) The woman is pregnant, if the test is positive.

    (d) The woman is not pregnant, if the test is positive (false positive).

18. **(Marbles)** Billy has 100 marbles, all of which are blue. Mary has 50 marbles: 10 blue, 20 white, and 20 red. Johnny has 75 marbles: 50 blue, 20 white, and 5 red. Sam has only 5 marbles, and all are red. A child is chosen at random, and that child gives you one of his or her marbles. Determine the probability that:

    (a) Billy was chosen. (b) You receive a white marble.

    (c) You receive a white marble, given that Mary was chosen.

    (d) You receive a white marble, given that Mary was not the one chosen.

    (e) Mary was the one chosen, given that you receive a white marble.

    (f) Mary was the one chosen, given that you did not receive a white marble.

    (g) Mary or Johnny was chosen, given that you receive a white marble.

    (h) Mary or Johnny was chosen, given that you receive a white or a blue marble.

19. **(Lactose Intolerance)** Lactose intolerance is the inability to digest significant amounts of lactose, the predominant sugar of milk. Between 30 and 50 million Americans are lactose intolerant, with some ethnic and racial populations being more widely affected than others. As many as 75% of all African-Americans and Native-Americans are lactose intolerant; 90% of Asian-Americans; and 50% of Hispanic-Americans are lactose intolerant. The condition is least common among persons of northern European descent, with only 20% of that population being lactose intolerant. Assume that a person is chosen at random from a group composed of 20 African-Americans, 10 Asian-Americans, 12 Hispanic-Americans, 3 Native-Americans, and 35 north-European-Americans. Determine the probability that the individual:
   (a) Is a Native-American. (b) Is lactose intolerant
   (c) Is a Hispanic-American, if the person is lactose intolerant.
   (d) Is a Native-American, given that the individual is lactose intolerant and is neither an Asian-American nor a north-European-American.

20. **(Two Cards)** A card is drawn from a standard deck, and then another card is drawn. If the first card is a face card, it is not replaced before drawing the second. If the first card is not a face card, it is replaced, and the deck re-shuffled before drawing the second card. Determine the probability that:
   (a) The second card is a face card.
   (b) The first card was a face card, given that the second card is a face card.

21. **(Marbles)** Joe has 100 marbles: 75 blue, and 25 white. Mary has 6 marbles: 1 blue, 2 white, and 3 red. Joe gives Mary a marble, after which Mary gives a marble to Joe. If all of this is randomly done, determine the probability that:
   (a) Joe gave Mary a blue marble. (b) Mary gives Joe a blue marble.
   (c) Joe gave Mary a blue marble, if Mary gives Joe a blue marble.

22. **(Card and Dice)** A card is drawn from a standard deck. If it is a face card, two dice are thrown; otherwise three dice are thrown. Determine the probability that:
   (a) The sum of the dice is less than 4.
   (b) You drew a face card, given that the sum of the dice is less than 4.

23. **(Urns)** Urn I contains 5 red marbles, 4 white marbles, and 3 blue marbles. Urn II contains 4 red marbles, 2 white marbles, and 5 blue marbles. Urn III contains 3 red marbles, 4 white marbles, and 5 blue marbles. A marble is drawn from Urn I: If it is red, it is placed in Urn III, if it is white, it is placed in Urn II, and if it is blue it is returned to Urn I. One of the modified three urns is then chosen at random, and a marble taken from that urn. What is the probability that the marble is red?

§5. The Fundamental Counting Principle

By now you worked your way through a number of problems, counting your way toward their solution:

$$Pr(\text{success}) = \frac{\text{number of successes}}{\text{number of possibilities}}$$

Fine, but what if the counting gets out of hand? What is the probability, for example, of being dealt four of a kind in a five card hand, or of winning the state lottery, or that at least two students in your class have the same birthday? The definition of probability will not change, but when dealing with large numbers, more fingers are called for. Those fingers are provided by the Fundamental Counting Theorem, which, for all of its awesome power, is rather apparent:

Suppose you are going to take a journey. You start off by choosing one of two paths; following which you can choose any of three paths [see Figure 1.10]. It is clear that a total of $2 \cdot 3 = 6$ journeys are available to you: **Aa**, **Ab**, **Ac**, **Bd**, **Be**, and **Bf.**

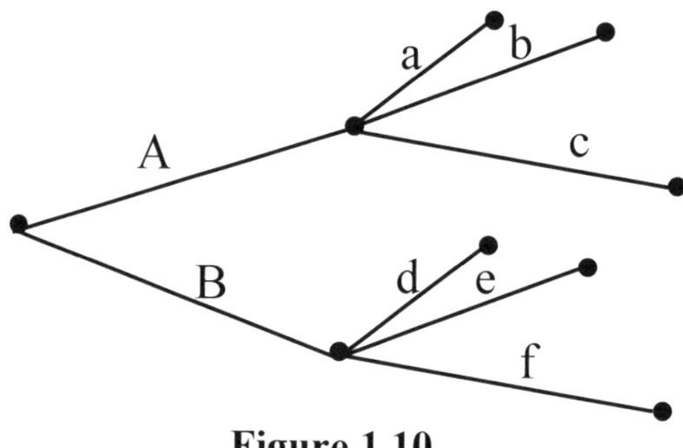

Figure 1.10

Generalizing, we have:

FUNDAMENTAL COUNTING PRINCIPLE

THEOREM 1.8 If there are n choices, each followed by m choices, then there is a total of $n \cdot m$ choices.

EXAMPLE 1.28
LETTER AND DIGIT
Choose a letter and follow it by a digit. How many different possibilities are there?

SOLUTION: Let's write down one of the things we are trying to count:

B 7

We chose the letter **B**, but could of chosen **any of 26** letters. We then chose the digit 7, but could have chosen **any of 10** digits: $0, 1, 2, 3, 4, 5, 6, 7, 8, 9$. Applying the Fundamental Counting Principle, we conclude that:

$$\text{Total number of choices} = \mathbf{26} \cdot \mathbf{10} = 260$$

As is illustrated in the following examples, the Fundamental Counting Principle can be used in "longer journeys:"

EXAMPLE 1.29
LICENSE PLATES

How many different license plates are possible if each is to consist of two capital letters, followed by four digits?

SOLUTION: When counting something, you may find it helpful to write down one of the things you are trying to count, say the plate:

ML 3423

Then step back and ask yourself how many choices you had along the way:

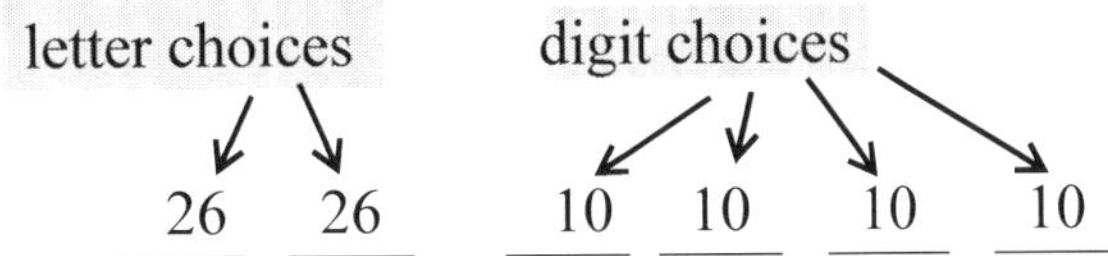

The Fundamental Counting Principle tells us that:

$$\text{Number of plates} = 26^2 \cdot 10^4 = 6{,}760{,}000$$

CHECK YOUR UNDERSTANDING 1.19

How many different license plates consisting of 2 letters followed by 4 digits are possible, if the two letters must be the same, and the 4 digits must be different.

Suggestion: Write down one of the things you are trying to count, and then step back and ask yourself how many choices you had along the way.

Answer: 131,040

EXAMPLE 1.30
SEATING ARRANGEMENT

Three couples (male and female) are to be randomly seated on one side of a rectangular table containing exactly six chairs. In how many ways can this be done if the male and female of each couple are to be seated next to each other?

SOLUTION: Here is a possible seating arrangement:

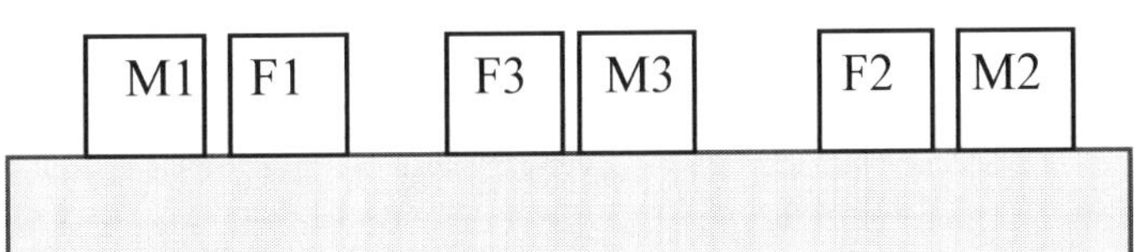

It is important to note that answers to these kind of problems don't "just happen." You really have to get personally involved; in this case: sit those bodies down!

We count the choices along the way:

Anyone can sit in the first chair:	A CHOICE OF 6
Followed by the other half of that couple:	A CHOICE OF 1
Any of the remaining 4 can then be seated in the third chair:	A CHOICE OF 4
Followed by the other half of that couple:	A CHOICE OF 1
Any of the remaining 2 can then be seated in the third chair:	A CHOICE OF 2
Followed by the other half of that last couple:	A CHOICE OF 1

Applying the Fundamental Principle of Counting, we conclude that the number of possible seating arrangements where couples are seated together is: $6 \cdot 1 \cdot 4 \cdot 1 \cdot 2 \cdot 1 = 48$.

AN ALTERNATE SOLUTION:

In seating the 3 couples, we might have written down a specific seating arrangement involving the couples as units, say:

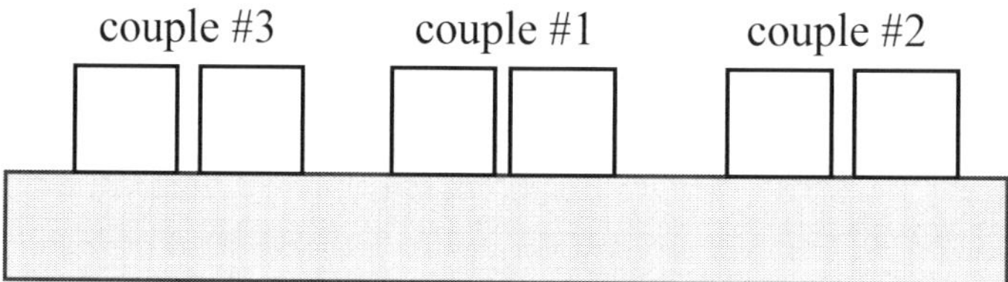

Stepping back, we note that we had a choice of 3 for the couple seated at the left, followed by a choice of 2 for the middle couple, and a choice of 1 for the remaining couple, for a total of $3 \cdot 2 \cdot 1 = 6$ choices. Having decided where the couples sit (6 choices), we still have to choose how each couple is to be seated in their two assigned seats: a choice of 2 for couple 1 (man at left, or woman at left); a choice of 2 for couple 2, and a choice of 2 for couple 3. Putting all of this together we have:

choices for seating the 3 couples (↓ 3, ↓ 2, ↓ 1)

$$3 \cdot 2 \cdot 2 \cdot 2 \cdot 1 \cdot 2 = 48$$

choices for seating the two individuals in the two chairs (↑ 2, ↑ 2, ↑ 2)

CHECK YOUR UNDERSTANDING 1.20

Referring to the previous example, how many seating arrangements are possible, if the only condition is that all of the men are to be seated next to each other at the table?

Answer: 144

BACK TO PROBABILITY

Actually, we're not going back too far, since probability involves counting two things: possibilities and successes.

EXAMPLE 1.31
VANITY PLATE

Assume that license plates consist of two letters followed by four digits. You purchase plates for your new car. What is the probability that the plates begin with the initial of your first name, followed by the initial of your family name?

SOLUTION: We already know the size of the sample space (see Example 1.29):

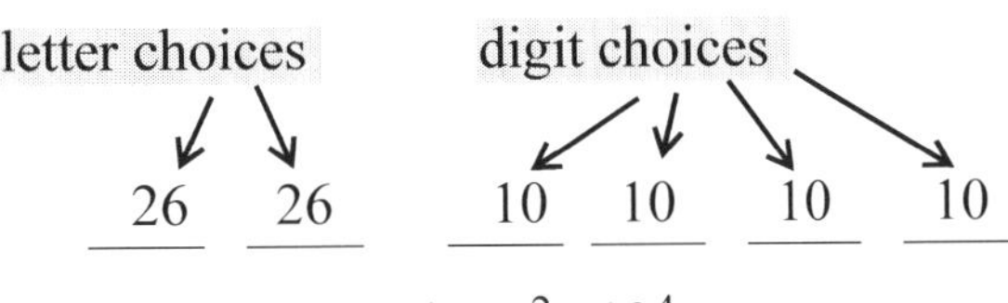

Total: $26^2 \cdot 10^4$

You now have to count successes, and it is important that you begin by writing down one of the things you are trying to count. If you happen to be John Smith, then you might write down:

JS 3414

We can't stress this enough:
JOT DOWN ONE OF THE THINGS YOU ARE TRYING TO COUNT
This will help you to better focus on the problem at hand.

Now stand back and count the number of choices along the way:

Number of Successes

J S 3 4 1 4

one choice you **had** to write down your initials → 1 1 10 10 10 10

Total: $1^2 \cdot 10^4 = 10^4$

Knowing the number of possibilities (sample space), and the number of successes, we arrive at the answer:

$$Pr(\text{initials}) = \frac{10^4}{26^2 \cdot 10^4} = \frac{1}{26^2} = \frac{1}{676} \approx 0.0015$$

CHECK YOUR UNDERSTANDING 1.21

License plates consist of 2 letters followed by 4 digits. What is the probability that the two letters in a randomly chosen plate are the same, and that the first and fourth digit of that plate are also the same?

Answer: $\frac{1}{260}$

EXAMPLE 1.32
PIZZA PALACE

Suppose you remember the first three digits and the last two digits of the (7 digit) phone number of the Pizza Palace. You also recall that none of the last four digits of their phone number are the same. Taking this into consideration, you hungrily cross you fingers and dial a number. What is the probability that you reach the Pizza Palace on your first try?

SOLUTION: We first determine the number of possible phone numbers (sample space). To better focus, we assume that the first three (known) digits of the phone number are 242; and that the last two (known) digits are 79 (they could not be 77, for example, since we are told that the last 4 digits are all distinct). With this in mind, we write down one of the numbers that might be dialed:

242- **62**79

Note that we picked these two numbers in such a way that none of the last four digits are the same (given information). We chose to write down a 6 and a 2, but know that our 6 could have been **any of 8** digits (any digit other than the last two chosen digits), and that our 2 could then have been **any of the remaining 7** digits. This leads us to:

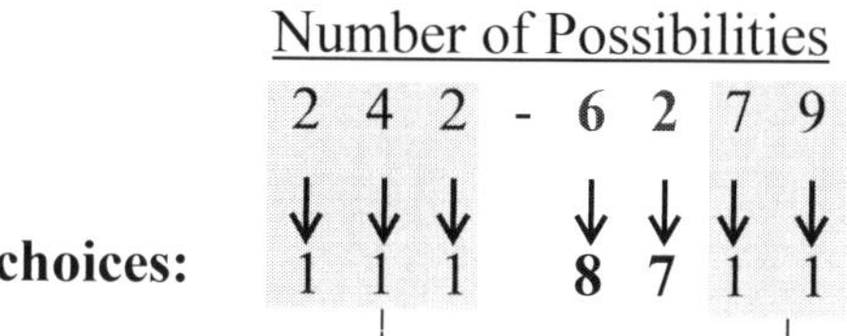

choices:

fixed--no choice

Total: $1 \cdot 1 \cdot 1 \cdot 8 \cdot 7 \cdot 1 \cdot 1 = 56$

Now we have to count the successes—all **one** of them, right? Thus:

$$Pr(\text{dial correctly}) = \frac{1}{56} \approx 0.018$$

CHECK YOUR UNDERSTANDING 1.22

There are four letter blocks: Block A, Block B, Block C, and Block D. Three of the blocks are randomly chosen and laid side by side. What is the probability that the blocks spell CAD, BAD, or DAB?

Answer: $\frac{3}{24}$

A Word of Caution

We have a question for you:

> A true/false quiz consists of 10 questions. In how many different ways can exactly two of the 10 questions be answered incorrectly?

Please try to answer the above before reading on.

How can it be both reasonable and incorrect? Easy, one can certainly analyze a problem in a rational manner, and simply forget to take something or other into account. This happens often in probability, and beyond.

Here is a "reasonable," **but incorrect**, approach:

Two of the 10 questions are going to be answered incorrectly; say *Question 5* and *Question 7*. We said "*Question 5*," but could have chosen any of the 10 questions (choice of 10); and then we said "*Question 7*," but could have chosen any of the remaining 9 questions. "*Choices followed by choices*," and we are lead to the following **INCORRECT** conclusion:

There are $10 \cdot 9 = 90$ different ways of answering 2 of the 10 questions incorrectly.

What's wrong with the above argument?

The point is that questions 5 and 7 could be marked incorrectly in two ways: *take the "5-road" before taking the "7-road,"* or *"take the 7-road before the "5 road"*:

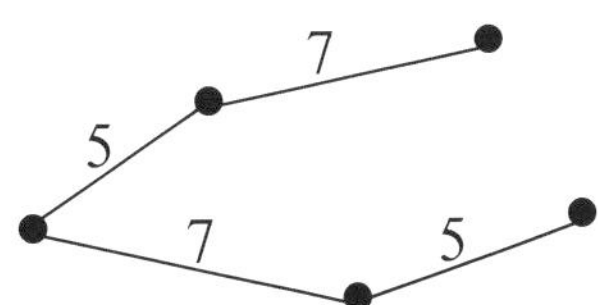

The Fundamental Principle of Counting counts these as two different journeys. So, our initial answer of $10 \cdot 9 = 90$ is incorrect by a factor of 2. The correct answer is $\frac{10 \cdot 9}{2} = 45$.

We are tempted to say: "*We **think** the correct answer is 45,*" for how do we know that we still haven't forgotten to take one thing or another into account? Exactly how does one go about concluding that one is reasoning incorrectly. Why, by reasoning, of course. Hmm...

If you need evidence that the original answer "*there are* $10 \cdot 9$ *ways of answering 2 of the 10 questions incorrectly*" is off by a factor of 2, try the same reasoning technique that lead us to that answer to the following question:

> A true/false quiz consists of 2 questions. In how many different ways can exactly two of the 2 questions be incorrectly answered?

We know that the answer must be **one**, but the previous reasoning process would lead to the incorrect answer of **two**: choice of 2 followed by a choice of 1.

EXERCISES

Exercises 1-11. (License Plates) How many different license plates consisting of two letters followed by four digits are possible if:

1. The two letters are different, and the four digits are all the same?
2. The two letters are the same, and the four digits are all odd?
3. The letter A appears first?
4. The letter A appears last?
5. The letter A appears exactly one time?
6. The letter A appears?
7. All of the digits are greater than 5, and the two letters are different?
8. The two letters are different and the first digit is a 0?
9. The two letters are the same and the last digit is a 0?
10. The two letters are the same, all the numbers are different, and the last digit is a 0?
11. The last digit is a four, and the digits increase as you read from left to right?

Exercises 12-18. (Phone Numbers) How many different 10-digit phone numbers are possible if:

12. The number's prefix (first 3 digits) is 201?
13. The number's prefix is either 201 or 914?
14. None of the first three digits is a 7?
15. Exactly two of the first three digits are 7?
16. At least two of the first three digits are 7?
17. None of the first three digits is even, and all of the remaining digits are divisible by 4?
18. The sum of the first three digits is odd?

19. **(Meal)** Each meal at Moe's restaurant comes with one of two soups, one of three salads with one of four salad dressings, one of nine entrées, and one of five desserts. How many different meals are available?

20. **(Barbara Dolly)**. Dolly has three pairs of shoes, four sweaters (one red), two skirts (one orange), and five hats. Determine the number of possible outfits consisting of:
    (a) A sweater, a skirt, a pair of shoes, and a hat.
    (b) A sweater, a skirt, a pair of shoes, and a hat or no hat at all.
    (c) A sweater, a skirt, a pair of shoes, and a hat; but not the red sweater with the orange skirt.

21. **(Social Security Number)** A typical social security number is of the form 986-87-9832. Assuming all such numbers are possible, and that one is drawn at random; what is the probability that:
    (a) The first three digits are all the same?
    (b) The digit 0 does not appear, and the last four digits are the same?
    (c) The digit 0 does not appear, and the last four digits are all different?
    (d) No digit is repeated, and the last digit is a 0?

22. (**Lottery**) In the pick-4 lottery, 4 digits are drawn randomly in order (drawing 1355 is not the same as drawing 1535). You buy one ticket. Determine the probability that:
    (a) You win the jackpot.
    (b) All but your first digit matches the winning four digit number.
    (c) Exactly three of your four digits matches those of the winning four digits number.
    (d) At least three of your four digits match those of the winning four digit number.

23. (**Molecules**) The "sub-molecules" adenine (A), cytosine (C), guanine (G), and thymine (T), link together to form larger molecules in genes. The same sub-molecule can appear more than once, and the order of linkage is a distinguishing factor (the three molecule link C-A-A, for example, is different than A-C-A). Assume that any linking is as likely to occur as any other. What is the probability that:
    (a) A 4-link molecule contains but one of the four sub-molecule types.
    (b) None of the four sub-molecules occurs more than once in a 4-link molecule.
    (c) One of the four sub-molecules occurs at least three times in a 4-link molecule.
    (d) The adenine sub-molecule occurs exactly twice in a 3-link molecule.
    (e) One of the four sub-molecules occurs at least twice in a 3-link molecule.

24. (**Disk Player**) Your friend has a disk collection consisting of 15 Jazz disks, 20 Blues disks, 18 Soft-Rock disks, and 22 Hard-Rock disks. At a party, he randomly chooses 5 of the disks and loads them in a 5-disk player. What is the probability that:
    (a) The first disk is Jazz?
    (b) None of the Hard-Rock disks is loaded?
    (c) All five disks are either Jazz or Blues
    (d) The first disk is a Blues disk, and the fifth disk is not a Blues disk?
    (e) The middle three disks are Soft-Rock?
    (f) Only the middle three disks are Soft-Rock?

25. (**Three Digit Integer**) A three digit positive integer is randomly chosen. (Don't forget that a three digit integer cannot start with a 0.) Determine the probability that:
    (a) All three digits are the same.
    (b) The integer is odd.
    (c) The digit 0 does not appear.
    (d) The digit 0 appears.
    (e) The digit 0 appears exactly one time.

26. (**Seating Arrangement**) Six individuals, consisting of three men and three women, are randomly seated on one side of a rectangular table containing exactly six chairs. What is the probability that:
    (a) Two but not three males are seated next to each other and two but not three females are seated next to each other?
    (b) A female is at each end of the table?
    (c) A female is at each end of the table, and no two females are seated next to each other?

27. **(Seating Arrangement)** Three couples (male and female) are randomly seated at a round table containing exactly six chairs. What is the probability that:
   (a) The male and female of each couple are seated next to each other?
   (b) The three males are seated next to each other?
   (c) No two males are seated next to each other?

§6. Permutations and Combinations

Some of the awesome power and versatility of the Fundamental Counting Principle was featured in the previous section. In this section, we focus on three important consequences of that principle: (1) Ordering n objects, (2) Selecting r objects from a collection of n objects, when order of selection plays a distinguishing role, and (3) Taking r objects from a collection of n objects, when order of selection does **not** matter.

ORDERING n OBJECTS (PERMUTATIONS)

EXAMPLE 1.33
BATTING ORDER

A little league baseball coach is to submit a batting order for the 9 children on her team. How many different batting orders are possible?

SOLUTION:

SEE THE ANSWER

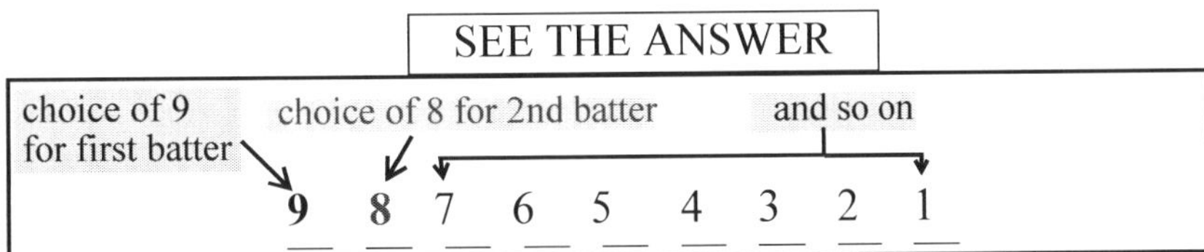

Applying the Fundamental Counting Principle, we find that:

Number of batting orders $= 9 \cdot 8 \cdot 7 \cdot 6 \cdot 5 \cdot 4 \cdot 3 \cdot 2 \cdot 1 = 362{,}880$

The above answer turned out to be the product of the first 9 positive integers, leading us to the following useful notation:

DEFINITION 1.6
FACTORIAL

For any positive integer n, the symbol $n!$, (read ***n*-factorial)** denotes the product of the integers from 1 to n, inclusive:

$$n! = 1 \cdot 2 \cdot 3 \cdots n$$

In addition: $0! = 1$

A justification for $0! = 1$ is offered at the end of the section.

Generalizing, we have:

THEOREM 1.9 There are $n!$ different ways of ordering n objects.

Any such ordering is said to be a **permutation of those n objects.** With this terminology, one says that there are $n!$ permutations of n objects.

For example:

There are 12! different ways of ordering 12 books on a shelf.
$(12! = 479{,}001{,}600)$
There are 35! different ways of ranking 35 individuals.
$(35! \approx 1.03 \times 10^{40})$

An important point of view: In probability, when you see a factorial expression, such as 6!, 15!, or 21!, you should not so much see it as a number, but rather as an abbreviated statement:

6!: The number of ways of ordering 6 objects.

15!: The number of ways of ordering 15 objects.

21!: The number of ways of ordering 21 objects.

CHECK YOUR UNDERSTANDING 1.23

In how many different ways can you arrange 7 books on a shelf?

Answer: 5,040

EXAMPLE 1.34
GIRLS BAT FIRST

A little league baseball coach is to submit a batting order for the 9 children on her team, exactly 4 of whom are girls. She decides to randomly order the players. What is the probability that all the girls will bat first?

SOLUTION:

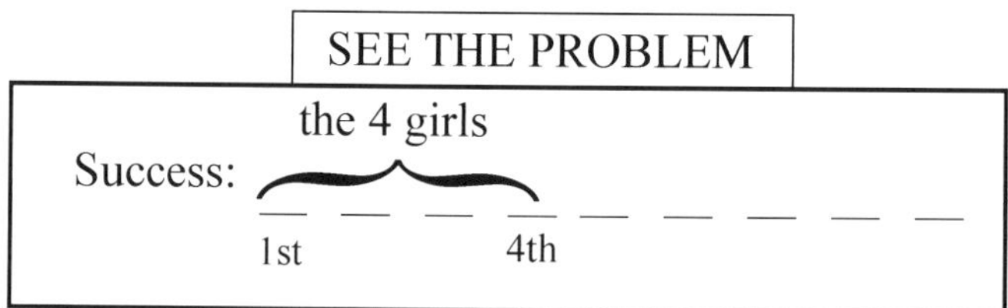

From Example 1.33 we know that there are 9! possible outcomes of the experiment (as many ways as you can order the 9 children). We now have to figure out the number of successes (number of batting orders with the 4 girls batting first). The answer is a blending of Theorem 1.9 and the Fundamental Counting Principle (Theorem 1.8, page 54):

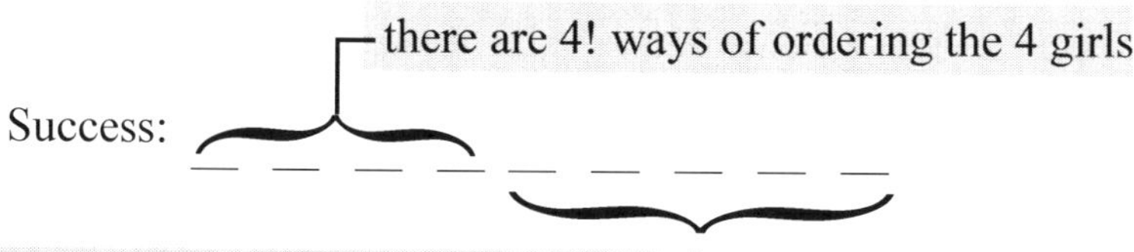

4! choices followed by 5! **Total choices: 4! times 5!**

Conclusion: $$Pr(\text{4 girls bat first}) = \frac{4!5!}{9!} \begin{array}{l}\leftarrow \text{successes}\\ \leftarrow \text{possibilities}\end{array}$$

$$= \frac{(1\cdot 2\cdot 3\cdot 4)(1\cdot 2\cdot 3\cdot 4\cdot 5)}{1\cdot 2\cdot 3\cdot 4\cdot 5\cdot 6\cdot 7\cdot 8\cdot 9}$$

"cancel like crazy:" $$= \frac{1}{126} \approx 0.008$$

The TI-84+ is featured throughout the text. You may be using a different calculator, and that's fine. Our purpose is simply to point out the capabilities of graphing calculators in general.

GRAPHING CALCULATOR GLIMPSE 1.1

Graphing calculators (and some non-graphing calculators) have factorial buttons. You can choose to have your answer appear in the default decimal form (*) or in fraction form (**).

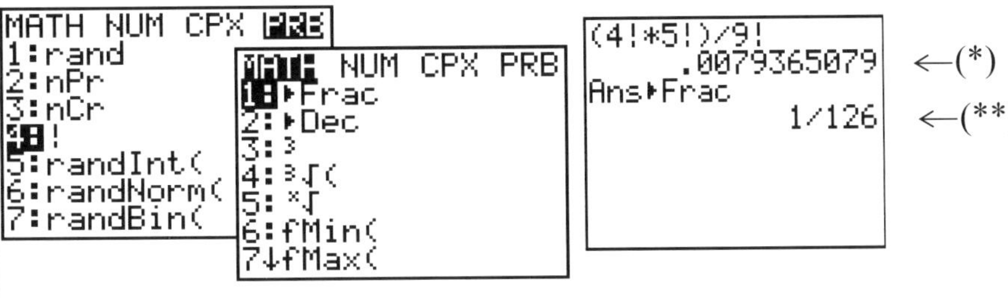

CHECK YOUR UNDERSTANDING 1.24

Fifteen children are randomly lined up, one after the other. What is the probability that the tallest child is first and the shortest child is last?

Answer: $\frac{1}{210}$

SELECTING *r* OBJECTS FROM *n*, WHEN ORDER COUNTS (PERMUTATIONS OF n OBJECTS TAKEN r AT A TIME)

EXAMPLE 1.35
BATTING ORDER

The baseball coach of Example 1.34 is again to submit a 9-player batting order. This time, however, she is to select the 9 players from 13 available players. In how many ways can this be done?

SOLUTION:

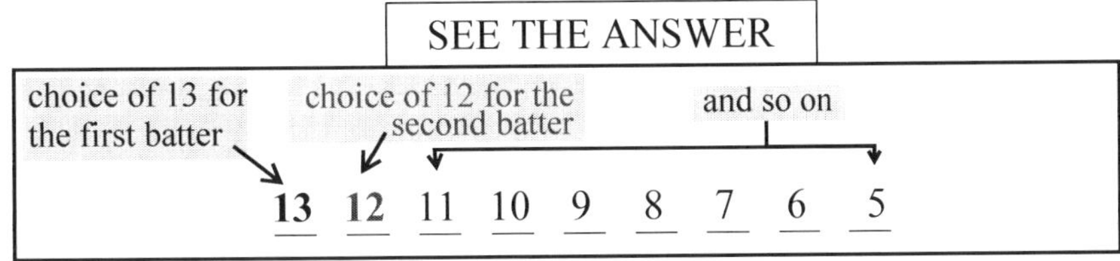

Applying the Fundamental Counting Principle, we have our answer:

$$\text{batting orders} = 13 \cdot 12 \cdot 11 \cdot 10 \cdot 9 \cdot 8 \cdot 7 \cdot 6 \cdot 5 = 259{,}459{,}200$$

It behooves us to write the above product in a different form, as it will lead us to a useful generalization:

ORDER COUNTS

$$\mathbf{13 \cdot 12 \cdot 11 \cdot 10 \cdot 9 \cdot 8 \cdot 7 \cdot 6 \cdot 5} = \frac{\mathbf{13 \cdot 12 \cdot 11 \cdot 10 \cdot 9 \cdot 8 \cdot 7 \cdot 6 \cdot 5 \cdot 4 \cdot 3 \cdot 2 \cdot 1}}{\mathbf{4 \cdot 3 \cdot 2 \cdot 1}}$$

$$= \frac{13!}{4!} = \frac{\mathbf{13!}}{\mathbf{(13-9)!}}$$

$$\frac{4 \cdot 3 \cdot 2 \cdot 1}{4 \cdot 3 \cdot 2 \cdot 1} = 1$$

Here is some notation that ties in with the above expression:

DEFINITION 1.7
PERMUTATIONS OF n OBJECTS TAKEN r AT A TIME

For integers r and n, with $0 \le r \le n$, the symbol $P(n, r)$ is that number given by:

$$P(n, r) = \frac{n!}{(n-r)!}$$

Returning to our result of Example 1.35, we can now say that:

There are $P(13, 9) = \frac{13!}{(13-9)!}$ ways of picking a 9- player batting order from 13.

Any such selection is said to be a **permutation of n objects taken r *at a time*.** With this terminology, one says that there are $P(n, r)$ permutations of n objects taken r at a time.

Generalizing, we have:

THEOREM 1.10 There are $P(n, r)$ ways of selecting, without replacement, r objects from a collection of n objects when order is a distinguishing factor.

For example:

There are $P(9, 4)$ different ways of picking 4 of 9 books and ordering them onto a shelf:

$$P(9, 4) = \frac{9!}{(9-4)!} = \frac{9!}{5!} = \frac{5! \cdot 6 \cdot 7 \cdot 8 \cdot 9}{5!} = 6 \cdot 7 \cdot 8 \cdot 9 = 3024$$

There are $P(21, 3)$ different ways of awarding a 1st, 2nd, and 3rd prize to 3 individuals from a group of 21:

$$P(21, 3) = \frac{21!}{(21-3)!} = \frac{21!}{18!} = \frac{18! \cdot 19 \cdot 20 \cdot 21}{18!} = 19 \cdot 20 \cdot 21 = 7980$$

CHECK YOUR UNDERSTANDING 1.25

Fifteen individuals are competing in a figure skating competition. In how many different ways can a Gold, Silver, and Bronze medal be awarded to the participants?

Answer: 2,730

EXAMPLE 1.36
EXECUTIVE BOARD

There are seven different colored signal flags which can be hoisted onto a mast, two of which are red and blue (red, white, blue, yellow, green, and black). Four of the flags are randomly selected and hoisted. What is the probability that both the red and the blue flags are hoisted and that the red is immediately above the blue?

SOLUTION: Applying Theorem 1.10 we easily find the number of possible ways four of the flags can be hoisted:

Pick 4 of the 7 — order count: $\boldsymbol{P(7, 4)}$

We now count successes and begin by focusing on one of them:

The "success journey"
We hoisted the red flag second, but had a **CHOICE OF 3** positions (could not have hoisted last since the blue flag must be below it).
Had to hoist the blue flag right after the red **CHOICE OF 1**.
We also had to hoist two **other** flags. Pick 2 of the remaining 5 flags (order counts): $P(5, 2)$

Applying the Fundamental Counting Principle we multiply the choices to arrive at the number of successes: $\mathbf{3 \cdot 1 \cdot} \boldsymbol{P(5, 2)}$.

So:

$$\text{Pr(R above B)} = \frac{\mathbf{3 \cdot 1 \cdot} \boldsymbol{P(5,2)}}{\boldsymbol{P(7,4)}} = \frac{1}{14} \text{ (margin)}$$

$$\frac{3P(5,2)}{P(7,4)} = \frac{3 \cdot \frac{5!}{(5-2)!}}{\frac{7!}{(7-4)!}} = \frac{3 \cdot \frac{5!}{3!}}{\frac{7!}{3!}} = 3 \cdot \frac{5!}{3!} \cdot \frac{3!}{7!} = 3 \cdot \frac{5!}{7!} = 3 \cdot \frac{5!}{5! \cdot 6 \cdot 7} = \frac{3}{6 \cdot 7} = \frac{1}{14} \approx 0.07$$

GRAPHING CALCULATOR GLIMPSE 1.2

You can use your graphing calculator to perform the above routine calculations. You can choose to have your answer appear in the default decimal form (*) or in fraction form (**).

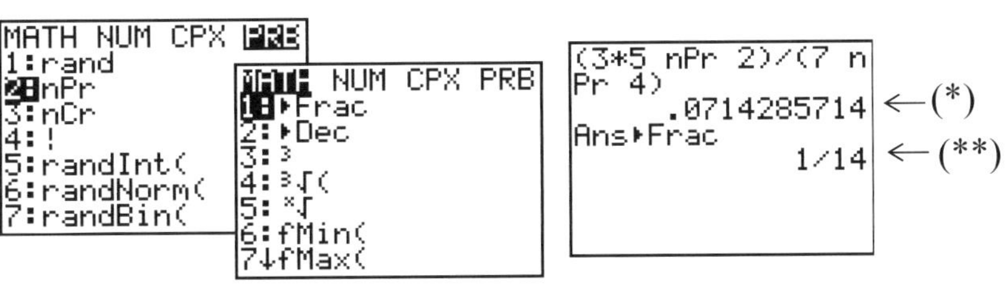

CHECK YOUR UNDERSTANDING 1.26

Referring to Example 1.36, what is the probability that red is above blue but not immediately above blue?

Answer: $\frac{22}{35}$

SELECTING r OBJECTS FROM n, WHEN ORDER DOES NOT COUNT (COMBINATIONS OF n OBJECTS TAKEN r AT A TIME)

In Example 1.36 we noted that there are $P(15,3)$ ways of electing a President, Vice President, and Secretary from a group of 15 (order counts). We now turn to a closely related question:

In how many ways can a committee of 3 be chosen from a group of 15? (Here, **order does not count**)

To arrive at the answer, we reason as follows:

order counts

In the President-VP-Secretary situation, choosing Mary as president, Johnny as vice-president, and Billy as secretary is not the same as choosing Johnny as president, Mary as vice-president, and Billy as secretary (order counts). Indeed, Mary, Johnny, and Billy can end up on the executive board in as many ways as those 3 individuals can be ordered: **3! ways**. The same can be said for any 3 of the individuals.

order does not count

In the committee situation, where order does not count, Mary, Johnny, and Billy (or any other group of 3) should be counted **once**— **not 3! times**. In other words, the "order counts" situation is **3! bigger** than the "order does not count situation." Consequently, to arrive at the number when order does not count, we simply take $P(15,3)$, and divide it by 3! (the "repetition factor"):

$$\text{committees of 3 from 15} = \frac{P(15,3)}{3!} = \frac{15!}{(15-3)!3!}$$

$$= \frac{12!\cdot 13\cdot 14\cdot 15}{12!\cdot 1\cdot 2\cdot 3} = \frac{13\cdot 14\cdot 15}{6} = 455$$

Some additional notation:

ORDER DOES NOT COUNT

The symbol $\binom{n}{r}$ is also used to denote $C(n,r)$

DEFINITION 1.8 **COMBINATIONS OF n OBJECTS TAKEN r AT A TIME** For integers r and n, with $0 \le r \le n$, the symbol $C(n,r)$ is that number given by:

$$C(n,r) = \frac{P(n,r)}{r!} = \frac{n!}{(n-r)!r!}$$

In particular, as we have seen: There are $C(15,3)$ committees of 3 that can be chosen from a group of 15.

Generalizing we have:

Any such selection is said to be a **combination of n objects taken r at a time.** With this terminology, one says that there are $C(n,r)$ combinations of n objects taken r at a time.

THEOREM 1.11 There are $C(n,r)$ ways of selecting, without replacement, r objects from a collection of n objects when order is **not** a distinguishing factor.

Here is a statement: $C(14,5)$. It reads:

The number of ways of selecting 5 objects from 14, when order does not count.

And here is how you can calculate that statement:

$14! = 1 \cdot 2 \cdot 3 \cdot 4 \cdot 5 \cdot 6 \cdot 7 \cdot 8 \cdot 9 \cdot 10 \cdot 11 \cdot 12 \cdot 13 \cdot 14$

$$C(14, 5) = \frac{14!}{(14-5)!5!} = \frac{14!}{9!5!} = \frac{9! \cdot 10 \cdot 11 \cdot 12 \cdot 13 \cdot 14}{9! \cdot 1 \cdot 2 \cdot 3 \cdot 4 \cdot 5} = \mathbf{11 \cdot 13 \cdot 14} = 2002$$

CHECK YOUR UNDERSTANDING 1.27

In draw-poker, you are dealt 5 cards (from 52) face down (so order does not count). How many different poker hands are possible?

Answer: 2,598,960

EXAMPLE 1.37

URN

An urn contains 9 red marbles, 7 white marbles, and 11 blue marbles. You reach in and grab 5 of the marbles. What is the probability that exactly 3 of the marbles you are holding are red?

SOLUTION:

You could use the tree-approach of the previous section to solve this problem, but it would have to be a very large tree, no?

SEE THE PROBLEM

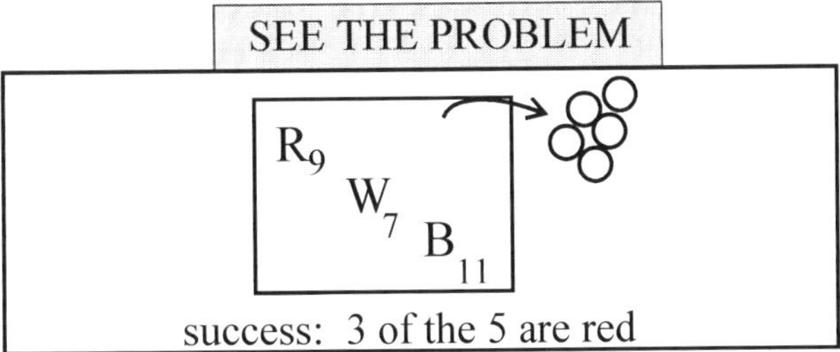

success: 3 of the 5 are red

There are as many possible outcomes of the experiment as there are ways of grabbing 5 objects from 27: $C(27, 5)$.

There are as many successes as there are ways of grabbing 3 of the 9 red marbles: $C(9, 3)$, and 2 of the 18 non-red marbles: $C(18, 2)$; for a total of $C(9, 3) \cdot C(18, 2)$ successes (Fundamental Principle of Counting).

Conclusion: $Pr(3 \text{ red}) = \dfrac{C(9, 3)C(18, 2)}{C(27, 5)} \approx 0.159$

GRAPHING CALCULATOR GLIMPSE 1.3

You can use your graphing calculator to perform the above routine calculations.

```
(9 nCr 3)*(18 nC
r 2)/27 nCr 5
            .159
```

EXAMPLE 1.38
FOUR OF A KIND

You are dealt 5 cards from a standard deck. What is the probability that you are dealt four of a kind (4 Kings, or 4 Aces, etc.)?

SOLUTION: There are as many possible hands as there are ways of grabbing 5 cards from the deck of 52 cards: $C(52, 5)$.

Time to count successes, and here is one of them:

Four Kings and a Queen

We chose 4 Kings, but could have chosen the four from any of the 13 types (Ace through Kings). Choice of 13, or, if you prefer:

$$C(13, 1) \leftarrow \text{one of the 13 types}$$

$$C(13, 1) = \frac{13!}{(13-1)!1!} = \frac{12! \cdot 13}{12! \cdot 1} = 13$$

We then went with a Queen, but that could have been any of the non-King cards. Choice of 48, or, if you prefer

$$C(48, 1) \leftarrow \text{one of the 48 non-Kings}$$

Using the Fundamental Counting Principle, we arrive at the number of successes: $C(13, 1) \cdot C(48, 1)$; and at our answer:

$$Pr(4 \text{ of a kind}) = \frac{C(13, 1) \cdot C(48, 1)}{C(52, 5)} \approx 0.00024$$

EXAMPLE 1.39
DEFECTIVE COMPUTERS

Computer Boutique receives a shipment of 25 computers, exactly five of which are defective. The manager will test four of the units, and return the entire shipment if more than one of the four units malfunctions. What is the probability that the shipment will be returned?

This is an urn problem consisting of 20 Green marbles (good computers) and 5 Black marbles (bad computers):

G_{20} B_5

SOLUTION: The formula tells the story; please read it carefully:

$$Pr(\text{returned}) = 1 - [\boldsymbol{Pr}(\mathbf{0\ bad}) + \boldsymbol{Pr}(\mathbf{1\ bad})]$$

$$= 1 - \left[\frac{C(5, 0) \cdot C(20, 4)}{C(25, 4)} + \frac{C(5, 1) \cdot C(20, 3)}{C(25, 4)}\right] \approx 0.166$$

You can use Definition 1.8 to see that $C(5, 0) = 1$, or simply observe that there is but 1 way of grabbing no objects from 5 (don't take any).

CHECK YOUR UNDERSTANDING 1.28

A bag contains 25 red jelly-beans, 7 black jelly-beans, 4 purple jelly-beans, and 9 white jelly-beans. You reach in and grab 8 of the jelly-beans. What is the probability that you end up with 2 of each color?

Answer: ≈ 0.0063

In Definition 1.6, we defined 0! to be 1. Here is one reason in support of that decision:

A justification for: **0! = 1**

In how many ways can you grab 5 marbles from a bag containing exactly 5 marbles? Clearly, **1** way.

Plugging into the expression of Theorem 1.11, with $n = r = 5$ we have:

$$\mathbf{1} = C(5, 5) = \frac{5!}{(5-5)!5!} = \frac{5!}{\mathbf{0}!5!}$$

The only way the above can pan out is if $\mathbf{0! = 1}$.

Figure 1.11 summarizes the Fundamental Counting Principle, along with some of its consequences. It is important that you see the first two columns as being "one and the same thing," and to be able to toggle back and forth between the words in column 2 and the mathematical expressions in column 1. The last column just tells you how to calculate the expression in column 1.

Symbol	Represents the number of:	Evaluate
$n \cdot m$	n-choices followed by m-choices	multiply
$n!$	ways of ordering of n objects	$n! = 1 \cdot 2 \cdot 3 \cdots n$
$P(n, r)$	ways of picking r from n, when order counts	$P(n, r) = \frac{n!}{(n-r)!}$
$C(n, r)$	ways of grabbing r from n, when order does not count	$C(n, r) = \frac{n!}{(n-r)!r!}$

Figure 1.11

EXERCISES

Exercises 1-9. Perform the indicated operation.

1. $4!$
2. $\frac{5 \cdot 3!}{5!}$
3. $P(10, 7)$
4. $P(7, 3)$
5. $C(7, 3)$
6. $\frac{C(7, 3)}{P(7, 3)}$
7. $P(7, 7)$
8. $C(7, 0)$
9. $C(7, 3) - C(7, 4)$

Exercises 10-13. (Ordering n objects)

10. In how many different ways can 9 objects be ordered?
11. In how many different ways can 5 children be seated at 5 desks?
12. In how many different ways can 5 different colored flags be hoisted on a mast?
13. In how many different ways can 10 pigeons roost in 10 pigeon holes, if pigeons like to sleep alone?

Exercises 14-17. (Picking r objects from n when order counts)

14. In how many different ways can 10 pigeons roost in 15 pigeon holes, if pigeons like to sleep alone?
15. In how many different ways can 5 children be seated at 11 desks?
16. In how many different ways can 5 of 11 children be seated at 5 desks?
17. In how many different ways can 5 of 7 different colored flags be hoisted on a mast?

Exercises 18-21. (Grabbing r objects from n when order does not count)

18. In how many different ways can 10 pigeons roost in 2 spacious pigeon holes, if 5 pigeons are to sleep in each of the 2 holes?
19. A high school student decides to apply to four of the 27 colleges in his state. In how many ways can this be done?
20. In how many different ways can you grab three coins from a box containing 12 coins?
21. A bridge hand consists of 13 cards (order not important). How many different bridge hands are there?

Exercises 22-25. (Pose a question leading to the given answer) (It should start with: "*In how many ways...*")

22. $4!$
23. $P(7, 5)$
24. $C(9, 4)$
25. $\frac{C(8, 2)C(7, 3)}{C(15, 5)}$

26. **(Permutations verses Combinations)** List all three letter combinations (without repetition) of the three letters A, B, C, when order counts (ABC and BCA are two possibilities). List all three letter combinations (without repetition) of those same three letters, when order does not count (ABC and BCA are to be considered to be the same).

27. **(Drawing two cards)** In how many different ways can you draw 2 cards from a standard deck if the order in which they are drawn:

    (a) Matters? (b) Does not matter?

    (c) The answer in (a) is how many times larger than that in (b)?

28. **(Drawing three cards)** In how many different ways can you draw 3 cards from a standard deck if the order in which they are drawn:

    (a) Matters? (b) Does not matter?

    (c) The answer in (a) is how many times larger than that in (b)?

29. **(Pizzas)** Pizza Shack offers 5 toppings for their regular cheese pizza. How many different pizzas are there which contain:

    (a) 3 toppings? (b) Less than 3 toppings (but at least one)?

Exercises 30-44. (Urn problems) (In many shapes and sizes)

30. **(Marbles)** In how many different ways can you grab 5 marbles from an urn containing 10 red, 8 white, and 4 blue marbles?
31. **(Marbles)** In how many different ways can you grab 3 red marbles and 2 white marbles from an urn containing 10 red, 8 white, and 4 blue marbles?
32. **(Marbles)** In how many different ways can you grab 3 red marbles, 2 white marbles, and 2 blue marbles from an urn containing 10 red, 8 white, and 4 blue marbles?
33. **(Marbles)** In how many different ways can you grab a red marble, a white marble, and a blue marble from an urn containing 10 red, 8 white, and 4 blue marbles?
34. **(Coins)** In how many different ways can you grab two dimes and a nickel from a box containing 6 dimes, 5 nickels, and 3 quarters?
35. **(Coins)** In how many different ways can you grab 3 dimes and 2 nickels from a box containing 6 dimes, 5 nickels, and 3 quarters?
36. **(Coins)** In how many different ways can you grab three quarters from a box containing 6 dimes, 5 nickels, and 3 quarters?
37. **(People)** In how many different ways can you select 3 men and 4 women from a group consisting of 7 men and 9 women?
38. **(People)** In how many different ways can you select 4 freshmen, 5 sophomores, 3 juniors, and 7 seniors from a group containing 15 students of each standing?
39. **(Classes)** In how many different ways can you select 2 math courses, 2 business courses, and 2 electives from a list of 5 math courses, 9 business courses, and 12 electives?
40. **(Ribbons)** In how many different ways can you select 5 red, 4 white, and 3 blue ribbons from a box that contains 10 red, 7 white, 8 blue, and 6 yellow ribbons?

41. **(Cards)** In how many different ways can you draw 2 Kings and 3 Queens from a standard deck?
42. **(Cards)** In how many different ways can you draw 2 Clubs and 3 Spades from a standard deck?
43. **(Cards)** In how many different ways can you draw 2 Clubs, 1 Spade, and 2 Hearts from a standard deck?
44. **(Cards)** In how many different ways can you be dealt 5 cards of the same suit from a standard deck?

Exercises 45-49. (Signals) There are six different colored signal flags which can be hoisted (order counts) onto a mast. One of the flags is red and another is yellow.

45. What is the total number of possible signals consisting of four flags, if the red flag is not to be hoisted?
46. What is the total number of possible signals consisting of four flags, if the red flag is to be hoisted?
47. What is the total number of possible signals consisting of four flags, if neither the red flag nor the yellow flag is to be hoisted?
48. What is the total number of possible signals consisting of four flags, if the red flag and the yellow flag are to be hoisted?
49. What is the total number of possible signals consisting of five or more flags?

Exercises 50-53. (Signals) There are six different colored signal flags which can be hoisted onto a mast (red, white, blue, yellow, green, and black). The six flags are randomly selected and hoisted. What is the probability that:

50. The red flag is at the top?
51. The red flag is at the top or at the bottom?
52. The red is at the top, followed by the white, and then the blue flag?
53. The red, white and blue flags are at the top, in any order?

Exercises 54-63. (Signals) There are six different colored signal flags which can be hoisted onto a mast (red, white, blue, yellow, green, and black). Four of the flags are randomly selected and hoisted. What is the probability that:

54. The red flag is hoisted?
55. The red flag is not hoisted?
56. Neither the red nor the green flag is hoisted?
57. The red flag is at the top?
58. The red flag is at the top or at the bottom?
59. The red is at the top, followed by the white, and then the blue flag?
60. The red, white and blue flags are at the top, in any order?
61. Both the red and the green flags are hoisted?
62. Either the red or the green flag is hoisted, but not both?
63. The red or the green flag is hoisted, possibly both?

Exercises 64-69. (Urn) An urn contains 9 red marbles, 7 white marbles, and 5 blue marbles. You grab 4 of the marbles. What is the probability that:

64. All are red?
65. None is red?
66. Two of the marbles are white and the other two are blue?
67. All are of the same color?
68. They are not all of the same color?
69. Each of the three colors is drawn?

Exercises 70-73. (Committee) A committee of 5 is randomly selected from a group consisting of 9 women and 6 men. What is the probability that:

70. None of the men are chosen?
71. Exactly 3 women are chosen?
72. At least 3 women are chosen?
73. At most 3 women are chosen?

Exercises 74-77. (Prizes) A first, second, third, fourth, and fifth prize are to be randomly awarded to 5 randomly selected children from a group consisting of 9 girls and 6 boys. What is the probability that:

74. None of the boys receives an award?
75. Exactly 3 of the girls receive an award?
76. Girls receive the top three prizes, and the other two prizes go to two of the boys?
77. Girls receive the top two and the fifth prize, and the other two prizes go to two of the boys?

Exercises 78-80. (Two Cards) Two cards are dealt from a standard deck. What is the probability that:

78. Both cards are Kings?
79. Neither card is a King?
80. Exactly one of the two cards is a King?

Exercises 81-89. (Poker) You are dealt a 5 card poker hand. What is the probability that:

81. All are clubs?
82. All are of the same suit (all hearts, all clubs, all spades, or all diamonds)?
83. You are dealt the full house consisting of 3 Kings and 2 Queens?
84. You are dealt a full house (3 of one kind and 2 of another—like 3 Kings and 2 Queens)?
85. You are dealt two pair [Two of one kind, and two of another (the remaining card of yet another kind)—like 2 Kings, 2 Queens, and a Five]?
86. You are dealt three-of-a-kind (three only of one kind and the remaining two cards are not a pair; for example: three Kings, a Jack and an Ace)?
87. You are dealt a royal flush (the 10, Jack, Queen, King, and Ace of the same suit)?

88. You are dealt a straight flush or a royal flush (five cards, of the same suit, in sequence—like the 3, 4, 5, 6, and 7 of clubs; or the Ace, 2, 3, 4, and 5 of spades)? (**Note:** an Ace can "start the straight," as in Ace, 2, 3, 4 and 5; or "end the straight," as in 10, Jack, Queen, King, and Ace; but Queen, King, Ace, 2, 3 is **not** a straight: straights do not "wrap-around.")

89. You are dealt a straight (five cards in sequence, but **not** of the same suit)? (See note in previous exercise.)

Exercises 90-94. (Bridge Hand) You are dealt a bridge hand (13 cards). What is the probability that:

90. Every card is a club?

91. All of the cards are of the same suit (all hearts, or all clubs, or all spades, or all diamonds)?

92. You are holding all of the face cards?

93. You are holding nothing but face cards and Aces?

94. No card is lower than a seven (Aces and face cards are higher than seven)?

95. **(Quality Inspection)** Your company ships boxes of calculators containing 24 units. Five calculators from each box are tested. If all 5 pass inspection the box is shipped. If 4 of the 5 pass inspection the box is again shipped (after replacing the defective unit). If more than 1 of the tested units fails inspection, then the entire box is sent back to production. What is the probability that a boxed is shipped if:
    (a) The box contains exactly 1 defective calculator?
    (b) The box contains exactly 5 defective calculators?
    (c) The box contains exactly 5 non-defective units?

96. **(PIN Number)** Your friend's PIN number for an automatic teller machine consists of 6 digits. You take a guess at it. What is the probability that:
    (a) You get it exactly right?
    (b) You get all but one of the digits right?
    (c) You get exactly half of the digits right?
    (d) You get none of the digits right?

97. **(Horse Race)** Seven horses are running in a race.
    (a) How many win-place-show combinations are possible? ("Win" is to come in first, "place" is to be second, and "show" is to be third).
    (b) You randomly place a bet on one of the seven horses. What is the probability that your horse will come in first?
    (c) You randomly place a bet on one of the seven horses. What is the probability that your horse will finish in the money (wins, places, or shows)?

98. **(Horse Race)** Seven horses are running in a race. You randomly place a bet on two of the horses. What is the probability that:

 (a) One of your horses comes in first?

 (b) Your two horses come in first and second?

 (c) Both of your horses finish in the money (first, second, or third)?

 (d) Neither of your horses finishes in the money?

 (e) Exactly one of your horses finishes in the money?

 (f) At least one of your horses finishes in the money?

99. **(Horse Race)** Seven horses are running in a race. You randomly place a bet on three of the horses. What is the probability that:

 (a) One of your horses comes in first?

 (b) Exactly two of your horses finish in the money?

 (c) Exactly one of your horses finishes in the money?

 (d) At least one of your horses finishes in the money?

 (e) At least two of your horses finish in the money?

100. **(Quinella and Trivecta)**

 (a) A Quinella is picking the two horses that will end up finishing first and second (in either order). What is the probability of winning a Quinella if you randomly choose two horses from a field of eight?

 (b) A Trivecta is picking the three horses that will end up finishing first, second, and third (in any order). What is the probability of winning a Trivecta if you randomly choose three horses from a field of eight?

101. **(Lottery)** In a pick-5 lottery, you are to pick 5 numbers from a card containing the numbers 1 through 50 (order is of no consequence). If the 5 numbers you selected match the 5 lottery numbers drawn, you win a million dollars. If exactly 4 of your numbers match, you win $10,000. You buy one card. What is the probability that you will win:

 (a) A million dollars? (b) $10,000? (c) Nothing?

102. **(Seating in a Van)** A van has twelve seats, including the driver's seat.

 (a) In how many ways can twelve people be seated, if each can drive?

 (b) In how many ways can seven people be seated, if each can drive?

 (c) In how many ways can twelve people be seated, if only 7 of them can drive?

 (d) In how many ways can seven people be seated, if only three of them can drive?

103. **(Matching)** The answer to each of 10-questions is given, but in a random order. You are to match the questions with their correct answer, and must assign a different answer to each question. You get 10 points for each correct answer. Assuming that you randomly pick an answer for each of the 10 questions, what is the probability that you end up with:

 (a) 100 points? (b) 90 points? (c) 80 points?

104. **(Birthday Problem)** What is the probability that in a class of 25 students at least 2 have the same birthday? (Assume that each year has 365 days.)

Suggestion: $P(\text{at least 2 have same birthday}) = 1 - P(\text{all have different birthdays})$

105. **(Urn)** An urn contains 5 red marbles, 4 white marbles, and 3 blue marble. A marble is drawn at random. The marble drawn is returned to the urn, along with two additional marbles of the same color. Two other marbles are then drawn from the resulting urn. What is the probability that those two marbles are both blue? Suggestion: Consider a tree.

106. **(Urns)** Urn I contains 5 red marbles, 4 white marbles, and 3 blue marbles. Urn II contains 4 red marbles, 2 white marbles, and 5 blue marbles. A marble is drawn from Urn I. If it is red, it is placed in Urn II; if it is white or blue it is returned to Urn I. A marble is then drawn from Urn II. If it is red it is placed in Urn I; if it is white or blue it is returned to Urn II. Finally, three marbles are drawn from Urn II. What is the probability that no two of the three marbles drawn are of the same color? Suggestion: Consider a tree.

107. **(Character Strings)** Determine the number of **different** character strings that can be formed utilizing all of the letters in the word:

(a) OHIO (b) CALIFORNIA (c) MISSISSIPPI

108. **(Character Strings)** Find a five letter word satisfying the property that its letters can be used to generate exactly:

(a) 30 distinct five character strings. (b) 20 distinct five character strings.

109. **(Theory)** Without resorting to any calculations, explain why the number. $C(100, 52)$ must equal the number $C(100, 48)$. Generalize this observation to any expression of the form $C(n, r)$.

110. **(Theory)** Express $C(100, 5)$ in terms of $C(99, 5)$ and $C(99, 4)$ without evaluating any of the three expressions. Explain your reasoning process. Generalize to $C(n, r)$.

Bernoulli Trials are also called **Binomial Trials.**

§7. Bernoulli Trials

Question A: Roll a die seven times. What is the probability that a **5** appears on the **first three** rolls only?

Answer: We are dealing with seven **independent trials** of an experiment. Applying Theorem 1.6, page 33, we have:

$$Pr(\mathbf{5\,5\,5}\not{5}\not{5}\not{5}\not{5}) = \frac{1}{6}\cdot\frac{1}{6}\cdot\frac{1}{6}\cdot\frac{5}{6}\cdot\frac{5}{6}\cdot\frac{5}{6}\cdot\frac{5}{6} = \left(\frac{1}{6}\right)^3\left(\frac{5}{6}\right)^4$$

Question B: Roll a die seven times. What is the probability that a **5** appears on the **first**, **fourth**, and **fifth** roll only?

Answer: Same as **A**:

$$Pr(\mathbf{5}\not{5}\not{5}\,\mathbf{5\,5}\not{5}\not{5}) = \frac{1}{6}\cdot\frac{5}{6}\cdot\frac{5}{6}\cdot\frac{1}{6}\cdot\frac{1}{6}\cdot\frac{5}{6}\cdot\frac{5}{6} = \left(\frac{1}{6}\right)^3\left(\frac{5}{6}\right)^4$$

Indeed, the answer will be $\left(\frac{1}{6}\right)^3\left(\frac{5}{6}\right)^4$ no matter where we position the three 5's. This fact, and the fact that there are precisely $\boldsymbol{C(7,3)}$ ways of positioning the three 5's, enables us to answer the following question:

Question C: Roll a die seven times. What is the probability that a **5** appears exactly three times?

Answer:

$$Pr(\text{three } 5's \text{ out of } 7) = \boldsymbol{C(7,3)}\cdot\left(\frac{1}{6}\right)^3\cdot\left(\frac{5}{6}\right)^4$$

probability of a success ($\frac{1}{6}$); number of successes (3); number of ways of grabbing 3 objects from 7 ($C(7,3)$); number of failures (4); probability of a failure ($\frac{5}{6}$)

Note that while a die can show six different faces, our concern was on whether a five was rolled (a success—with probability $\frac{1}{6}$); or a five was not rolled (a failure—with probability $\frac{5}{6}$). In effect, we were conducting what is called a Bernoulli (or binomial) trial:

After the Swiss mathematician Jakob Bernoulli (1654-1705).

DEFINITION 1.9
BERNOULLI TRIAL

A **Bernoulli trial** is an experiment with two outcomes; one termed a success, and the other a failure.

Generalizing the above dice-rolling situation, we come to:

THEOREM 1.12
BERNOULLI FORMULA

If n **independent Bernoulli trials**, each with probability of success p are conducted, then the probability of obtaining exactly r out of n successes is given by:

$$Pr(r \text{ successes out of } n) = C(n, r) \cdot p^r \cdot (1-p)^{n-r}$$

Analyzing the Bernoulli Formula:

$$Pr(r \text{ successes out of } n) = C(n, r) \cdot p^r \cdot (1-p)^{n-r}$$

- grab r of the n to be successes (points to $C(n, r)$)
- probability of a successes and there are r of them (points to p^r)
- probability of a failure and there are $n-r$ of them (points to $(1-p)^{n-r}$)

EXAMPLE 1.40
CARDS FROM SEPARATE DECKS

Each of nine individuals holds a deck of cards, and each draws a single card from his deck. What is the probability that exactly five of them draw a face card?

SOLUTION: We suggest that as soon as you realize that you are in a Bernoulli trial situation, that you write down: $\boldsymbol{n}$ $\boldsymbol{p}$ $\boldsymbol{r}$

And then jot down their values: 9 $\frac{12}{52}$ 5

probability of drawing a face card ↑ (points to $\frac{12}{52}$)

All that remains is to substitute our n-p-r values into the Bernoulli Formula of Theorem 1.12:

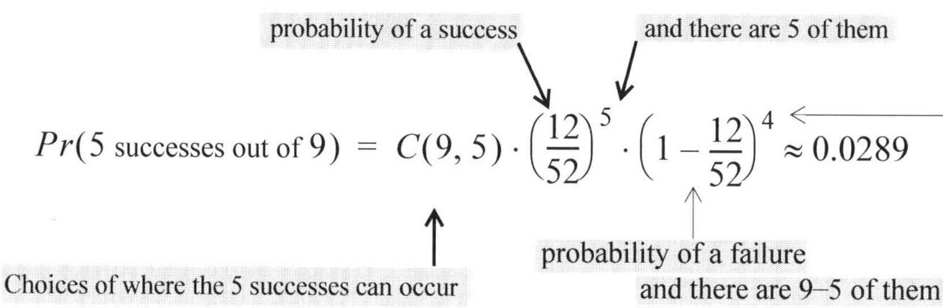

$$Pr(5 \text{ successes out of } 9) = C(9, 5) \cdot \left(\frac{12}{52}\right)^5 \cdot \left(1 - \frac{12}{52}\right)^4 \approx 0.0289$$

GRAPHING CALCULATOR GLIMPSE 1.4

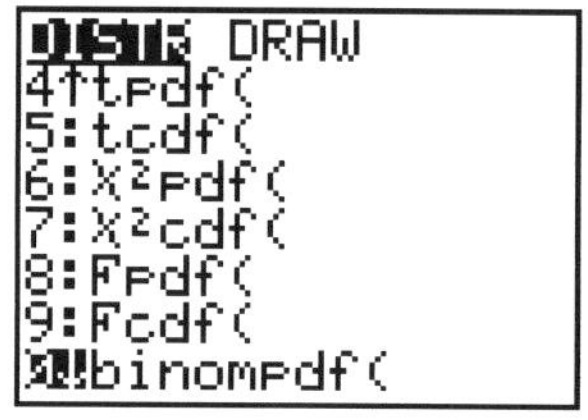

binompdf(9,12/52
,5)
.0289

EXAMPLE 1.41
CARDS FROM SAME DECK

Each of nine individuals draws a card from the same deck, without replacement. What is the probability that exactly five of them draw a face card?

This is **NOT** a Bernoulli trial situation—we are not dealing with independent trials.

SOLUTION: Be careful. There is a big difference between this example and the previous one. In the previous one, the nine events were independent. Now however, the events are dependent: the second person cannot draw the card drawn by the first, and so on. We can not apply Theorem 1.12 in this (dependent) setting. We can, however, employ the counting techniques of the previous section to arrive at the answer:

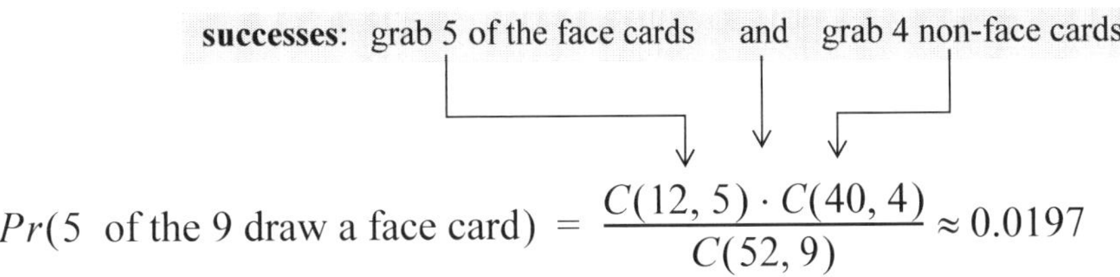

$$Pr(5 \text{ of the 9 draw a face card}) = \frac{C(12, 5) \cdot C(40, 4)}{C(52, 9)} \approx 0.0197$$

possibilities: grab any 9 from the deck

CHECK YOUR UNDERSTANDING 1.29

The probability that a marksman will hit the bull's-eye is 0.84. She fires 20 shots at the target. What is the probability that she will hit the bull's-eye exactly 17 times?

Answer: ≈ 0.24

EXAMPLE 1.42
PASSING AN EXAM

An exam consists of 10 multiple choice questions, with five answers for each question. An unprepared student randomly guesses an answer to each question. What is the probability that the student will pass the test, given that 70% is a passing grade and that all 10 questions are of equal weight?

SOLUTION: This is a Bernoulli situation, with:

n	p	r
10	$\frac{1}{5}$	7 **or** 8 **or** 9 **or** 10 The "**or**'s" break the answer into the sum of four parts:

Bringing us to:

$$Pr(\text{Passing}) = C(10, 7)\left(\frac{1}{5}\right)^7\left(\frac{4}{5}\right)^3 + C(10, 8)\left(\frac{1}{5}\right)^8\left(\frac{4}{5}\right)^2$$
$$+ C(10, 9)\left(\frac{1}{5}\right)^9\left(\frac{4}{5}\right)^1 + C(10, 10)\left(\frac{1}{5}\right)^{10}\left(\frac{4}{5}\right)^0 \approx 0.000864$$

GRAPHING CALCULATOR GLIMPSE 1.5

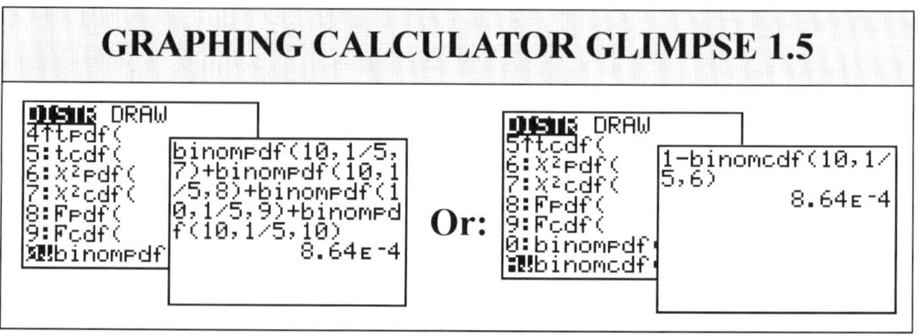

CHECK YOUR UNDERSTANDING 1.30

Referring to the situation of Example 1.42, determine the probability that the student answers at least 3 questions correctly.

Suggestion: Use Theorem 1.1, page 15.

Answer: ≈ 0.32

EXAMPLE 1.43
FIRECRACKERS

The HAPPY-FOURTH company sells cases consisting of 102 firecrackers. The company has a policy that allows a purchaser to test 2 of the firecrackers in a case and will replace that case if one or both of those firecrackers do not go off. Production is such that 7% of the firecrackers are bad. A retailer purchases 37 cases and tests each case. Determine the probability that exactly 2 cases will be returned to the company.

SOLUTION: There are a lot of numbers floating around, but the essence of the problem is that you are dealing with 37 cases and you want the probability that (exactly) two of those cases will be returned. In short:

n	*P*	r
37	?????	2

(where $\boldsymbol{P}$ is the probability that a case will be returned)

At this point, we have a partial answer:

$$Pr(2\text{ cases returned}) = C(37,2)\boldsymbol{P}^2(1-\boldsymbol{P})^{35} \quad (*)$$

But what is the value of $\boldsymbol{P}$? It is the probability that a case will be returned; which is to say that at least one of the two tested firecrackers are duds; bringing us to another Bernoulli situation:

n	p	r
2	.07	1 or 2

(where p is the given probability that a firecracker is bad)

Here, then, is the value of $\boldsymbol{P}$:

$$\begin{aligned} \boldsymbol{P} &= C(2,1)(.07)^1(.93)^1 + C(2,2)(.07)^2(.93)^0 \\ &= 2(.07)(.93) + 1(.07)^2 \cdot 1 \approx 0.14 \end{aligned}$$

And here is our final answer [see (*) above]:

$$Pr(2 \text{ cases returned}) = C(37,2)(0.14)^2(1-0.14)^{35} \approx 0.067$$

CHECK YOUR UNDERSTANDING 1.31

You roll a die 5 times. If you roll at least four sixes (out of the five rolls), you win a penny. You play the game three times. What is the probability that you will win exactly two cents?

Answer: 0.000033

Exercises

1. **(Rolling a Die)** A die is rolled 6 times. Find the probability that:
   (a) The number 2 is rolled on the first 3 rolls, and not on the remaining 3 rolls.
   (b) The number 2 is rolled exactly 3 times.
   (c) The number 2 is rolled at least 5 times.
   (d) The number 2 is rolled at most 5 times.
   (e) A number less than 3 is rolled exactly 4 times.
   (f) A number less than 3 is rolled the first and fifth time only.
   (g) The number 1 is rolled exactly 2 times, with one right after the other.
   (h) The same number is rolled each time.

2. **(Rolling a Pair of Dice)** A pair of dice are rolled 8 times. Find the probability that:
   (a) A sum of 7 is rolled exactly 2 times.
   (b) A sum of 7 is rolled at least 2 times.
   (c) A sum of 7 is rolled at most 2 times.
   (d) A sum of 7 or 11 is rolled exactly 5 times.
   (e) A sum of 7 or 11 is never rolled.
   (f) A sum of 7 or 11 is rolled the third and fifth time, only.

3. **(Drawing from a Hat)** A hat contains 9 slips of paper marked 1 through 9.
   (a) You draw a slip and then replace it. You do this 9 times. Determine the probability that you draw the number 5 twice.
   (b) You draw 9 slips without replacement. Determine the probability that you draw the number 5 twice.

4. **(EZ-PASS)** Sixty-eight percent of cars on a toll-road have EZ-pass. Five cars are approaching the toll-gates. What is the probability that:
   (a) None have EZ-pass.
   (b) All have EZ-pass.
   (c) All but 2 have EZ pass.
   (d) At least 2 have EZ pass.

5. **(Graduate School)** Due to your impressive scholastic work, there is a 0.67 probability that you will be accepted at any of the 7 graduate schools to which you have applied. What is the probability that:
   (a) All will accept you.
   (b) None will accept you.
   (c) Exactly one will accept you.
   (d) At least one will accept you.

6. **(Negative Reaction)** Experience shows that 0.03% of individuals will have a negative reaction to a certain medication. Eleven people begin taking the medication. What is the probability that:
   (a) None will have a negative reaction.
   (b) At most 2 will have a negative reaction.
   (c) At least 2 will have a negative reaction.

7. (**Skating**) Gliding Glenn, the world famous figure skater, can properly execute a triple axle 95% of the time. His program involves 4 triple axles. What is the probability that he will properly execute:
   (a) All four of them. (b) None of them. (c) Exactly 2 of them. (d) At least 2 of them.

8. (**Grades**) Past experience has shown that 6% of the students in Math 101 will receive an A. What is the probability that in a class of 25 students:
   (a) Exactly 6 will receive As.
   (b) At most 2 will receive As.
   (c) At most 2 will receive As or Bs, given that 9.5% of the students in Math 101 receive Bs.

9. (**Snow Days**) Past experience has shown that there is a 0.57 probability that there will be at least 4 school closings per year due to inclement weather in the Pine Bush school district. What is the probability that:
   (a) Exactly 4 of the next five years will have at least 4 school closings.
   (a) At least 4 of the next five years will have at least 4 school closings.
   (c) At most 4 of the next five years will have at least 4 school closings.

10. (**Poker**) The highest hand in poker is a royal flush: ten, jack, queen, king, and ace, all of the same suit. One hundred poker hands are dealt. Determine the probability that:
    (a) Exactly one royal flush is dealt.
    (b) No royal flush is dealt.
    (c) At least one royal flush is dealt.

11. (**Bridge Hand**) A bridge hand consists of 13 cards. One hundred bridge hands are dealt. Determine the probability that:
    (a) Exactly one contains all four aces.
    (b) None contain all four aces.
    (c) At least one contains all four aces.

12. (**Urns**) Billy draws three marbles from an urn containing 5 red marbles, 3 white marbles and 2 blue marbles, and Judy draws three marbles from an urn containing 4 red marbles, 2 white marbles, and 1 blue marble. If Billy draws three different colored marbles and Judy does not, then Billy wins a penny. They play this game 10 times. Determine the probability that Billy will win at least one penny.

13. (**Slips in Hat**) You draw a slip from a hat which contains 9 slips of paper marked 1 through 9. If the slip is the number 5, you win a penny. You play the game 12 times (always with the hat containing the 9 slips of paper). Determine the probability that:
    (a) You win a total of 3 pennies.
    (b) You win less than 3 pennies.

14. **(Slips in Hats)** Hats A and B both contain 9 slips of paper marked 1 through 9. You draw a slip from hat A and put it in hat B, and then draw a slip from B. You then draw another slip from hat B. If that slip is the number 5, you win a penny. You do this 12 times. Determine the probability that:
    (a) You win a total of 3 pennies. [Compare your answer with Exercise 13(a).]
    (b) You win less than 3 pennies. [Compare your answer with Exercise 13(b).]

15. **(Slips in Hats)** Hats A and B both contain 9 slips of paper marked 1 through 9. You draw a slip from hat A and put it in hat B, and then draw a slip from B and place it in Hat A. You then draw another slip from hat A. If that slip is the number 5, you win a penny. You do this 12 times. Determine the probability that:
    (a) You win a total of 3 pennies. [Compare your answer with Exercises 13(a).]
    (b) You win less than 3 pennies. [Compare your answer with Exercises 13(b).]

16. **(Cupid)** It is well known that Cupid hits his mark 99% of the time, and that he fires exactly 99 arrows each day.
    (a) What is the probability that none of the arrows will hit their mark on any given day?
    (b) What is the probability that exactly 97 of his arrows will hit their mark on any given day?
    (c) What is the probability that at least 97 of his arrows will hit their mark on any given day?
    (d) If, during the course of a week, at least 97 of his arrows hit the mark on at least 5 of the days, then the Flaming Firefly gives Cupid a marigold. What is the probability that Cupid earns a marigold on any given week? (Cupid, of course, never takes a day off.)
    (e) At the end of each 13-week period, Cupid will be awarded a ringlet of gold, if during that 13 week period he earns at least 12 marigolds. What is the probability that he will receive a ringlet at the end of a 13 week period?
    (f) What is the probability that he will get 13 golden ringlets in the next thirteen 13-week periods?

17. **(Insurance)** An insurance company will not renew an auto policy if the driver received 2 (or more) moving traffic violations during the one year policy period. Assume that 0.01% of drivers will receive a moving violation ticket on any given day (and that no driver receives more than one such ticket on any given day). What is the probability that:
    (a) The company will be willing to renew the policy of a randomly chosen policy holder.
    (b) The company will be willing to renew the policy of at least 92 of the 93 policies which will expire on April 15.

18. **(Watches)** A wholesale distributor for Tic-Toc watches guarantees that a box of 50 watches will contain at most two defective units. Past experience has shown that 1.3% of the watches are defective?
    (a) What is the probability that a box of watches will satisfy the guarantee?
    (b) A customer buys ten boxes from the distributor. What is the probability that at least two of those boxes will not satisfy the guarantee?

§8. EXPECTED VALUE AND DECISION MAKING

We recall the equiprobable sample space for the rolling of two dice:

$$S = \begin{Bmatrix} (1,1) & (1,2) & (1,3) & (1,4) & (1,5) & (1,6) \\ (2,1) & (2,2) & (2,3) & (2,4) & (2,5) & (2,6) \\ (3,1) & (3,2) & (3,3) & (3,4) & (3,5) & (3,6) \\ (4,1) & (4,2) & (4,3) & (4,4) & (4,5) & (4,6) \\ (5,1) & (5,2) & (5,3) & (5,4) & (5,5) & (5,6) \\ (6,1) & (6,2) & (6,3) & (6,4) & (6,5) & (6,6) \end{Bmatrix}$$

Yes, the sample space contains 36 elements, but if you are hoping to roll a seven, then you really don't care if you get it by rolling a 2 and a 5, or by rolling a 4 and a 3. Your concern is with the sum of the two dice — you are interested in the function, let us call it X, which assigns the number 7 to $(4, 3)$, the number 5 to $(3, 2)$, and so on:

The letter X, rather than f, is typically used in this probability-setting.

$$X[(4, 3)] = 4 + 3 = 7, \ X[(3, 2)] = 3 + 2 = 5, \text{ and so on.}$$

In general:

In spite of its name, a random variable is neither random nor is it a variable—it is what it is: **a function**.

DEFINITION 1.10
RANDOM VARIABLE

A **random variable**, X, is a function that assigns a numerical value to each element of a sample space.

The concept of mathematical expectation first appears in Gerolamo Cardano's (1501-1576) book *Liber de Ludo Aleas* (Book on Games and Chance), which many consider to be the first book on probability. One can say that Cardano was truly dedicated to the discipline: Having predicted the date of his death, and realizing on that fatal day that he must have erred somewhere along the line, he resolved the problem and saved his reputation by committing suicide.

EXPECTED VALUE OF A RANDOM VARIABLE

The following table gives the number of children per family, for 1000 families surveyed[1]:

Number of Families	Number of Children
150	0
350	1
325	2
125	3
50	4

To determine the average number of children per family you would simply divide the total number of children by the number of families:

$$\text{Average \# of Children} = \frac{0 \cdot 150 + 1 \cdot 350 + 2 \cdot 325 + 3 \cdot 125 + 4 \cdot 50}{1000}$$

Which can be written in the form:

$$\text{Average \# of Children} = 0 \cdot \frac{150}{1000} + 1 \cdot \frac{350}{1000} + \mathbf{2} \cdot \mathbf{\frac{325}{1000}} + 3 \cdot \frac{125}{1000} + 4 \cdot \frac{50}{1000}$$

Let's look closely at a piece of the above expression, say the term:

1. We are concerned with the random variable that assigns the number of children to a given family.

$$2 \cdot \frac{325}{1000}$$

The **2** is a possible number of children, and the $\frac{325}{1000}$ is the probability that a randomly chosen family has 2 children: (325 of the 1000 families have 2 children)

In probability, the term **expected value** is used instead of average (or mean). Replacing the word "Average" with "Expected," we find that:

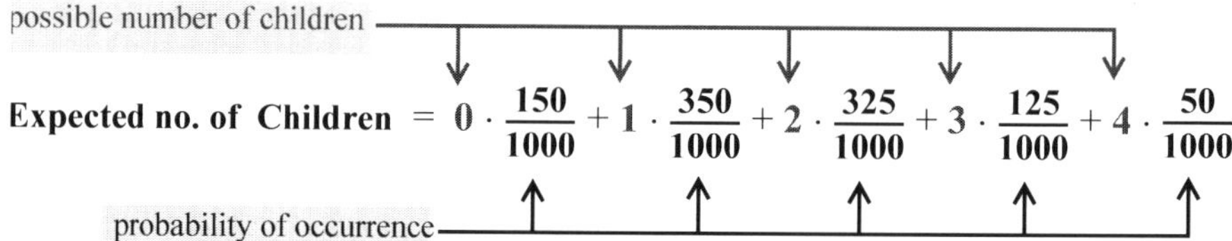

Generalizing::

DEFINITION 1.11
EXPECTED VALUE

Let the random variable X assume values x_1 through x_n. The **expected value of X**, denoted by $E(X)$, is given by:

$$E(X) = x_1 Pr(X = x_1) + \ldots + x_n Pr(X = x_n)$$

IN WORDS: The expected value is the sum of the possible values times their probabilities of occurrence.

EXAMPLE 1.44
URN

Three marbles are drawn, without replacement from an urn containing 3 red marbles, 4 white marbles, and 6 blue marbles. What is the expected number of white marbles drawn?

SOLUTION: We suggest that you begin by replacing the word "expected" with the word "**possible**," and ask yourself "what are the **possible** number of white marbles drawn?" This brings you to:

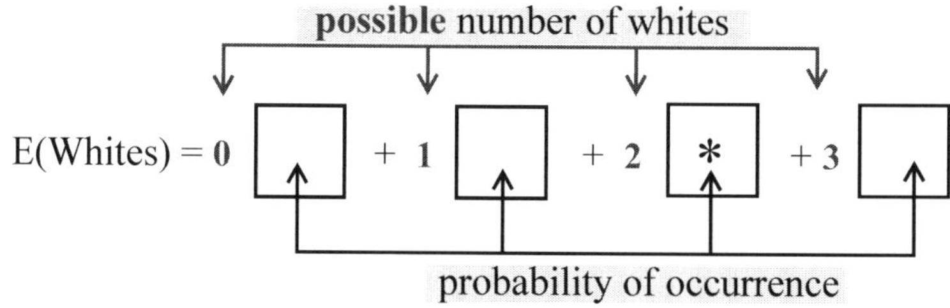

It remains to fill in the four probability boxes. In particular, the $*$-box is to contain the probability of drawing 2 white:

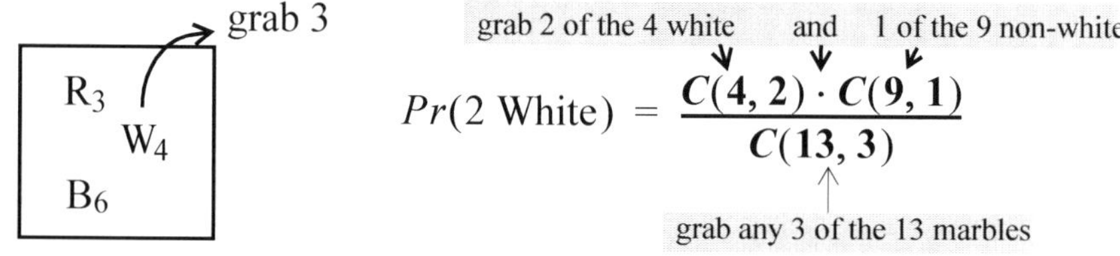

The other boxes are filled in the same way, leading us to:

While on any given draw you cannot end up with a fraction of a white marble, on the **average** you can expect to draw 0.92 white marbles. For example, if you conduct the experiment 1000 times, then you can expect to end up with about 920 white marbles

$$E(\text{Whites}) = 0 \cdot \frac{C(4,0) \cdot C(9,3)}{C(13,3)} + 1 \cdot \frac{C(4,1) \cdot C(9,2)}{C(13,3)} + 2 \cdot \frac{C(4,2) \cdot C(9,1)}{C(13,3)} + 3 \cdot \frac{C(4,3) \cdot C(9,0)}{C(13,3)} \approx 0.92$$

CHECK YOUR UNDERSTANDING 1.32

You are dealt 5 cards from a standard deck. What is the expected number of aces drawn?

Answer: ≈ 0.385

EXAMPLE 1.45

CHECK OUT TIME

The following table records the time, rounded to the nearest minute, that it took for a number of customers on a grocery line to check out. (In particular, from that table we see that 33 customers had a check-out-time of 4 minutes.)

A table which displays the number of times each possible value occurs is said to be a **frequency distribution table**.

Time	1	2	3	4	5	6	7
Frequency	12	21	35	33	18	5	1

Find the expected waiting time.

SOLUTION: One can easily go from the above frequency distribution table to a probability distribution table. The first step is to determine the total number of customers processed (total frequency):

$$12 + 21 + 35 + 33 + 18 + 5 + 1 = 125$$

Dividing the frequency of a value by the total frequency, yields the probability of occurrence of that value:

For example, since 33 of the 125 individuals are assigned the value of 4 (minutes), the probability of occurrence of that value is:

$$Pr(4 \text{ minutes}) = \frac{33}{125}$$

values →

Time	1	2	3	4	5	6	7	Sum
Fequency	12	21	35	33	18	5	1	125
Probability	$\frac{12}{125}$	$\frac{21}{125}$	$\frac{35}{125}$	$\frac{33}{125}$	$\frac{18}{125}$	$\frac{5}{125}$	$\frac{1}{125}$	1

probability of occurrence →

Note: being a probability distribution the sum of the probabilities must add up to 1

With the probability distribution at hand, we can calculate the expected value of checkout time (average amount of minutes one has to wait at the cash register):

possible values

$$E(\text{Time}) = 1 \cdot \frac{12}{125} + 2 \cdot \frac{21}{125} + 3 \cdot \frac{35}{125} + 4 \cdot \frac{33}{125} + 5 \cdot \frac{18}{125} + 6 \cdot \frac{5}{125} + 7 \cdot \frac{1}{125} \approx 3.34$$

probability of occurrence

Answer: In Example 1.45 the expected waiting time was found to be 3.34 minutes. Here, it is 3.28 minutes, so the situation improved a bit.

CHECK YOUR UNDERSTANDING 1.33

The manager of Example 1.45 tried out a different scanning machine at the checkout line, and tabulated the following data:

Time	1	2	3	4	5	6	7	8	9
Frequency	15	32	25	19	16	9	3	0	1

Did the checkout time situation improve from that of Example 1.45?

EXPECTED WINNINGS

Monetary expected values in games of chance are often called expected winnings. Consider the following examples.

EXAMPLE 1.46
DRAW A CARD

Maverick is to draw a card from a standard deck. If it is an ace, he wins \$5; if it is a face card, he wins \$2; if he doesn't draw either a face card or an ace, then he loses \$1. Should he play the game?

SOLUTION: The possible values (winnings) are \$5, \$2, and $-\$1$, with probability of occurrence:

no. of aces → $P(\$5) = \frac{4}{52}$ no. of face cards → $P(\$2) = \frac{12}{52}$ no. of non-aces and non-face cards → $P(-\$1) = \frac{52-4-12}{52} = \frac{36}{52}$

Putting this together, we have:

possible winnings

$$E(\text{Winnings}) = \$5 \cdot \frac{4}{52} + \$2 \cdot \frac{12}{52} + (-\$1) \cdot \frac{36}{52} \approx \$0.15$$

probability of occurrence

Maverick should play the game, since **on the average** he will win fifteen cents per game.

EXAMPLE 1.47
RAFFLE

A church sells 1000 raffle tickets at a dollar each. You buy one ticket. Two tickets are drawn without replacement. If yours is drawn first, you win \$350; if it is drawn second, you win \$225. What are your expected winnings?

Here is a "common sense" approach to the problem: The church takes in \$2000 and gives back \$1500. It follows that the 1000 ticket holders end up losing \$500—suggesting that, on the average, each ticket holder (including you) will end up losing \$0.50. This, as you will see, turns out to be the correct answer.

SOLUTION: Since the church is not going to give you back the \$1 you paid for the ticket if you win, the actual possible winnings are \$349, \$224, and $-\$1$.

It is easy to see that the probability of winning the first prize is $\frac{1}{1000}$, and the probability of not winning any prize is $\frac{998}{1000}$. Please try to figure out the probability that you will win the second prize before reading on.

$$E(\text{Winnings}) = \$349 \cdot \frac{1}{1000} + \$224 \cdot \boxed{?} + (-\$1) \cdot \frac{998}{1000}$$

Note also that the number in the above box must be $1/1000$ since the sum of the probabilities has to sum to 1.

If you said $\frac{1}{1000}$, you are correct. The probability that you are holding the ticket for any given number (including the number that will be called for the second prize) is $\frac{1}{1000}$.

Conclusion:

$$E(\text{Winnings}) = \$349 \cdot \frac{1}{1000} + \$224 \cdot \frac{1}{1000} + (-\$1) \cdot \frac{998}{1000} = -\$0.425$$

An alternate approach: $E(\text{Winnings}) = \$350 \cdot \frac{1}{1000} + \$225 \cdot \frac{1}{1000} + (-\$2) = -\$0.425$

How so?

CHECK YOUR UNDERSTANDING 1.34

In a moment of unbridled benevolence, in addition to the \$1000 first prize, and \$500 second prize, the church of Example 1.47 decides to award two third prizes of \$100. You purchase one ticket. What are your expected winnings?

Answer: -\$0.20.

EXPECTED VALUES FOR BERNOULLI TRIALS

Roll a die 600 times. What is the expected number of fives rolled? One hundred, right? After all, the probability of rolling a five is $\frac{1}{6}$, and since you have 600 shots at it, you would certainly expect to throw (about) $600 \cdot \frac{1}{6} = 100$ fives in 600 rolls. By the same token, since the probability of flipping a Head with a coin is $\frac{1}{2}$, if you flip a coin 1000 times you would expect to flip about $1000 \cdot \frac{1}{2} = 500$ heads. In either case, intuition tells us that if n independent trials are conducted, and if the probability of success in any of the trials is p, then the expected number of successes is given by the product np. Intuition is right on target:

BERNOULLI TRIALS

THEOREM 1.13 If n independent Bernoulli trials are conducted, each with probability of success p, then the expected number of successes is given by pn.

(You are invited to establish the above result in Exercise 31.)

EXAMPLE 1.48
GRADUATING WITH HONORS

Assuming that 7.2% of graduating students graduate with honors, find the expected number of students that will receive honors in a graduating class of 735 students?

SOLUTION: This is a Bernoulli trials situation, with $n = 735$ and $p = 0.072$. Applying Theorem 1.13, we find that:

Expected number receiving honors $= pn = (0.072)(735) = 52.92$

CHECK YOUR UNDERSTANDING 1.35

An individual, X, is placed in an isolated room with the five cards below.

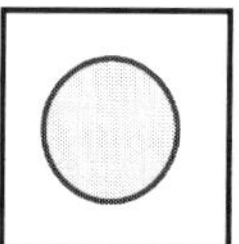 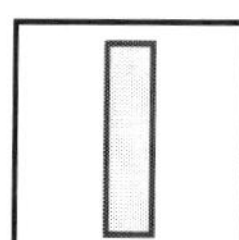 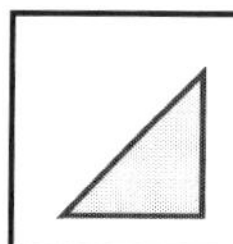 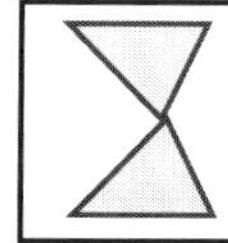 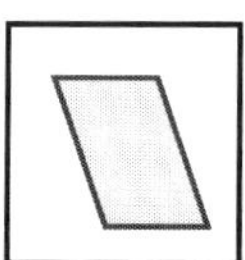

Another individual, Y, has five identical cards in another room, and concentrates on one of them. The hope is that through some sort of mental telepathy, the thoughts of Y can be transmitted to X, and that X will be able to identify the card. The experiment is conducted 100 times, and will be labeled *successful*, if the number of matches equals or exceeds the expected number by at least 25%. How many matches are needed for the experiment to be labeled *successful*?

Answer: At least 25.

DECISION MAKING

Probability is neat and exact when throwing a die or drawing a card, but gets murkier and murkier as more and more non-quantitative factors come into play. We can, for example, calculate the probability that Maverick will be dealt a Jack-high flush, and can also calculate the probability that it would be the strongest of six poker hands dealt. One cannot, however, so easily calculate that it will be the winning hand; the betting may force Maverick to fold. Is someone bluffing?

Should I do this, or should I do that? Expected values can influence the decision process. Consider the following example:

EXAMPLE 1.49
GAMES OF CHANCE

You have a choice of playing two games.
Game 1: Draw a card from a standard deck. If it is an ace, you win \$5; if it is a face card, you win \$2; if neither, then you lose \$1.
Game 2: Draw two cards from a standard deck. If it is a pair, you win \$75; else you lose \$4.
Which game should you play?

SOLUTION: From Example 1.46:

$$E(\text{Win-Game1}) = \$5\cdot\frac{4}{52}+\$2\cdot\frac{12}{52}+(-\$1)\cdot\frac{36}{52}\approx\$0.15$$

Now for Game 2:

choice of 13 types followed by two cards of that type

$$E(\text{Win-Game2}) = \$75\cdot\frac{13\cdot C(4,2)}{C(52,2)}+(-\$4)\cdot\left(1-\frac{13\cdot C(4,2)}{C(52,2)}\right)$$

$$\approx \$4.41-\$3.76=\ \$0.65$$

Conclusion: Play Game 2.

CHECK YOUR UNDERSTANDING 1.36

You have a choice of playing Game 2 of Example 1.49 or you can **pay** \$1.50 to play the following Game 3:

Roll a dice three times. If you roll a one exactly once you will get your \$1.50 back. If you roll a two exactly twice you will get \$10. If you roll a three exactly three times you will get \$25.

Should you play Game 2 or Game 3?

Answer: Game 2.

EXAMPLE 1.50
GAS STATIONS

Mr. Goodfellow is going to open a gas-station in one of two developing regions of the county. Currently, there is little competition in either region. There is a 0.7 probability that heavy competition will soon emerge at site A, and a 0.6 probability that it will soon emerge at site B. With heavy competition, he anticipates an annual profit of \$50,000 at site A, and an annual profit of \$60,000 at site B. Without heavy competition, profits would be \$45,000 and \$40,000 at sites A and B, respectively. Which site should he choose?

SOLUTION:

SEE THE PROBLEM

	Probability of heavy competition	Profit with heavy competition	Profit without heavy competition
Site A	0.7	\$50,000	\$45,000
Site B	0.6	\$60,000	\$40,000

We determine the expected annual profit at both sites:

$$E(\text{profit at A}) = \$50{,}000(0.7) + \$45{,}000(0.3) = \$48{,}500$$
$$E(\text{profit at B}) = \$60{,}000(0.6) + \$40{,}000(0.4) = \$52{,}000$$

Conclusion: Site B should be chosen.

CHECK YOUR UNDERSTANDING 1.37

An oil company is considering drilling in one of two locations. It is anticipated that there is a 0.25 probability that oil will be discovered at site A, yielding revenues of \$45 million. The probability that oil will be discovered at site B is 0.17, and it would yield revenues of \$60 million. The cost for developing sites A and B are \$4.2 and \$5.5 million, respectively. Which site should be chosen to maximize expected profit?

Answer: Site A.

EXAMPLE 1.51

FISH MARKET

Past experience at the Friendly Fish Market gave rise to the following probability distribution for the number of calls for one salmon on any given day:

no. of calls	0	1	2	3	4
probability	0.1	0.2	0.3	0.3	0.1

On the average, each salmon costs the market $17, and the average profit per salmon sold is $11. Salmon at the end of a day must be discarded. How many salmon should be ordered? (Assume that at least one fish will be ordered).

SOLUTION: Friendly has four options: Order 1, 2, 3, or 4 salmon. We calculate the expected profit, for each of the four options:

Ordering 1 Salmon		
number sold	probability	profit
0	0.1	$1(-\$17)+0(\$11) = -\$17$
1	$1-0.1 = 0.9$	$0(-\$17)+1(\$11) = \$11$
Expected Profit: $-\$17(0.1)+\$11(0.9) = \$8.20$		

Ordering 2 Salmon		
number sold	probability	profit
0	0.1	$2(-\$17)+0(\$11) = -\$34$
1	0.2	$1(-\$17)+1(\$11) = -\$6$
2	$1-(0.1+0.2) = 0.7$	$0(-\$17)+2(\$11) = \$22$
Expected Profit: $-\$34(0.1)-\$6(0.2)+\$22(0.7) = \$10.80$		

Ordering 3 Salmon		
number sold	probability	profit
0	0.1	$3(-\$17)+0(\$11) = -\$51$
1	0.2	$2(-\$17)+1(\$11) = -\$23$
2	0.3	$1(-\$17)+2(\$11) = \$5$
3	0.4	$0(-\$17)+3(\$11) = \$33$
Expected Profit: $-\$51(0.1)-\$23(0.2)+\$5(0.3)+\$33(0.4) = \$5$		

Ordering 4 Salmon		
number sold	probability	profit
0	0.1	$4(-\$17)+0(\$11) = -\$68$
1	0.2	$3(-\$17)+1(\$11) = -\$40$
2	0.3	$2(-\$17)+2(\$11) = -\$12$
3	0.3	$1(-\$17)+3(\$11) = \$16$
4	0.1	$0(-\$17)+4(\$11) = \$44$
Exp. Profit: $-\$68(0.1)-\$40(0.2)-\$12(0.3)+\$16(0.3)+\$44(0.1) = -\$9.20$		

Conclusion: Friendly should order 2 salmon.

EXAMPLE 1.52

QUALITY CONTROL

You are considering instituting one of two quality control procedures on an assembly line. Past experience has shown that 4% of the units on the line are defective. Procedure **A** is 97% accurate in detecting a defective unit, and procedure **B** is 94% accurate. It will cost $2/unit to initiate procedure **A**, and $1/unit to initiate procedure **B**. It costs the company $19 to manufacture each unit. The company realizes a $23 profit on a sold non-defective unit, and a total loss of $55 (including cost of production) on a sold defective unit. Should you introduce one of the quality control procedures, and if so, which one?

SOLUTION: The main challenge is to sort through all of the given information and to compress it in some manageable form:

Probability unit is good: 0.96			**PROFIT**		
Quality Procedure	Probability that unit is bad and is not shipped	Probability that unit is bad and is shipped	Profit on each good unit shipped	Loss due to shipment of bad unit	Loss due to non-shipment of bad unit
None	0	0.04	$23	$55	N/A (no quality control—all units are shipped)
A	(.04)(.97) = 0.0388	(.04)(.03) =0.0012	$(23 - 2) =$21	$(55 + 2) = $57	$(19 + 2) = $21
B	(.04)(.94) = 0.0376	(.04)(.06) =0.0024	$(23 - 1) = $22	$(55 + 1) = $56	$(19 + 1) = $20

The rest is fairly straight forward; we simply calculate the expected dollar profit per unit in each of the three cases:

$$E(\text{None}) = \mathbf{23}(0.96) - \mathbf{55}(0.04) = 19.88$$
$$E(\text{A}) = \mathbf{21}(0.96) - \mathbf{57}(0.0012) - \mathbf{21}(0.0388) \approx 19.28$$
$$E(\text{B}) = \mathbf{22}(0.96) - \mathbf{56}(0.0024) - \mathbf{20}(0.0376) \approx 20.23$$

possible profits

probability of occurrence

Conclusion: To optimize profit, procedure B should be initiated.

CHECK YOUR UNDERSTANDING 1.38

The Yankees are in the pennant race, and Beauford decides to order 1000 Yankee World Series shirts, at $6 a shirt. Beauford will put the same price tag of either $21, or $15, or $7, on each of the 1000 shirts, and being a very stubborn individual, will not alter that price under any circumstance.

At $7: All 1000 shirts will sell, regardless.

At $15: All 1000 shirts will sell if the Yankees get into the World Series, but none if they don't.

At $21: All 1000 shirts will sell if the Yankees win the World Series; only 500 will sell if the Yankees get into the series but lose the series; and none will sell if the Yankees do not get into the series.

Assume that there is a 35% chance that the Yankees will get into the World Series; and, if they do, a 60% chance that they will win the series. Which of the three prices will maximize expected profit.

Answer: $7

EXERCISES

1. **(Children per Family)** The following table records the number of children per family in a certain town:

Children	0	1	2	3	4	5	6	7
Frequency	25	53	71	18	5	2	0	1

Determine, to two decimal places, the expected number of children per family.

2. **(Cumulative Grade)** The following table records the grade distribution for last year's Math 101 sections.

Grade	A	B	C	D	F
Frequency	25	47	341	62	36

Determine, to two decimal places, the expected cumulative grade for the course, given the corresponding numerical values: $A = 4, B = 3, C = 2, D = 1, F = 0$.

3. **(Coins)** Flip three coins. What is the expected number of Tails tossed?
4. **(Dice)** Roll a pair of dice. What is the expected sum rolled?
5. **(Delegation)** A delegation of three is to be randomly selected from a group of 5 men and 7 women. Determine, to two decimal places, the expected number of women on the delegation.
6. **(Five Cards)** Draw, without replacement, five cards from a standard deck. Determine, to two decimal places, the expected number of aces drawn.
7. **(Three Cards)** Draw, without replacement, three cards from a standard deck. Determine, to two decimal places, the expected number of clubs drawn.
8. **(Transistors)** A box of 100 transistors contains (exactly) 3 faulty units. Ten of the transistors are randomly chosen. Determine, to two decimal places, the expected number of faulty units chosen.
9. **(Dice)** You roll a pair of dice. If the same number shows on both dice (doubles), you win \$10, otherwise you lose \$1.50. What are your expected winnings?
10. **(Die)** You roll a die. If you roll an odd number, you will receive as many dollars as the number rolled. If you roll an even number, you will lose as many dollars as the number rolled. What are your expected winnings?
11. **(Urn)** You draw a marble from an urn containing 5 Red marbles, 3 Blue marbles, and 2 White marbles. If you draw a white marble, you win \$5; if you draw a blue marble, you win \$2; if you draw a red marble you lose \$3.20. What are your expected winnings?
12. **(Urn)** You draw two marbles from an urn containing 5 Red marbles, 3 Blue marbles, and 2 White marbles. If both are white, you win \$5; if both are blue, you win \$2; otherwise you lose \$3.20. What are your expected winnings?

13. **(Cards)** You draw two cards without replacement from a standard deck. If you draw a pair, you win \$10; if both cards are of the same suit, you win \$1; otherwise, you lose \$1. What are your expected winnings?
14. **(Church Fair)** You pay \$1 for one of 1000 tickets at a church fair. Four numbers are drawn. The holder of the first number drawn will win \$500, the holders of the remaining three numbers will each receive \$100. What are your expected winnings?
15. **(Church Fair)** You pay \$2 for one of 1000 tickets at a church fair. There is a first prize of \$500, two second prizes of \$250 each, and three third prizes of \$100 each. What are your expected winnings?
16. **(Lottery)** In a lottery, you are to pick 5 numbers from a card containing the numbers 1 through 50 (order is of no consequence). To win, your 5 numbers must match the 5 lottery numbers drawn. You purchase a ticket for \$1. What are your expected winnings, if the winning prize is \$1,000,000?
17. **(Lottery)** In a lottery, you are to pick 5 numbers from a card containing the numbers 1 through 50 (order is of no consequence). If your 5 numbers match the 5 lottery numbers drawn you win a million dollars. If exactly 4 of your numbers match you win \$10,000. You purchase a ticket for \$2. What are your expected winnings?
18. **(Dice)** A pair of dice are rolled 75 times. What is the expected number of times a sum of five is rolled?
19. **(Kicking the Smoking Habit)** Assume that 37% of individuals enrolled in SSS (Stop Smoking Seminar) will quit smoking within a month's time. What is the expected number of smokers, out of 100, that will quit smoking within a month's time?
20. **(Negative Reaction)** The probability that a patient will have a negative reaction to a certain medication is 0.03. The medication is administered to 75 patients. What is the expected number of negative reactions?
21. **(Coins)** Flip two coins 40 times. What is the expected number of times you will flip two heads?
22. **(Multiple Choice)** A student guesses blindly on each of ten 4-choice multiple choice questions. What is his expected (percentage) grade?
23. **(Grade)** The probability that a student receives an A in English 101 is 0.14. Twenty-five students are in the class. What is the expected number of A's?
24. **(Multiple Choice)** On a 10 question multiple-choice test, each question has 4 possible answers. A correct answer earns 3 points, an incorrect answer earns minus-one point, and a blank receives 0 points.
    (a) What is the expected point-value earned if a student guesses randomly on each of the 10 questions?
    (b) Would the answer to part (a) be any different if the test consisted of 25 questions? Justify your answer.
25. **(Weighted Die)** With reference to the following probability distribution

Face rolled	1	2	3	4	5	6
Probability	1/12	2/12	3/12	3/12	1/12	2/12

    (a) Determine, to two decimal places, the expected roll of the die.
    (b) The die is rolled 100 times. What is the expected number of times a 3 is rolled?

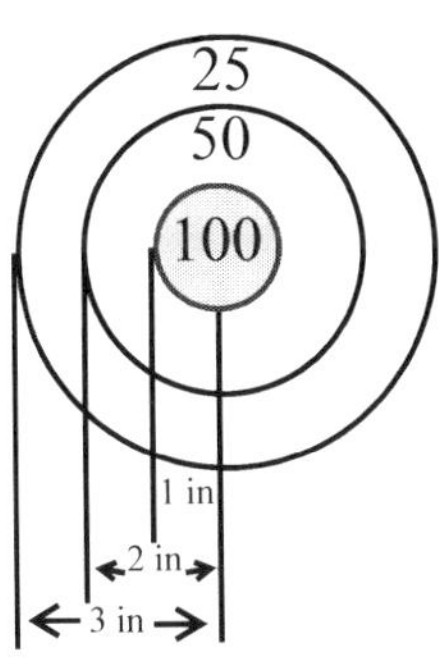

26. **(Dart)** A dart is thrown at the adjacent board and will randomly end up somewhere on the board. What is the expected point score on the throw? Note: The area of a circle of radius r is πr^2

27. (a) **(Three Darts)** Three darts are thrown at the adjacent board and will randomly end up somewhere on the board. What is the expected combined point score of the three tosses?

    (b) **(Thirty Darts)** Thirty darts are thrown at the adjacent board and will randomly end up somewhere on the board. What is the expected number of times that the bull's-eye will be hit?

28. **(Dice)** You roll a pair of dice. If the same number shows on both dice (doubles), you win \$10, otherwise you lose a certain amount. Determine the amount in order for the game to be fair (0 expected winnings).

Exercises 29-32. (Sum Theorem for Random Variables) If several random variables, $X_1, X_2, \ldots, X_n$ are defined on the same sample space, then their sum $S = X_1 + X_2 + \cdots + X_n$ is also a random variable, and it can be shown that:

$$E(S) = E(X_1) + E(X_2) + \ldots + E(X_n)$$

In words: The expected value of the sum is the sum of the expected values.

To illustrate, consider the rolling of a die. Let X_1 be the random variable (function) that assigns twice the number rolled, let X_2 be the random variable that assigns the number 6 to any odd roll and the number 10 to any even roll, and let S denote the random variable that is the sum of the random variables X_1 and X_2:

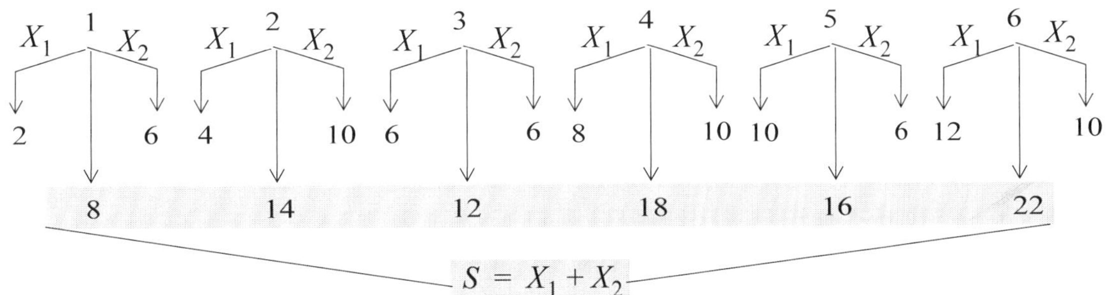

Evaluating the expected values for the variables X_1, X_2, and $X_1 + X_2$, we see that $E(X_1 + X_2) = E(X_1) + E(X_2)$:

$$E(X_1) = 2 \cdot \frac{1}{6} + 4 \cdot \frac{1}{6} + 6 \cdot \frac{1}{6} + 8 \cdot \frac{1}{6} + 10 \cdot \frac{1}{6} + 12 \cdot \frac{1}{6} = 7$$

$$E(X_2) = 6 \cdot \frac{3}{6} + 10 \cdot \frac{3}{6} = \mathbf{8}$$

$$E(X_1 + X_2) = 8 \cdot \frac{1}{6} + 14 \cdot \frac{1}{6} + 12 \cdot \frac{1}{6} + 18 \cdot \frac{1}{6} + 16 \cdot \frac{1}{6} + 22 \cdot \frac{1}{6} = \mathbf{15}$$

29. **(Keys)** Each of 100 people drops a key in a hat and then draws a key from the hat. Show that the expected number of people that will end up drawing his or her key from the hat is one? Suggestion: Let X_i be the random variable that assigns the value 1 if person i draws his or her key from the hat, and the value 0 if not.

30. **(Keys)** Will the answer to Exercise 29 change if one person drops his or her key in the hat, or if a million people drop their keys in a (large) hat? Justify your answer.

31. (**Theory**) Show that if n independent Bernoulli trials of an experiment, with probability of success p, are conducted, then the expected number of successes is given by np.
Suggestion: Let X_i be the random variable that assigns the value 1 if trial i results in a success, and 0 if not.

32. (**Blood Tests**) One thousand individuals are to be subjected to a blood test which returns a positive result with probability 0.01. One could administer the test to each individual, requiring 1000 tests. Instead, suppose you partition individuals into 10 groups of 100, pool the blood of each group, and then test the 10 vessels of combined blood. If a test is negative, then that one test suffices for 100 individuals. If the test is positive, then each of the 100 individuals will be tested separately, and a total of 101 tests are required for those 100 individuals. What is the expected number of tests to be administered?
This problem is based on a technique developed during World War II for testing recruits for syphilis (see R. Dorfman, *The detection of defective members of large populations,* Annals of Mathematical Statistics, vol. 14 (1943), pp. 436-440).

33. (**Petersburg Paradox**) Two players A and B play a game in which A tosses a coin. The game continues until the first head appears. Player B will give player A a coin if a head appears on the first throw, two coins if a head appears for the first time on the second throw, four coins if on the third throw, eight if on the fourth, and so on. What amount should A pay B before the start of the game to make it a fair game (zero expected winnings)?
Suggestion: To make the "arithmetic easier," assume that A will receive \$2 if a head comes up on the first toss (with probability 1/2); \$4 if on the second (with probability 1/4); \$8 if on the third (with probability 1/8); and so on.

DECISION MAKING

34. (**Three Games**) Which of the following three games of chance would you play?
    I. Draw a card from a standard deck. If it is an Ace, you win \$5; if it is Red, you win \$1 (a Red Ace wins \$6); else, you lose \$2.
    II. Draw a card from a standard deck. If it is a club or a face card, you win \$2.50; else, you lose \$1.
    III. Draw two cards, without replacement, from a standard deck. If they are of the same suit, you win \$5; if it is a pair, you win \$10; else, you lose \$2.
    Which of the three games should you play?

35. (**Advertising)** A company has budgeted \$50,000 for advertising on TV, or on the radio, or in magazines. A study has yielded the following probability distribution for the number of potential customers (in thousands) that will be reached for the \$50,000 in each of the three media:

TV	**100**	**90**	**80**	**70**	**60**
Probability	**0.15**	**0.20**	**0.3**	**0.25**	**0.10**
Radio	75	70	60	50	40
Probability	0.25	0.25	0.25	0.20	0.05
Magazines	**50**	**40**	**30**	**20**	**10**
Probability	**0.50**	**0.30**	**0.10**	**0.06**	**0.04**

Which of the three mediums should be selected to reach the most customers?

36. (**Farming**) There is a 0.25 probability that the temperature will drop below freezing, and Best Orange Company has to decide on whether or not to fire up its smudge pots to protect its \$9500 crop investment. It will cost the company \$2000 to employ the smudge pots. If there is a freeze the crop would realize a revenue of \$23,000 at market, as opposed to a revenue of \$17,000 if there is no freeze (a matter of supply and demand). The crop will not survive a freeze without the smudge pots. Should the pots be used?

37. (**Political Campaign**) You are a political campaign manager, and are to choose one of three fund raising strategies: a door-to-door campaign, a mailing campaign, or a phone campaign. Which option should you take, given the following data:

Door-To-Door Average Contribution: \$15.70 per person		**Mailing** Average Contribution: \$5.10 per person		**Phone** Average Contribution: \$2.50 per person	
Number of contributors	Probability	Number of contributors	Probability	Number of contributors	Probability
3000	0.6	6000	0.4	11000	0.7
2500	0.3	5500	0.4	8000	0.2
2000	0.1	2500	0.2	7000	0.1
Operating cost: \$17,000		Operating cost: \$12,000		Operating cost: \$6,000	

38. (**Investment**) You plan to invest in one of three options: Safe, Not-So-Bad, and Cross-Your-Fingers. Which way should you go if:

SAFE % return on investment	3.2				
Probability	1				
NOT-SO-BAD % return on investment	5	4	3.2	1.2	0
Probability	0.15	0.25	0.20	0.25	0.15
CROSS-YOUR-FINGERS % return on investment	25	20	5	0	
Probability	0.02	0.10	0.2	0.68	

39. (**Airline**) An airline can designate one of two aircraft for flights from New York to Chicago. Aircraft A has a seating capacity of 170, and aircraft B has a capacity of 200. Tickets on either aircraft will go for \$350. Total fixed operating expenses (including craft-depreciation) are \$40,000 per flight for aircraft A, and \$45,000 for flight B. There would also be a \$75 per passenger cost to the airline, independent of aircraft. Which aircraft should the airline choose, to maximize profit subject to the following probability distribution of the number of ticket requests:

Ticket Requests	200	190	180	170	160	150	140
Probability	0.30	0.20	0.17	0.15	0.10	0.05	0.03

40. (**Horse Race**) Blue Velvet, Big Bucks, and Lost Cause are running in the derby. Blue Velvet pays 2.5 to 1 if she wins, 1.7 to 1 if she at least places (comes in first or second), and 1.3 to 1 if she at least shows (comes in first, second, or third). Big Bucks pays 3.5 to 1, 2.1 to 1, and 1.7 to 1 to win, place, or show, respectively. Lost Cause pays 13 to 1, 9 to 1, and 8.2 to 1 to win, place, or show, respectively. You figure that there is a 0.23 probability that Blue Velvet will win, a 0.37 probability that she will place, and a 0.39 probability that she will show; a 0.17, 0.19, and 0.23 probability that Big Bucks will win, place or show, respectively; and a 0.098, 0.12, and 0.13 probability that Lost Cause will win, place or show, respectively. Deep Pockets is going to place a two dollar bet on one of the three horses. Which of the three horses should he bet on to: (a) Win (b) Place (c) Show

41. **(Insurance)** Your friend asks you to help her determine whether or not to carry collision insurance on her car. The premiums on a $100-deductible collision insurance are $250 for the year. You impress your friend when you inform her that the average collision repair cost is $1200. Wanting to impress you in return, she tells you that the probability of her being involved in an accident during the year is 0.07, and that there is virtually no chance of her being involved in more than one collision. It is your turn to impress her back, by mathematically determining whether or not she should carry the collision insurance.

42. **(Rock Group)** You are given the opportunity to invest money to bring the Sliding Pebble rock group to town. If at least $70,000 is raised, then their lead singer, Donald Sharp, will perform, and you can expect to double your investment. If less than $70,000 but more than $50,000 is raised, then all but Donald will appear, and you can expect to make a profit of 10% on your investment. If $50,000 or less is raised, only the drummer will show up, and you will probably end up losing all of your investment. You decide to liquidate one of three savings bonds valued at $10,000, $20,000, and $30,000 and to invest all of the money in the concert. After many sleepless nights, and many hours of contemplation, you come up with the following probability distribution for the amount of concert money (in thousand dollars) that will be raised, excluding your investment. Which bond should you sell?

Amount raised (in thousands)	$\$ < 20$	$20 \le \$ < 30$	$30 \le \$ < 40$	$40 \le \$ < 50$	$50 \le \$ < 60$	$60 \le \$ < 70$	$\$ \ge 70$
Probability	0.25	0.25	0.20	0.10	0.10	0.07	0.03

43. **(Car Rental)** Rent-4-Wheels makes a profit of $12 per day for each car rented, and loses $4 per day for each car that is not rented. How many cars should the company have on hand to maximize profit, subject to the following probability distribution:

Anticipated Number of Rental Requests	12	11	10	9	8	7	6
Probability	0.09	0.11	0.22	0.24	0.12	0.12	0.10

44. An urn containing 10 red marbles, 8 blue marbles and two white marbles. You have three options:
    A. Draw two marbles without replacement from the urn and: receive $3 for each red marble drawn, but will have to pay $4 for each blue marble drawn along with $7 for each white marble drawn.
    B. You can pay $2 to replace one of the white marbles in the original urn with a blue marble, and then proceed as in option A.
    C. You can pay $3 to add three red marbles to the original urn, and then proceed as in option A.

    Should you play the game? If so, which option should you choose?

45. **(Game Show)** You are in a game show. There are three closed doors. Behind one of the doors is $100,000. Sacks of potatoes lie beyond the other two doors. You get to open one door and keep whatever is beyond it.

    You choose door 1. Before it is opened the host of the show, aware of the location of the check, opens door 3, revealing a sack of potatoes. The host then gives you the opportunity to stick with door 1, or to switch to door 2. Should you switch?

PART 1 SUMMARY

DEFINITIONS AND THEOREMS

EQUIPROBABLE SAMPLE SPACE	An **equiprobable sample space** for an experiment is a set that represents all of the possible outcomes of the experiment, with each outcome **as likely** to occur as any other. Let S be an equiprobable sample space (*possible outcomes of an experiment*), and let E be a subset of S (the event, or *successes*). Then:
PROBABILITY	$Pr(E) = \frac{\#(E)}{\#(S)}$ ← number of successes / ← number of possibilities
ODDS	Odds (in favor of E) $= \frac{\#(E)}{\#(E^c)}$ ← number of successes / ← number of failures Odds (against E) $= \frac{\#(E^c)}{\#(E)}$ ← number of failures / ← number of successes
COMPLEMENT OF AN EVENT	The probability that something does not occur is one minus the probability that it does occur. Equivalently: the probability that something does occur is one minus the probability that it does not occur. 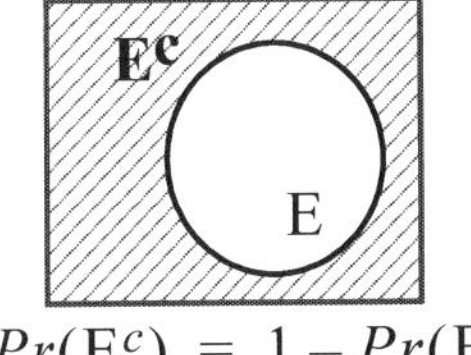 $Pr(E^c) = 1 - Pr(E)$
UNION OF EVENTS	The probability that one or another thing happens (possibly both), is the probability that one happens, plus the probability that the other happens, minus the probability that they both happen. 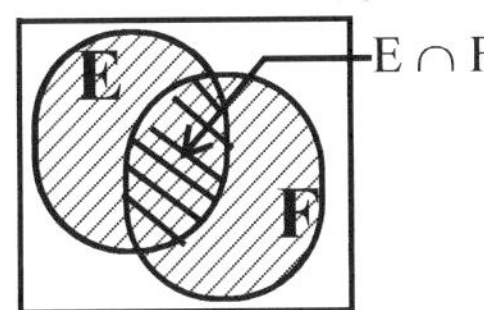 $Pr(E \cup F) = Pr(E) + Pr(F) - Pr(E \cap F)$
MUTUALLY EXCLUSIVE EVENTS	If E and F are **mutually** exclusive ($E \cap F = \varnothing$), then the probability that one or the other happens is the probability that one happens plus the probability the other happens. 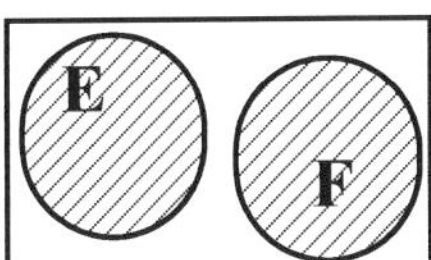 $Pr(E \cup F) = Pr(E) + Pr(F)$

CONDITIONAL PROBABILITY	The probability of a success, given a condition that restricts the possible outcomes of the experiment; as in: $\text{Pr(King} \mid \text{Face Card)} = \frac{4}{12}$ ↑ read: **given**
THEOREM	For E and F events, with $Pr(F) \neq 0$: $Pr(E\vert F) = \frac{Pr(E \cap F)}{Pr(F)}$
PROBABILITY OF TWO EVENTS OCCURRING	The probability of two events occurring is the probability that one occurs times the probability that the other occurs, **given** that the one has occurred: For events E and F: $Pr(E \cap F) = Pr(F)Pr(E\vert F)$
INDEPENDENT EVENTS	Two events E and F are independent if the occurrence of one has no influence on the occurrence of the other: $Pr(E\vert F) = Pr(E)$ and $Pr(F\vert E) = Pr(F)$. If E and F are independent events, then: $Pr(E \cap F) = Pr(E)Pr(F)$
BAYES' THEOREM	Let $U_1, U_2, \ldots, U_n$ be n mutually exclusive events, each with positive probability, whose union is the sample space S. Let E be an event in S such that $Pr(E) \neq 0$. Then for any $1 \le i \le n$: $Pr(U_i\vert E) = \frac{Pr(U_i \cap E)}{Pr(U_1 \cap E) + Pr(U_2 \cap E) + \ldots + Pr(U_n \cap E)}$
TO APPLY:	Construct an appropriate probability tree and divide the probability of ending up at E via U_i, by the sum of the probabilities of ending up at E in any manner.

THE FUNDAMENTAL PRINCIPLE OF COUNTING AND 3 OF ITS CONSEQUENCES

Symbol	Represents the number of:	Evaluate
$n \cdot m$	n-choices followed by m-choices	multiply
$n!$	ways of ordering of n objects	$n! = 1 \cdot 2 \cdot 3 \cdots n$
$P(n, r)$	ways of picking r from n, when order counts	$P(n, r) = \frac{n!}{(n-r)!}$
$C(n, r)$	ways of grabbing r from n, when order does not count	$C(n, r) = \frac{n!}{(n-r)!r!}$

BERNOULLI TRIAL	An experiment with two possible outcomes; one termed a success, and the other, a failure.
THEOREM	If n **independent Bernoulli trials** with probability of success p are conducted, then the probability of obtaining exactly r out of n successes $(0 \le r \le n)$ is given by: $$Pr(r \text{ successes out of } n) = C(n, r) \cdot p^r \cdot (1-p)^{n-r}$$
RANDOM VARIABLE	A random variable is a function that assigns a numerical value to each element in a sample space.
EXPECTED VALUE	Let the random variable X assume values x_1 through x_n. The **expected value of X**, denoted by $E(X)$, is given by: $$E(X) = x_1 Pr(X = x_1) + \ldots + x_n Pr(X = x_n)$$
IN WORDS:	Expected value is the sum of the possible values times their probabilities of occurrence.
EXPECTED VALUE FOR BERNOULLI TRIALS	If n independent Bernoulli trials are conducted, each with probability of success p, then the expected number of successes is given by pn.

Review Exercises
Brief

1. A card is drawn from an ordinary deck of 52 cards. Find the probability that the card is:
   (a) A club.
   (b) A red face card
   (c) A red card or a face card.
   (d) A red card that is not a face card.
   (e) An 8 or a heart, but not the eight of hearts.
2. A marble is drawn from an urn containing two red marbles numbered 1 and 2, and three white marbles numbered 1, 2, and 3. A marble is drawn at random.
   (a) What is the probability that the marble is not numbered 2?
   (b) What are the odds in favor of the marble being red?
   (c) What are the odds against the marble displaying an odd number?
3. A high school has 550 students, 56 of whom are in band, 34 in chorus, and 50 in orchestra. 3 students are in the band, the chorus, and the orchestra. Two are in the orchestra and band but not in chorus. Four are in the band and chorus, and eight are in the orchestra and chorus. What is the probability that a student chosen at random is:
   (a) Neither in the band nor the chorus?
   (b) In the orchestra but not in the band?
   (c) In at least two of the three groups, given that the student is in at least one group?
4. Two marbles are drawn without replacement from an urn containing two red, three white, and four blue marbles. Determine the probability that you end up with:
   (a) No white marbles.
   (b) Both marbles of the same color.
   (c) No white marbles, given that neither is red.
   (d) Two different colors, given that neither is red.
5. 25% of a population smoke, and 20% of those have a breathing disorder. Only 5% of the non-smokers have a breathing disorder. A person is chosen at random from the population.
   (a) What is the probability that the person has a breathing disorder?
   (b) What is the probability that the person smokes, given that he or she has a breathing disorder?
6. How many different license plates consisting of three letters followed by four digits are possible if:
   (a) The three letters are different, and the four digits are all the same and greater than 5?
   (b) Exactly two of the letters are the same and none of the digits is greater than 5?
7. A house has four doors and eleven windows. In how many ways can a burglar enter and exit if the burglar:
   (a) Enters through a window and leaves through a door?
   (b) Enters through a window and leaves through either a window or a door?
8. A president, vice-president, and secretary is to be randomly selected from a group of six women and four men. Determine the probability that:
   (a) The president and secretary are women and the vice-president is a man.
   (b) The president is not a woman.
   (c) The president is a woman and the secretary is not a man.

9 Five cards are drawn from a standard deck. Determine the probability that:
(a) None is a spade. (b) All are of the same suit. (c) Exactly three are face cards.
(d) At least one is a face card. (e) At least one is a face card, given that they are all black.

10. A die is flipped twelve times. What is the probability that:
(a) Exactly two fives appear?
(b) A number greater than 4 is tossed at most once.
(c) A number greater than four is tossed at least once?

11. Two cards are to be drawn from a standard deck of 52. What is the expected number of aces drawn?

12. You are to flip a coin while Caterina flips two. If your coin is different from either of Caterina's, then she will give you two dollars. If not, you will give her one dollar. Should you play?

13. Maverick removes the twelve face cards from a deck to which you may choose to add **up to** three aces. The first ace you add will cost you \$8, the second: \$6, and the third: \$5. He will then draw a card from the augmented stack of cards (that will now consist of 13, 14, or 15 cards). If he draws an ace, then you will win \$100. Should you play? If so, how many aces should you buy?

Answers

1. (a) $\frac{13}{52}$ (b) $\frac{6}{52}$ (c) $\frac{32}{52}$ (d) $\frac{20}{52}$ (e) $\frac{15}{52}$ 2. (a) $\frac{3}{5}$ (b) 2:3 (c) 2:3 3. (a) $\frac{464}{550}$ (b) $\frac{45}{550}$ (c) $\frac{11}{126}$

4. (a) $\frac{C(6,2)}{C(9,2)}$ (b) $\frac{C(2,2)+C(3,2)+C(4,2)}{C(9,2)}$ (c) $\frac{C(4,2)}{C(7,2)}$ (d) $\frac{C(3,1)C(4,1)}{C(7,2)}$

5. (a) $(.25)(.20)+(.75)(.05)$ (b) $\frac{(.25)(.20)}{(.25)(.20)+(.75)(.05)}$

6. (a) $26\cdot 25\cdot 24\cdot 4$ (b) $3\cdot 26\cdot 25\cdot 6^4$ 7. (a) $11\cdot 4$ (b) $11\cdot 15$

8. (a) $\frac{6\cdot 4\cdot 5}{P(10,3)}$ (b) $\frac{4\cdot 9\cdot 8}{P(10,3)}$ (c) $\frac{6\cdot 8\cdot 5}{P(10,3)}$

9. (a) $\frac{C(39,5)}{C(52,5)}$ (b) $\frac{4C(13,5)}{C(52,5)}$ (c) $\frac{C(12,3)C(40,2)}{C(52,5)}$ (d) $1-\frac{C(40,5)}{C(52,5)}$ (e) $1-\frac{C(20,5)}{C(26,5)}$

10. (a) $C(12,2)\left(\frac{1}{6}\right)^2\left(\frac{5}{6}\right)^{10}$ (b) $C(12,0)\left(\frac{2}{6}\right)^0\left(\frac{4}{6}\right)^{12}+C(12,1)\left(\frac{2}{6}\right)^1\left(\frac{4}{6}\right)^{11}$

(c) $1-C(12,0)\left(\frac{2}{6}\right)^0\left(\frac{4}{6}\right)^{12}$ 11. $0\cdot\frac{C(4,0)}{C(52,2)}+1\cdot\frac{C(4,1)C(48,1)}{C(52,2)}+2\cdot\frac{C(4,2)}{C(52,2)}$

12. No. On the average, you can expect to lose a quarter if you play.
13. Buy three aces for an expected winning of \$1.

Review Exercises

1. (**Drawing a Card**) A card is drawn from a standard deck. Determine the probability that it is:
   (a) A five or a seven.
   (b) A five and a heart.
   (c) A five or a heart.
   (d) A five or a heart, but not the five of hearts.

2. (**Choosing a Calendar Date**) A date is chosen at random. Assuming it is not leap year, determine the probability that it is:
   (a) April 15.
   (b) The thirtieth of a month.
   (c) In December.
   (d) Not in December.

3. (**Choosing a Number**) A number is selected at random from the integers 2 through 16. Determine the probability that it is:
   (a) Even.
   (b) Greater than 12.
   (c) Divisible by 4.
   (d) Divisible by 4 and 5.
   (e) Divisible by 4 or 5.
   (f) Not divisible by either 4 or 5.

4. (**Nothing but Counting**) *A word is to be selected at random from this sentence.* What is the probability that it will contain less than three letters?

5. (**Odds**) What are the odds in favor of rolling a sum of 7 with a pair of dice?

6. (**Odds**) What are the odds against drawing an ace from a standard deck?

7. (**From Odds to Probability**) What is the probability that you will win the door prize, if the odds are 100 to 1 that you will not win the prize?

8. (**From Probability to Odds**) There is a 35% chance that it will rain tomorrow. What are the odds that it will rain?

9. (**Summer Plans**) In a survey of 100 college students, 26 indicated that they planned to travel in the summer, 34 indicated that they planned to go to school, and 64 said that they planned to work. Eight said that they planned to travel and go to school, 5 said that they planned to travel and work, 20 planned to go to school and work, and 3 planned to do all three.
   (a) Draw a suitable Venn Diagram.
   (b) Determine the probability that a student is planning to travel and not work.
   (c) Determine the probability that a student is planning to travel or work.
   (d) Determine the probability that a student is planning neither to work nor to go to school.

10. (**Students**) In a class of 125 students, 60 are females. Seventy of the students are taking math, and 60 of the students live on campus. Twenty of the males in the class are taking math and live on campus, while only 15 of the females are taking math and live on campus. Thirty-five of the females live on campus, and 27 of the females are taking math. A student is chosen at random from the class.
    (a) Draw a suitable Venn diagram.
    (b) Determine the probability that a student chosen at random does not live on campus.
    (c) Determine the probability that a student chosen at random is a female not taking math.
    (d) Determine the probability that a student chosen at random is a male who is not taking math and is not living on campus.

11. (**Circuit**) Pushing button A activates both switches in the adjacent figure to enable current to flow. There is a 0.05 probability that each of the switches will malfunction. Button A is pushed. What is the probability that the light will go on?

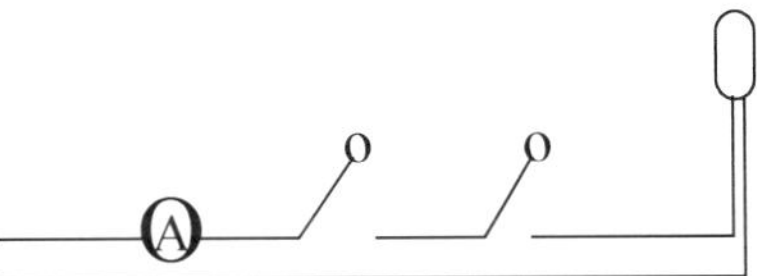

12. (**Circuit**) Pushing button A activates all three switches in the adjacent figure to enable current to flow. Each switch has a probability of malfunctioning of 0.05. Button A is pushed. What is the probability that the light will go on?

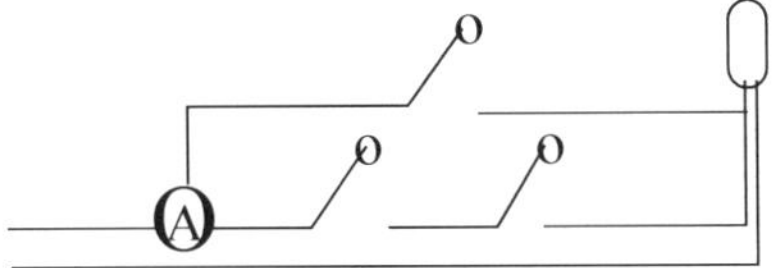

13. (**Left-Handed Math Majors**) Five of 20 students in a class are left-handed, and exactly 3 of those five are math majors. Exactly 3 of the right-handed students are math majors. A student is selected at random from the class. What is the probability that:
    (a) The student is a math major?
    (b) The student is not a math major if the student is left-handed?
    (c) The student is left-handed, given that the student is a math major?
14. (**Defective Chips**) Machines X, Y, and Z produce transistor chips. Machine X produces 40% of the chips, Y produces 35%, and Z produces the remaining 25%. It is known that X, Y, and Z produce 1%, 2%, and 3% defective chips, respectively.
    (a) Determine the probability that a randomly selected chip is defective.
    (b) Determine the probability that a randomly selected chip came from machine X, given that the chip is not defective.
    (c) Determine the probability that a randomly selected chip did not come from machine X, given that the chip is defective.
15. (**Trolls and Elves**) In a faraway land, trolls make up 70% of the population, and elves the remaining 30%. It is well known that trolls lie 90% of the time and elves tell the truth 90% of the time. Being a stranger in the land, you don't know the difference between a troll and an elf. You come to a fork in the road and ask an inhabitant "which is the way to La-La Land." He tells you to take the road to the right, and you do.
    (a) What is the probability that the road will take you to La-La?
    (b) The road does take you to La-La. What is the probability that the inhabitant was a troll?
16. (**Meal**) A complete meal at a restaurant consists of a selection of one of four appetizers, one of two soups, two choices of salad, one of seven entrees, and one of five desserts. How many different complete meals are available?
17. (**Card Hands**) Determine the number of five card hands which:
    (a) Do not contain an Ace.
    (b) Contain neither an Ace nor a King.
    (c) Contain 2 Aces and 3 Kings.
    (d) Contain exactly 2 Aces and at least 2 Kings.

18. (**Stacking a Deck**) How many different ways can a deck of cards be stacked, if each of the 13 different types are to be together (the four kings, for example, can not be separated by another card).

19. (**Seating Arrangements**) Determine the number of ways four couples (male and female) may be seated in a row of eight chairs if:
    (a) No couple is to be separated.
    (b) All the women are to be seated next to each other.
    (c) The first three chairs are occupied by men, and the last chair is occupied by a man.

20. (**Integer Less than 1000**) (Note: the first digit of any positive integer cannot be a 0.) Determine the number of positive integers less than 1000 which contain:
    (a) Only odd digits. (b) Only even digits.
    (c) At least two fives. (d) At least two zeros.

21. (**Socks**) A drawer contains 5 different pairs of socks, all in disarray. You reach in and grab some socks. What is the probability that you end up with at least one matching pair if you grab:
    (a) 2 socks (b) 3 socks (c) 6 socks

22. (**Horse Race**) Three horses are in a race. Big Boy is two times as likely to win as Lucky who is three times as likely to win as Sweet Lady.
    (a) What is the probability that Sweet Lady will win?
    (b) What are the odds against Sweet Lady winning?
    (c) What is the probability that Big Boy will win?
    (d) What are the odds in favor of Big Boy winning?

23. (**Books**) You brought along five novels, four fiction books, and three poetry books on your vacation, and randomly grab three of them to take down to the beach. What is the probability that:
    (a) You have only poetry books?
    (b) You have one of each of the three types?
    (c) You will be able to read some poetry?

24. (**Books**) A box contains 4 math books, 5 art books, and 3 history books. You randomly grab the books and line them up on a shelf. Determine the probability that:
    (a) The math books occupy the first four positions on the shelf.
    (b) The math books are together.
    (c) The math books are together, the art books are together, and the history books are together.

25. (**Urn**) An urn contains 5 Red, 4 White, 3 Blue, and 2 Yellow marbles. Four marbles are chosen at random without replacement. Determine the for the following event:
    (a) None are White. (b) All are white. (c) At least one is White.
    (d) None are White, given that none are Yellow.
    (e) As many White marbles as Yellow marbles are drawn.
    (f) None are White, given that exactly one is Red.
    (g) All are White, given that at least three are White.
26. (**Coins**) You grab two coins from a box containing 2 quarters, 5 dimes, and 3 nickels. Determine the probability that you end up with a total of:
    (a) Fifty cents. (b) Thirty cents. (c) At least 30 cents. (d) More than 10 cents.
    (e) At least 30 cents, given that exactly one of the coins is a dime.
    (f) Less than 30 cents, given that exactly one of the coins is a dime.
27. (**Coins**) You grab four coins from a box containing 2 quarters, 5 dimes, and 3 nickels. Determine the probability that you end up with a total of:
    (a) Fifty cents. (b) Sixty cents. (c) Fifty-five cents, given that you do not grab a nickel.
28. (**Delegation**) Bill, Jill, and Marvin are three of ten members of an environmental club. A delegation of three is randomly chosen to attend a state convention. Determine the probability that:
    (a) Bill and Jill will attend. (b) Bill or Jill, but not both, will attend.
    (c) Bill or Jill (possibly both) will attend. (d) Bill and Jill and Marvin will attend.
    (e) Exactly two of the three will attend.
29. (**Three Digit Number**) An integer between100 and 999 inclusive is randomly selected. Determine the probability that:
    (a) All 3 digits are the same. (b) The digit 0 does not appear.
    (c) The number is odd. (d) No digit is repeated.
30. (**Card and Urn**) An urn contains 5 Red, 4 White, and 3 Blue marbles. A card is dealt. If it is a spade, one marble is drawn from the urn, otherwise two marbles are drawn without replacement. Determine the probability that:
    (a) At least one red marble is drawn. (b) Two red marbles are drawn.
    (c) A spade was dealt, given that no red marble was drawn.
31. (**Marbles in Pockets**) Alexandra has four blue and five red marbles in her left pocket, and three blue and four white marbles in her right pocket. She transfers one marble at random from her right to her left pocket, and then draws a marble at random from her left pocket. Determine the probability that:
    (a) The marble drawn from the left pocket is blue.
    (b) The marble drawn from the left pocket is white.
    (c) The marble drawn from the left pocket is white, given that she drew a blue marble from her right pocket.
    (d) She drew a blue marble from her right pocket, given that she draws a red marble from the left pocket.

32. (**Matching Cards**) Individual X, and individual Y are in different rooms. Both have the jack, queen, king, and ace of hearts in front of them. Individual Y arranges the four cards in a row and concentrates on the arrangement in an attempt to communicate the arrangement to individual X by mental telepathy. Eventually, individual X arranges his four cards. Determine the probability that:
    (a) All of the cards are a match, as read from left to right.
    (b) Exactly 3 of the cards match.
    (c) Exactly 2 of the cards match.
    (d) Exactly 1 of the cards matches.
    (e) None of the cards match.

33. (**Weighted Coin**) A coin is weighted in that a head is twice as likely to occur as a tail. It is tossed five times. What is the probability that:
    (a) Exactly three heads occur.
    (b) At least four heads occur
    (c) At least one head occurs.
    (d) More heads than tails occur.

34. (**Cards**) Ten people draw a card from 10 different decks. Determine the probability that:
    (a) Exactly 4 of them draw an Ace.
    (b) At least 1 of them draws an Ace.
    (c) No more than 2 of them draw an Ace.

35. (**Flu**) One hundred individuals are given a flu shot that is 80% effective. There is a 10% probability that an individual will be exposed to the flu virus. Determine the probability that:
    (a) Exactly ten of the individuals will be exposed to the flu virus.
    (b) Exactly ten of the individuals will come down with the flu.
    (c) At least one of the individuals will come down with the flu.
    (d) At most one of the individuals will come down with the flu.

36. (**Defective Switches**) Seven switches are randomly chosen from a box containing 100 switches; exactly 5 of which are defective. Determine the probability that, of the switches chosen:
    (a) Exactly 3 are defective.
    (b) At least one is defective.
    (c) At most 6 are defective.

37. (**Two Cards**) Five people draw two cards (without replacement) from 5 different decks. Determine the probability that:
    (a) Exactly 4 of them draw two clubs.
    (b) At least 1 of them draws two clubs.
    (c) No more than 1 of them draws two clubs.

38. (**Urn**) Ten people draw 3 marbles, without replacement, from 10 urns; each containing 3 red marbles, 4 white marbles, and 5 blue marbles. Determine the probability that:
    (a) Exactly four of them draw 3 blue marbles.
    (b) At least one of them draws 3 blue marbles.
    (c) Exactly two of them draw 3 differently colored marbles.
    (d) At least nine of them draw at least 2 white marbles.
39. (**Urn**) An urn contains 3 red marbles, 4 white marbles, and 5 blue marbles. Five marbles are drawn without replacement. Determine the probability that:
    (a) All of the red marbles are drawn.
    (b) At least 2 red marbles are drawn.
    (c) Less than 3 red marbles are drawn.
    (d) All of the red marbles are drawn, given that exactly 1 white marble is drawn.
40. (**Defective Switches**) There are 6 boxes, each containing 95 good switches and 5 defective switches. Two switches are randomly taken from each of the boxes. Determine the probability that:
    (a) None of the switches chosen from any of the 6 boxes are defective.
    (b) At most one of the two switches taken from each of the 6 boxes is defective.
    (c) Exactly two of the 6 boxes have two defective switches chosen.
41. (**Disney Land**) William is taking four courses, and has a fifty-fifty chance of passing any one of them. "Tell you what I'm going to do," says William's concerned father, "if you pass every course, then I will take you to Disney Land. If you pass three of them, then I'll flip a coin and take you to Disney if it turns out Heads. If you pass two of them, then I'll roll a die, and will take you only if a 6 comes up. If you don't pass more than one of your courses, then you can just forget all about Mickey and his friends.
    (a) What is the probability that William gets to go to Disney Land?
    (b) What is the probability that William passed all four courses if he went to Disney Land?
42. (**Die**) A die is rolled and you will win an amount, in dollars, shown on the die if the number rolled is even and will lose the amount shown if the number is odd. What is the expected winnings of the game?
43. (**Commission**) A door-to-door vacuum-cleaner salesman receives a \$100 commission on the deluxe model, a \$75 commission on the super model and a \$50 commission on the standard model. Experience shows that 3% of the people he solicits buy the deluxe model, 5% buy the super model, and 4% buy the standard model. What is his expected commission on the next encounter?
44. (**Draw 3 Cards**) You draw 3 cards, without replacement, from a standard deck. What is the expected number of Aces drawn?
45. (**Flip a Coin**) You flip a coin 4 times. If three consecutive heads (or more) are thrown, you win \$10, otherwise you lose \$2. Should you play the game? Justify your answer.

46. **(Investing)** You plan to invest \$5000 in Companies A, B, or C, and will sell your share in whatever company you invest in at the end of the year. There is a 0.75 probability that company A will fail; but if not, you will be able to sell your share in the company for \$25,000 (a profit of \$20,000). There is only a 5% chance that company B will go belly up; a 55% chance that you will lose half of your investment, a 10% chance that you will break even, and a 30% chance that you will double your money. There is a guaranteed profit of \$500 if you go with company C. What should you do, and why?

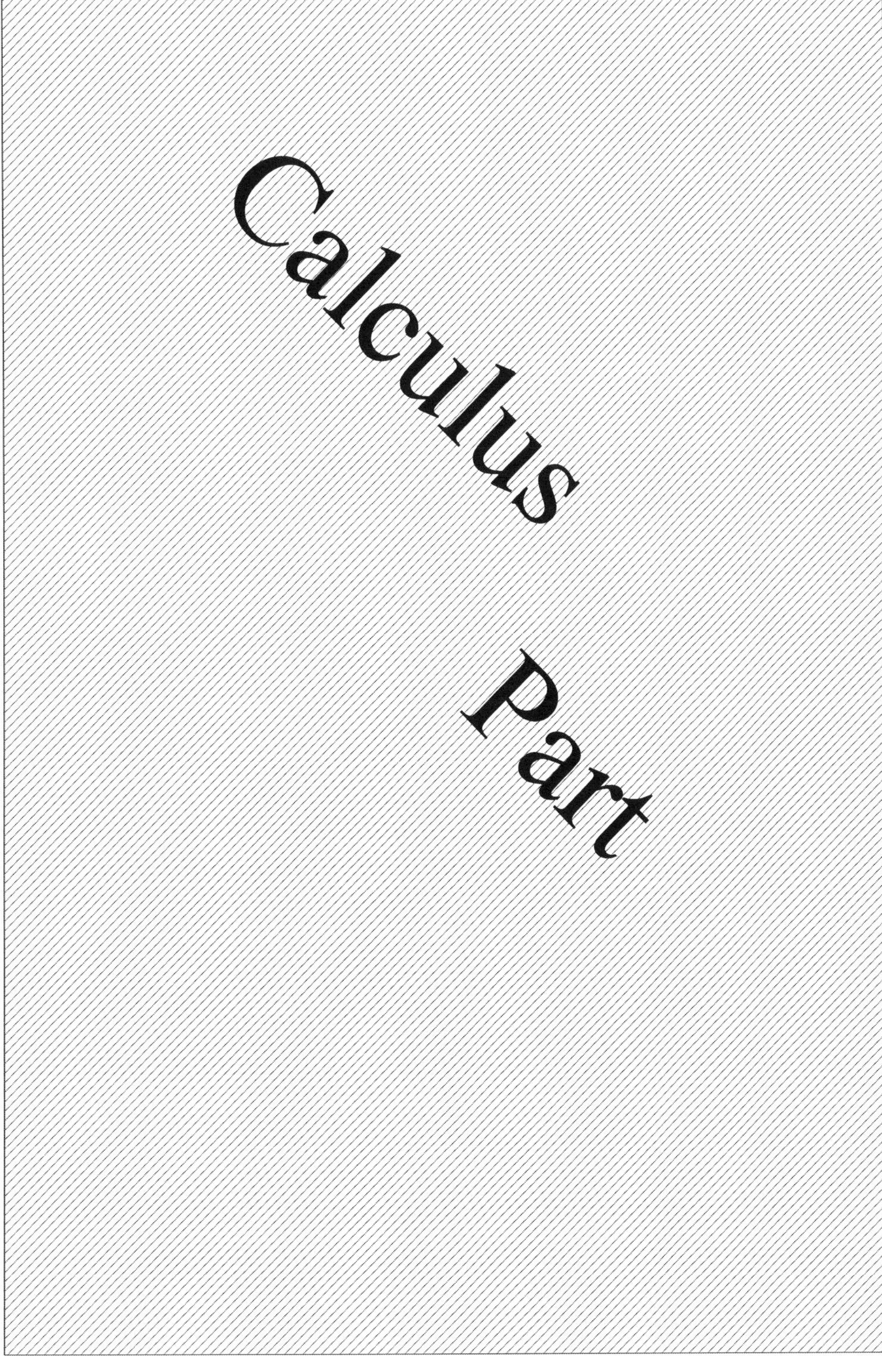
Calculus
Part

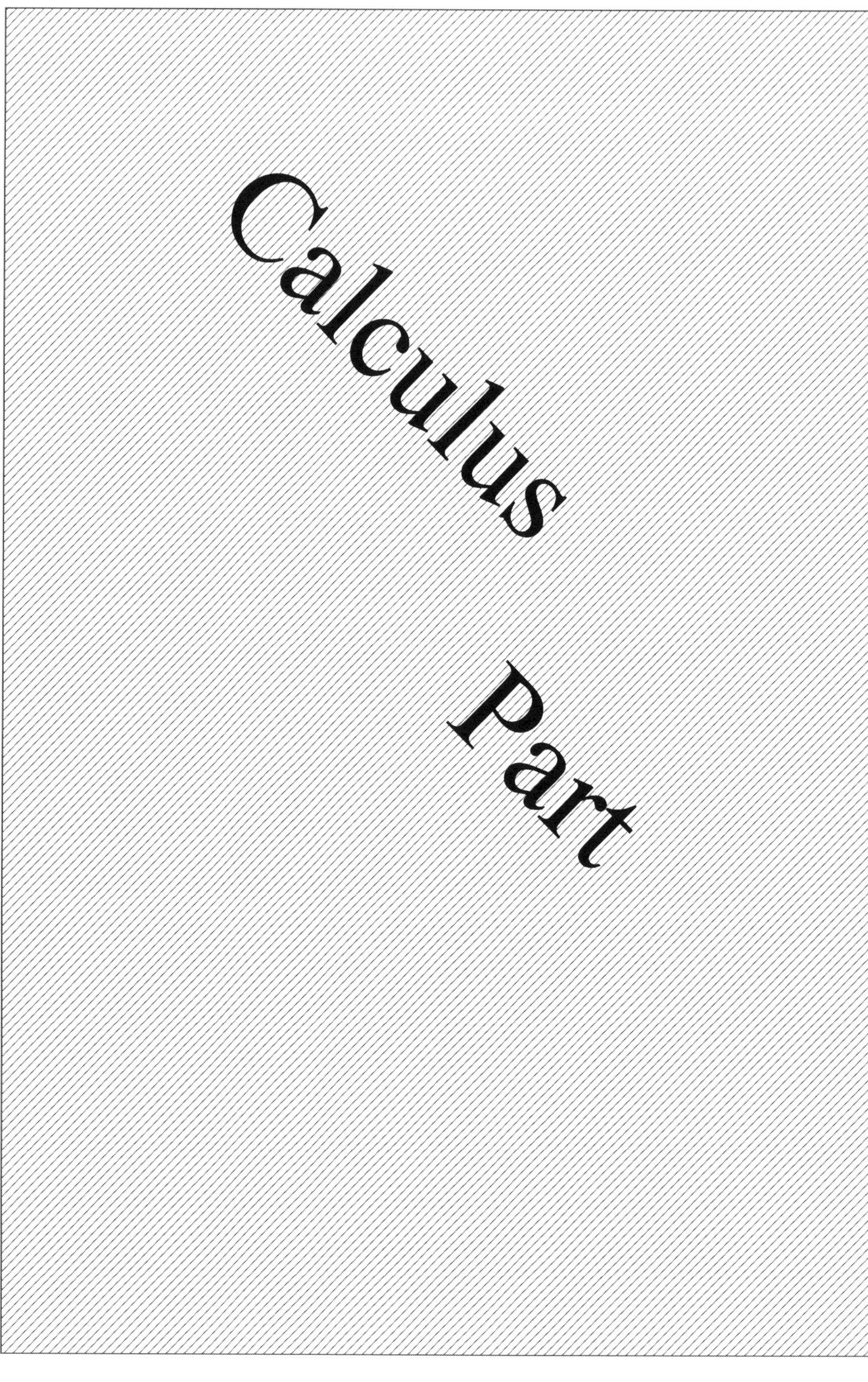
Calculus Part

§1. PRELIMINARIES

Mathematics is, in part, a visual art, and much of its power stems from the fact that a great deal of information can be represented compactly. An example of this is the familiar exponent form:

DEFINITION 2.1

INTEGER EXPONENTS

For any positive integer n and any number a:

$$a^n = \underbrace{a \cdot a \cdot \ldots \cdot a}_{n\text{-times}}$$

a raised to the nth power

From its very definition, we see that:

$$9^2 \cdot 9^3 = (9 \cdot 9)(9 \cdot 9 \cdot 9) = 9^5 = 9^{2+3}$$

$$\frac{9^5}{9^2} = \frac{9^2 \cdot 9^3}{9^2} = 9^3 = 9^{5-2} \text{ (see margin)}$$

$$(9^2)^5 = 9^2 \cdot 9^2 \cdot 9^2 \cdot 9^2 \cdot 9^2 = 9^{10} = 9^{2 \cdot 5}$$

$$(2 \cdot 5)^3 = (2 \cdot 5)(2 \cdot 5)(2 \cdot 5) = (2 \cdot 2 \cdot 2)(5 \cdot 5 \cdot 5) = 2^3 \cdot 5^3$$

$$\left(\frac{2}{5}\right)^3 = \frac{2 \cdot 2 \cdot 2}{5 \cdot 5 \cdot 5} = \frac{2^3}{5^3}$$

CANCELLATION LAW

For $c \neq 0$

$$\frac{ac}{bc} = \frac{a}{b}$$

Note: The cancellation law is often abused. You can only cancel a common **factor** in the numerator and denominator of a fraction. **DON'T** do this:

$$\frac{2a + b}{a} = 2 + b$$

WRONG!

In general:

THEOREM 2.1

EXPONENT RULES

Let m and n be positive integers. When all expressions are defined, we have:

(i) $a^m a^n = a^{m+n}$ — When multiplying add the exponents

(ii) $\dfrac{a^m}{a^n} = a^{m-n}$ — When dividing subtract the exponents

(iii) $(a^m)^n = (a^n)^m = a^{mn}$ — A power of a power: multiply the exponents

(iv) $(ab)^n = a^n b^n$ — A power of a product equals the product of the powers

(v) $\left(\dfrac{a}{b}\right)^n = \dfrac{a^n}{b^n}$ — A power of a quotient equals the quotient of the powers

It is important to note that these exponent rules deal with products, and not with sums. In particular, though it is true that:

$$(3 \cdot 4)^2 = 3^2 \cdot 4^2$$

it is **NOT TRUE** that

$$(3+4)^2 \neq 3^2 + 4^2$$

WRONG:

$(3+4)^2 = 49$ while $3^2 + 4^2 = 25$.

In (ii) we stated *When dividing, subtract the exponents*. But if we try this with the quotient $\frac{9^2}{9^5}$, we end up with an expression $\frac{9^2}{9^5} = 9^{2-5} = 9^{-3}$, which at this point if meaningless. Observing that $\frac{9^2}{9^5} = \frac{9^2}{9^2 \cdot 9^3} = \frac{1}{9^3}$, we breathe meaning into the expression 9^{-3} by defining it to be $\frac{1}{9^3}$. By the same token, since $\frac{9^2}{9^2} = 1$, if we want the "*subtract exponent*" rule $\frac{9^2}{9^2} = 9^{2-2} = 9^0$ to hold, we are forced to define 9^0 to be 1.

In general:

Under the assumption that Theorem 2.1 holds for all positive integers n and m, one can show that it indeed holds for all integers: positive, negative, or zero (see Exercise 117).
Indeed, we note that Theorem 2.1 holds for all exponents. For example:
$$2^{\pi} \cdot 2^{\sqrt{3}} = 2^{\pi+\sqrt{3}}$$

DEFINITION 2.2 For any positive integer n and any $a \neq 0$:

$$a^0 = 1 \text{ and } a^{-n} = \frac{1}{a^n}$$

EXAMPLE 2.1 Simplify:

(a) $\frac{(2a)^2b^3}{(a^{-3}b^2)^5}$ (b) $\frac{(ab^2c^{-3})^4\left(\frac{a}{b}\right)^3}{(a^{-2}b^2)^3}$

SOLUTION: There are many ways of simplifying the given expressions. Our approach is designed to highlight the exponent rules of Theorem 2.1. Please make sure that you see a justification for each step along the way.

(a) $$\frac{(2a)^2b^3}{(a^{-3}b^2)^5} = \frac{2^2a^2b^3}{a^{-15}b^{10}} = 4a^{2+15}b^{3-10} = 4a^{17}b^{-7} = \frac{4a^{17}}{b^7}$$

(b) $$\frac{(ab^2c^{-3})^4\left(\frac{a}{b}\right)^3}{(a^{-2}b^2)^3} = \frac{a^4b^8c^{-12}(ab^{-1})^3}{a^{-6}b^6}$$
$$= \frac{a^4b^8c^{-12}a^3b^{-3}}{a^{-6}b^6}$$
$$= a^{4+3+6}b^{8-3-6}c^{-12} = a^{13}b^{-1}c^{-12} = \frac{a^{13}}{bc^{12}}$$

CHECK YOUR UNDERSTANDING 2.1

Simplify:

(a) $(-2ax^2)^4$ (b) $\left(\frac{-x^2}{2}\right)^3$ (c) $(-x)^2\left(\frac{x}{2y}\right)^3$

Answers: (a) $16a^4x^8$ (b) $-\frac{x^6}{8}$ (c) $\frac{x^5}{8y^3}$

LINEAR EQUATIONS

Equations with the same solution sets (any solution of one is also a solution of the other) are said to be **equivalent**. It is apparent that:

Adding (or subtracting) the same number to both sides of an equation will result in an **equivalent equation**.

For example:

The equation: $4x+3 = -2x-9$ is equivalent to: $4x+\mathbf{3}-\mathbf{3} = -2x-9-\mathbf{3}$ $4x+0 = -2x-9-\mathbf{3}$ $4x = -2x-9-\mathbf{3}$	**EFFECT:** That **3** which was previously on the left side is now on the right side, but with its sign changed. $4x+\mathbf{3} = -2x-9$ $4x = -2x-9-\mathbf{3}$
By the same token, the equation: $4x+3 = -2x-9$ is equivalent to: $4x+3+2x = -\mathbf{2x}-9+2x$ $4x+3+\mathbf{2x} = -9+0$ $4x+3+\mathbf{2x} = -9$	**EFFECT:** That $-\mathbf{2x}$ which was previously on the right side is now on the left side, but with its sign changed. $4x+3 = -\mathbf{2x}-9$ $4x+3+\mathbf{2x} = -9$

The above should convince you of the fact that:

When solving an equation, you may bring over any term from one side of the equation to the other side, as long as you <u>change its sign</u>.

For example: $4x+\mathbf{3} = -\mathbf{2x}-9$

bring over and change signs: $4x+\mathbf{2x} = -\mathbf{3}-9$

$6x = -12$

see margin: $x = -2$

Multiplying (or dividing) both sides of an equation by the same nonzero number (in this case "6") results in an equivalent equation.

EXAMPLE 2.2 Solve:

$$3x-x+2-9 = 4x+6+x$$

SOLUTION:

$$3x-x+2-9 = 4x+6+x$$

Combine like terms on both sides of the equation: $2x-7 = 5x+6$

Move all the variable terms to one side of the equation and all the constant terms to the other remembering to change signs: $2x-5x = 6+7$

Combine terms once more: $-3x = 13$

Divide both sides by -3: $x = -\frac{13}{3}$

You are invited to verify that $-\frac{13}{3}$ is indeed a solution to the given equation.

CHECK YOUR UNDERSTANDING 2.2

Solve:

(a) $3-2x+5-x = -4x-2+1$

(b) $\frac{-3x}{5}-\frac{x}{2}+1 = \frac{2x+1}{10}$

Suggestion: Start off by multiplying both sides of the equation by 10.

Answers: (a) -9 (b) $\frac{9}{13}$

LINEAR INEQUALITIES

One solves linear inequalities in exactly the same fashion as one solves linear equations, with one notable exception:

If you multiply both sides of the inequality $-2 < 3$ by the positive number 2, then the inequality sign remains as before:

$$-2 < 3$$

multiply by 2: $-4 < 6$

But if you multiply both sides by the negative quantity -2 then the sense of the inequality is reversed:

$$-2 < 3$$

multiply by -2 : $4 > -6$

WHEN MULTIPLYING OR DIVIDING BOTH SIDES OF AN INEQUALITY BY A **NEGATIVE** QUANTITY, ONE MUST **REVERSE** THE DIRECTION OF THE INEQUALITY SIGN.

To illustrate:

Equation	Inequality
$3x - 5 = 5x - 7$	$3x - 5 < 5x - 7$
$3x - 5x = -7 + 5$	$3x - 5x < -7 + 5$
$-2x = -2$	$-2x < -2$
$x = 1$	reverse dividing by a negative number
	$x > 1$

CHECK YOUR UNDERSTANDING 2.3

Solve:

$$\frac{3x}{5} - \frac{2-x}{3} + 1 < \frac{x-1}{15}$$

Suggestion: Begin by multiplying both sides of the inequality by 15.

Answer: $x < -\frac{6}{13}$

LINES

The following definition attributes a measure of "steepness" to any nonvertical line in the plane.

DEFINITION 2.3
SLOPE OF A LINE

For any nonvertical line L and any two distinct points (x_1,y_1) and (x_2,y_2) on L, we define the **slope** of L to be the number m given by:

$$m = \frac{y_2 - y_1}{x_2 - x_1} = \frac{\text{change in } y}{\text{change in } x}$$

It can be shown that the slope of a line does not depend on the two points chosen.

For Example:

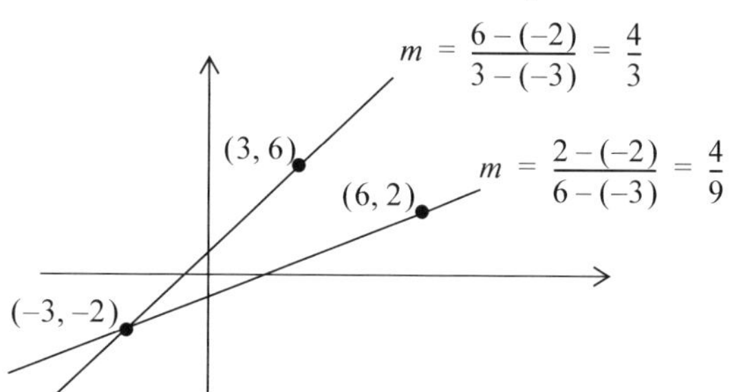

The steeper the climb, the more positive the slope.

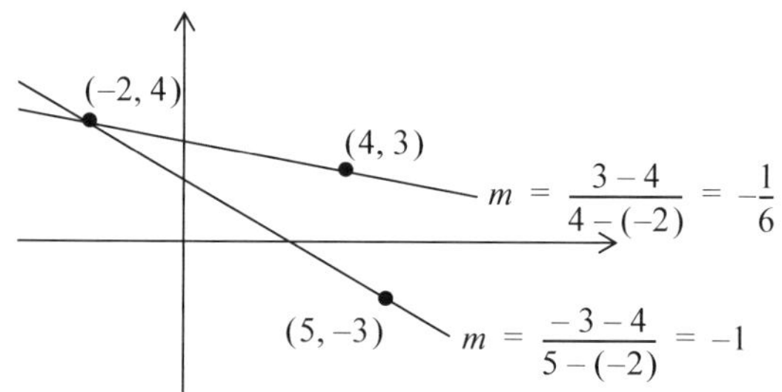

The steeper the fall, the more negative the slope.

Note:

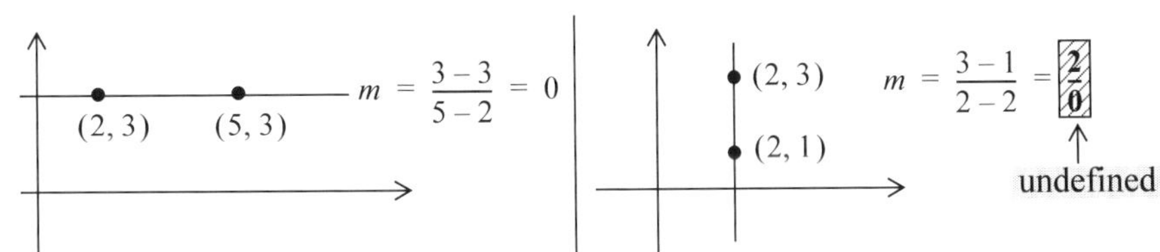

The slope of a horizontal line is 0 | **No slope is associated with a vertical line**

Answer: $-\frac{7}{4}$

CHECK YOUR UNDERSTANDING 2.4

Determine the slope of the line L that passes through (or contains) the points $(-1, 5)$ and $(3, -2)$.

SLOPE-INTERCEPT EQUATION OF A LINE

***Y*-INTERCEPT**

Consider the line L of slope m in Figure 2.1. Being nonvertical, it must intersect the y-axis at some point $(0, b)$. The number b, where the line intersects the y-axis, is called the ***y*-intercept** of the line.

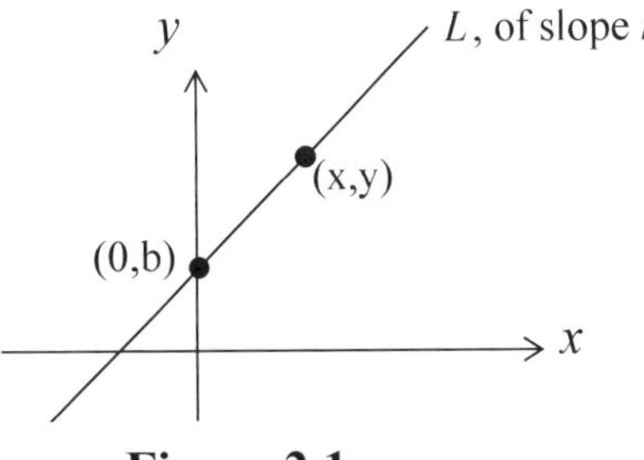

Figure 2.1

Now let (x, y) be any point on L other than $(0, b)$. From the fact that any two distinct points on a line determine its slope, we have:

$$m = \frac{y-b}{x-0}$$

$$m = \frac{y-b}{x}$$

multiply both sides of the equation by x:

$$y - b = mx$$

add b to both sides of equation: $y = mx + b$

Direct substitution shows that the above equation also holds at the point $(0, b)$; thus:

The Point-Slope form is introduced in the exercises.

THEOREM 2.2 Slope-Intercept Equation of a line A point (x, y) is on the line of slope m and y-intercept b if and only if its coordinates satisfy the equation $y = mx + b$.

It is a good idea to sketch the line in question:

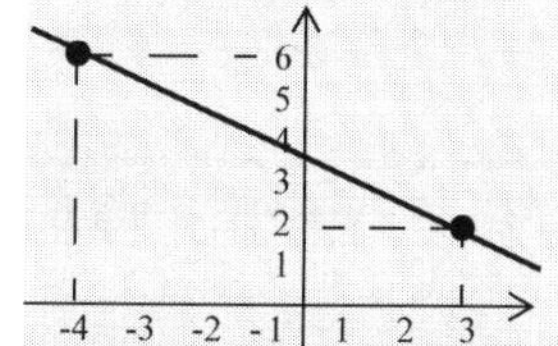

We see that the line has a negative slope, and that its y-intercept is about 4.

The other point, $(-4, 6)$ could be used instead of $(3, 2)$, and it would lead to the same result (try it).

EXAMPLE 2.3 Find the slope-intercept equation of the line that contains the points $(3, 2)$ and $(-4, 6)$.

SOLUTION: Using the given points, we find that:

$$m = \frac{6-2}{-4-3} = -\frac{4}{7}$$

We now know that the equation is of the form:

$$y = -\frac{4}{7}x + b$$

Since the point $(3, 2)$ is on the line, the above equation must hold when 3 is substituted for x and 2 for y, and this enables us to solve for b:

$$2 = -\frac{4}{7}(3) + b$$

$$b = 2 + \frac{12}{7} = \frac{26}{7}$$

Leading us to the equation: $y = -\frac{4}{7}x + \frac{26}{7}$

CHECK YOUR UNDERSTANDING 2.5

Find the slope-intercept equation of the line
(a) Of slope 7 and y-intercept 3.
(b) Of slope 3 which passes through the point $(2, 5)$.
(c) Which passes through the points $(2, 4)$ and $(3, -5)$.

Answers: (a) $y = 7x + 3$
(b) $y = 3x - 1$
(c) $y = -9x + 22$

FACTORING POLYNOMIALS

Factoring is the reverse process of multiplication. By reading the following equation from right to left:

$$(a+b)(a-b) = a^2 - b^2 \text{ (see margin)}$$

we obtain the formula for factoring a difference of two squares:

$(a+b)(a-b)$
$= a(a-b) + b(a-b)$
$= a^2 - ab + ba - b^2$
$= a^2 = b^2$

THEOREM 2.3 $a^2 - b^2 = (a+b)(a-b)$

For example: $4x^2 - 9 = (2x+3)(2x-3)$

And: $4x^2 - 5 = (2x + \sqrt{5})(2x - \sqrt{5})$

CHECK YOUR UNDERSTANDING 2.6

Factor:
(a) $25x^2 - 1$ (b) $3x^2 - 2$ (c) $x^4 - 16$

Answers:
(a) $(5x+1)(5x-1)$
(b) $(\sqrt{3}x + \sqrt{2})(\sqrt{3}x - \sqrt{2})$
(c) $(x^2+4)(x+2)(x-2)$

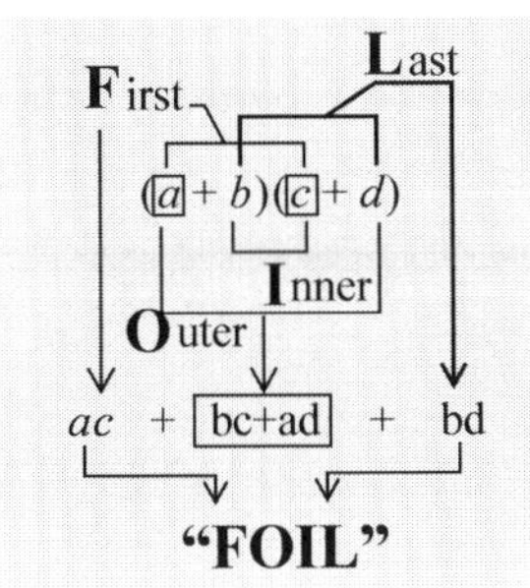

Some trinomials (three terms) can be factored by reversing the so-called "**FOIL**" multiplication method (see margin). To gain a factorizations of $5x^2 + 27x - 18$, for example, we start with the template

$$(5x \qquad)(x \qquad) \qquad (*)$$

which gives us the **F**irst term $5x^2$. We then envision pairs of integers in the template whose product equals the magnitude of the **L**ast term 18, and position them in (*) in the hope that the sum of the **O**uter-**I**nner terms has a magnitude of 27:

$(5x \quad 9)(x \quad 2)$	No luck: "in 9 and 10 does not sit a 27"
$(5x \quad 2)(x \quad 9)$	No luck: "in 2 and 45 does not sit a 27"
$(5x \quad 6)(x \quad 3)$	No luck: "in 6 and 15 does not sit a 27"
$(5x \quad 3)(x \quad 6)$	Some luck: "27 does sit in 3 and 30," and the signs can also be accommodated: To end up with -18 one of the numbers 3 and 6 has to be negative and the other positive. In addition, the "OI" combination must be a positive 27. Success: $(5x^2 + 27x - 18) = (5x - 3)(x + 6)$

CHECK YOUR UNDERSTANDING 2.7

Factor:

(a) $2x^2 + 7x - 4$ (b) $18x^4 - 6x^3 - 60x^2$

Answers: (a) $(2x - 1)(x + 4)$
(b) $6x^2(3x + 5)(x - 2)$

Polynomial Equations

In the following examples we illustrate how factorization can be used to solve certain polynomial equations.

EXAMPLE 2.4 Solve:

$$x^4 = 16$$

Solution: Move the 16 to the left side of the equation — factor, and set the factored form to zero:

$$x^4 - 16 = (x^2 + 4)(x^2 - 4) = (x^2 + 4)(x + 2)(x - 2) = 0$$

Since **a product is zero if and only if one of its factors is zero**, the problem reduces to solving the three equations:

$x^2 + 4 = 0$	$x + 2 = 0$	$x - 2 = 0$
$x^2 = -4$	$x = -2$	$x = 2$
no solution		

The two solutions $x = 2$ and $x = -2$ can also be expressed in the form $x = \pm 2$.

EXAMPLE 2.5 Solve:

$$x^3 + x^2 - 6x = 0$$

SOLUTION: The solution again hinges on the important fact that a product is zero **if** and **only if** one of its factors is zero:

$$x^3 + x^2 - 6x = 0$$

Pull out the common factor, x: $x(x^2 + x - 6) = 0$

factor further: $x(x+3)(x-2) = 0$

solutions: $x = 0$ or $x = -3$ or $x = 2$

Answers: (a) $x = \pm\frac{1}{5}$

(b) $x = \pm\sqrt{\frac{2}{3}} = \pm\frac{\sqrt{6}}{3}$

(c) $x = \frac{1}{2}, x = -4$

CHECK YOUR UNDERSTANDING 2.8

Solve:

(a) $25x^2 - 1 = 0$ (b) $3x^2 - 2 = 0$ (c) $2x^2 + 7x - 4 = 0$

INTERVAL NOTATION

As you will see, the following notation will enable us to compactly represent the solution set of certain inequalities.

Description	Interval Notation	Geometrical Representation
All real numbers strictly between 1 and 5 (not including 1 or 5)	$(1, 5)$	-3 -2 -1 0 1 2 3 4 5 6
All real numbers between 1 and 5, including both 1 and 5.	$[1, 5]$	-3 -2 -1 0 1 2 3 4 5 6
All real numbers between 1 and 5, including 1 but not 5.	$[1, 5)$	-3 -2 -1 0 1 2 3 4 5 6
All real numbers between 1 and 5, including 5 but not 1.	$(1, 5]$	-3 -2 -1 0 1 2 3 4 5 6
All real numbers greater than 1.	$(1, \infty)$	-3 -2 -1 0 1 2 3 4 5 6
All real numbers greater than or equal to 1.	$[1, \infty)$	-3 -2 -1 0 1 2 3 4 5 6
All real numbers less than 5.	$(-\infty, 5)$	-3 -2 -1 0 1 2 3 4 5 6
All real numbers less than or equal to 5.	$(-\infty, 5]$	-3 -2 -1 0 1 2 3 4 5 6
The set of all real numbers.	$(-\infty, \infty)$	-3 -2 -1 0 1 2 3 4 5 6

The solution set of an inequality may consist of several intervals. In general, for any two sets A and B, the **union** of A and B is defined to be that set, denoted by $A \cup B$, which consists of all elements that are either in A or in B. For example:

$(-2, 0) \cup [3, 5]$	All numbers between -2 and 0, along with those between 3 and 5, including 3 and 5.	-2 0 3 5
$(-\infty, 1) \cup [2, 5] \cup (7, \infty)$	All numbers less than 1, along with those between 2 and 5, inclusive, along with all numbers greater than 7.	1 2 5 7

POLYNOMIAL INEQUALITIES

The four-step procedure used to solve the polynomial inequality:

$$(x+3)^3(x+1)^2(4-x) < 0 \qquad (*)$$

in the next example is based on the following two observations:

(i) The sign of a polynomial can only change from positive to negative or from negative to positive about a zero of the polynomial.
(Where the graph of a polynomial function crosses the x-axis.)

The sign of $(x+3)$ changes as one moves from one side of its zero -3 to the other side. That sign-change is not "erased" by raising $(x+3)$ to the third (or any odd) power.
The sign of $(x+1)$ also changes as one moves from one side of its zero -1 to the other side. But that sign-change will be "erased" by raising $(x+3)$ to the second (or any even) power.

(ii) If a factor of the polynomial is raised to an odd power, as is the case with $(x+3)^3$ in (*), then the sign of the polynomial will change about the zero of that factor (see margin). On the other hand, if the factor is raised to an even power, as is the case with $(x+1)^2$, then the sign will not change (see margin).

EXAMPLE 2.6 Solve:

$$(x+3)^3(x+1)^2(4-x) < 0$$

SOLUTION:

Step 1. Locate the zeros of the polynomial on the number line:

Place the letter ***c*** above the zeros of factors which are raised to an odd power [the factors $(x+3)^3$ and $(4-x)^1$] to remind you that the sign of the polynomial will **c**hange about the odd-zeros -3 and 4. Place the letter ***n*** above the zero of the factor $(x+1)^2$ to remind you that the sign of the polynomial will **n**ot change about that even-zero -1 :

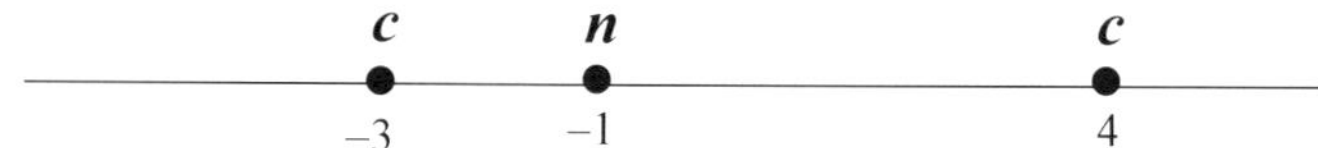

Determining the sign of the polynomial to the left of -3 or within the interval $(-3,-1)$ [or $(-1,4)$] can also serve as starting points.

Step 2. Determine the sign of $(x+3)^3(x+1)^2(4-x)$ to the right of its last zero. In this example, for $x>4$ the product is easily seen to be **negative** [$(4-x)$ is the only negative factor], thus:

Step 3. (Walk the sign to the left.) The ***c*** above 4 indicates that the sign will change about 4 (from negative to positive):

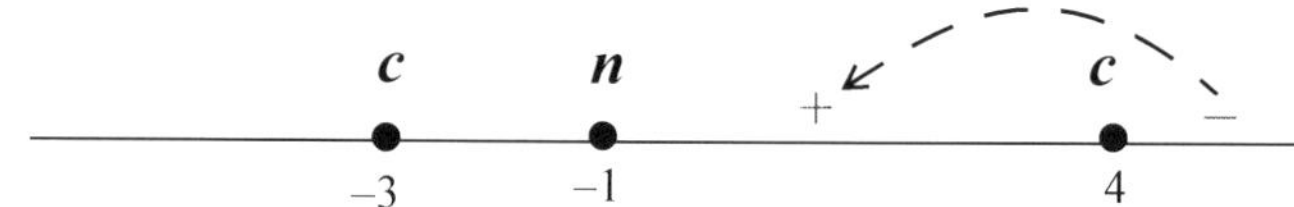

The ***n*** above -1 indicates that the sign will not change as you traverse -1 (will remain positive):

Finally, the ***c*** above -3 indicates a sign change:

Here is the end result:

SIGN $(x+3)^3(x+1)^2(4-x)$

Figure 2.2

Step 4. Since we are solving $(x+3)^3(x+1)^2(4-x)<0$ we read off the intervals where the polynomial is negative (the "**-**" intervals): $(-\infty,-3)\cup(4,\infty)$.

NOTE: The information in Figure 2.2 also enables you to solve the inequalities

$$(x+3)^3(x+1)^2(4-x)>0: \quad (-3,-1)\cup(-1,4)$$

$$(x+3)^3(x+1)^2(4-x)\ge 0: \quad [-3,4]$$

$$(x+3)^3(x+1)^2(4-x)\le 0: \quad (-\infty,-3]\cup[4,\infty)\cup\{-1\}$$

Answers:
(a) $(-2,3)\cup(5,\infty)$
(b) $(-\infty,-2]\cup\{-1,4\}$

CHECK YOUR UNDERSTANDING 2.9

Solve:

(a) $(x-3)(x+2)(-x+5)<0$ (b) $(x+1)^2(x+2)^3(x-4)^2\le 0$

FUNCTIONS

We will be concerned exclusively with functions, f, which assign a (real) number $f(x)$ to a given (real) number x. We want to emphasize the fact that the variable x is a placeholder; a "box" that can hold any meaningful expression. For example:

$$f(\mathbf{x}) = 2\mathbf{x}+5$$

$$f\boxed{} = 2\boxed{}+5$$

So:

$$\begin{aligned} f(\mathbf{3}) &= 2\cdot\mathbf{3}+5 = 11\\ f(\mathbf{c}) &= 2\cdot\mathbf{c}+5 = 2c+5\\ f(\mathbf{3t}) &= 2\cdot\mathbf{3t}+5 = 6t+5\\ f(\mathbf{x^2+3}) &= 2(\mathbf{x^2+3})+5 = 2x^2+11 \end{aligned}$$

EXAMPLE 2.7 For $f(x) = -3x^2+6x-1$, determine:

(a) $f(x+h)$ (b) $\dfrac{f(x+h)-f(x)}{h}$, for $h\ne 0$

SOLUTION:

(a)

$$\begin{aligned} f(\mathbf{x+h}) &= -3(\mathbf{x+h})^2+6(\mathbf{x+h})-1\\ \text{(margin)}\quad &= -3(x^2+2xh+h^2)+6x+6h-1\\ &= -3x^2-6xh-3h^2+6x+6h-1 \end{aligned}$$

You want to remember the following two formulas:

$$(a+b)^2 = a^2+2ab+b^2$$
$$(a-b)^2 = a^2-2ab+b^2$$

Proof of the top formula:

$$\begin{aligned} (a+b)^2 &= (a+b)(a+b)\\ &= a(a+b)+b(a+b)\\ &= a^2+ab+ba+b^2\\ &= a^2+ab+ab+b^2\\ &= a^2+2ab+b^2 \end{aligned}$$

(b) $\dfrac{f(x+h)-f(x)}{h} = \dfrac{[-3(x+h)^2+6(x+h)-1]-(-3x^2+6x-1)}{h}$

$$= \frac{[-3(x^2+2xh+h^2)+6x+6h-1]+3x^2-6x+1}{h}$$

$$= \frac{-3x^2-6xh-3h^2+6x+6h-1+3x^2-6x+1}{h}$$

$$= \frac{-6xh-3h^2+6h}{h} = \frac{h(-6x-3h+6)}{h} = -6x-3h+6$$

CHECK YOUR UNDERSTANDING 2.10

For $f(x) = 3x^2-5$, determine:

(a) $f(-2)$ (b) $f(x+h)$ (c) $\dfrac{f(x+h)-f(x)}{h}$, for $h \neq 0$

Answer: (a) 7
(b) $3x^2+6hx+3h^2-5$
(c) $6x+3h$

Finally, we note that the (understood) **domain** of a function consists of all numbers for which the function is defined. The domain of the function $f(x) = \dfrac{x}{x-1}$, for example, consists of all numbers except 1; or: $(-\infty, 1) \cup (1, \infty)$. One may, however, restrict the domain of a function, as in:

The monthly cost (in dollars) of a company is given by

$$C(x) = 125x - \frac{x^2}{1500} + 2500 \text{ for } \boxed{0 \le x \le 500}$$

↑ the specified domain

EXERCISES

Exercises 1-9. Evaluate each of the following.

1. $-2^2(2+3)^2$
2. $(-2)^2(2\cdot 3)^2$
3. $\dfrac{(2+3)^2}{3^2}$
4. $\dfrac{\left(\frac{1}{2}\right)^2+\frac{5}{4}}{2}$
5. $\dfrac{\left(\frac{1}{2}+\frac{1}{4}\right)^2}{2^3}$
6. $2^3\left(\frac{1}{2}+\frac{1}{4}\right)^2$
7. $\dfrac{2^{-1}+2}{9\cdot 3^{-1}}$
8. $\dfrac{3(3-\frac{1}{2})^2}{4^0\cdot 3^2}$
9. $\dfrac{3^{-2}-2^{-3}}{-3^2}$

Exercises 10-12. Simplify.

10. $\dfrac{\left(\frac{a}{b}\right)^2}{(2ab)^2}$
11. $\dfrac{(ab)^2}{a^3(-b)^3}$
12. $\dfrac{(-ab)^2a^{-3}}{(2ab)^3}$

Exercises 13-18. Solve the given linear equation.

13. $2x+8 = 7x-5$
14. $5x+2 = -7x-12$
15. $-3x+7 = 4x-5+\frac{x}{2}$
16. $-(-2y-4)+\frac{y}{3}+3y = 5$
17. $\frac{x-2}{3} = 2(-x+1)$
18. $7x+\frac{2x-4-3x}{2} = \frac{7x-5+2x}{4}$

Exercises 19-24. Solve the given linear inequality.

19. $3x-5>2x+4$
20. $-4x+3>x+2$
21. $\frac{x}{2}-3x\geq 4x+5$
22. $-(-2x+4)\leq 5x-5$
23. $\frac{x+2}{4}+x-1<x+2$
24. $\frac{x+4}{-4}+x-1<x+2$

Exercises 25-30. Determine the slope of the line containing the given points.

25. $(2, 5), (3, 7)$
26. $(-2, 5), (3, 7)$
27. $(2, -5), (3, 7)$
28. $(-2, -5), (3, -7)$
29. $(3, 6), (7, 6)$
30. $(0, 0), (2, -7)$

31. On the same set of axes, sketch the four lines that pass through the origin and the additional point: $L_1: (1,1)$; $L_2: (1,-1)$; $L_3: (2,4)$; $L_4: (5,2)$. Calculate the slope for each of the lines.

32. On the same set of axes, sketch two distinct lines of slope 4, and two distinct lines of slope -4.

33. Which of the given points lie on the line $y = 4x + 3$: (7,1), (1,7), $(-2, 5)$, $(-2, -5)$?

34. Find the slope-intercept equation of two different lines, each of slope $\frac{2}{3}$.

35. Find the slope-intercept equation of two different lines, each of y-intercept 9.

Exercises 36-38. Find the slope-intercept equation of the line of given slope, m, which passes through the given point P.

36. $m = 3, P = (1, 2)$ 37. $m = 0, P = (1, 2)$ 38. $m = -1, P = (-1, \frac{1}{2})$

Exercises 39-41. Find the slope-intercept equation of the line passing through the two points.

39. $(2, 1), (1, 2)$ 40. $(1, -3), (0, 2)$ 41. $(-1, 3), (5, 3)$

Exercises 42-48. (Slope-Intercept Form) Find the slope-intercept equation of the line which:

42. Has slope $-\frac{1}{2}$ and contains the point (1,2).
43. Has slope 3 and contains the point (1,3).
44. Has slope 3 and y-intercept 4.
45. Has slope 3 and passes through the point (3,5).
46. Contains the points (3,5), (2,4).
47. Passes through the points $(2, -4)$, $(-3, 5)$.
48. Passes through the origin and the point (3,5).

Exercises 49-50. (Parallel Lines) Two lines are said to be **parallel** when they have the same slope, or when they are both vertical.

49. Determine the slope-intercept equation of the line that passes through the point $(3, -4)$, and is parallel to the line $y = 3x + 4$.

50. Determine the slope-intercept equation of the line that passes through the point $(2, 3)$, and is parallel to the line $2x + 3y = 1$.

Exercises 51-52. (Perpendicular Lines) Two lines that intersect at right angles are said to be **perpendicular**. It can be shown that two lines with nonzero slopes m_1 and m_2 are perpendicular if and only if the slope of one is the negative reciprocal of the slope of the other: $m_1 = -\frac{1}{m_2}$

51. Determine the slope-intercept equation of the line that passes through the point $(2, 1)$, and is perpendicular to the line $y = -2x + 1$.

52. Determine the slope-intercept equation of the line that passes through the point $(-1, -5)$, and is perpendicular to the line $x - 4y = 2$.

Exercises 53-54. (Point-Slope Equations of Lines) A point (x, y) is on the line of slope m which passes through the point (x_0, y_0) if and only if it satisfies the equation:

$$y - y_0 = m(x - x_0)$$

Example: Find a point-slope equation of the line of slope 3 passing through the point $(4, -9)$.

Solution: Since $m = 3$, $x_0 = 4$, and $y_0 = -9$, we have:

$$y - (-9) = 3(x - 4)$$
$$y + 9 = 3(x - 4)$$

53. (a) Find a point-slope equation of the line of slope 3 passing through the point $(-2, 1)$.

    (b) Find the slope-intercept equation of the line in (a).

    (c) Show that the two equations represent the same line by expressing the point-slope equation in (a) in the form $y = mx + b$.

54. (a) Find a point-slope equation of the line of slope 2 passing through the point $(1, -3)$.

    (b) Find the slope-intercept equation of the line in (a).

    (c) Show that the two equations represent the same line by expressing the point-slope equation in (a) in the form $y = mx + b$.

Exercises 55-72. Factor the polynomial.

55. $9x^2 - 4$
56. $-x^2 + 4$
57. $4x^2 - 25$
58. $100x^2 - 1$
59. $-4x^2 + 1$
60. $4x^2 - 5$
61. $2x^2 - 5$
62. $x^2 + 3x + 2$
63. $x^2 + 7x + 12$
64. $x^2 - x - 12$
65. $x^2 - 7x + 12$
66. $2x^2 + 5x + 2$
67. $6x^2 + 7x + 2$
68. $6x^2 + 13x + 6$
69. $6x^2 - 7x - 5$
70. $9x^3 - x$
71. $6x^3 + 31x^2 + 40x$
72. $-6x^3 + x^2 + 12x$

Exercises 73-90. (Equations) Solve the polynomial equation.

73. $2x^2 + 9x - 35 = 0$
74. $x^2 + 2x - 35 = 0$
75. $25x^2 - 16 = 0$
76. $x^2 - 25 = 0$
77. $5x^2 - 6 = 0$
78. $x^2 - 5 = 0$
79. $x^2 + 10x + 25 = 0$
80. $4x^2 - 4x + 1 = 0$
81. $\frac{x^2}{3} - \frac{4x}{3} - 7 = 0$
82. $-x^2 + x - \frac{1}{4} = 0$
83. $3x^3 - 14x^2 - 5x = 0$
84. $3x^3 + 16x^2 = -5x$
85. $2x^3 - 5x^2 - 3x = 0$
86. $3x^3 + 2x^2 = 5x$
87. $(2x^2 + 5x - 25)(5x^2 + 9x - 2) = 0$
88. $(x^3 - 9x)(2x^2 + 7x - 15) = 0$
89. $(4x^2 - 25)(x^2 - 2) = 0$
90. $x^4 - 4 = 0$

Exercises 91-100. (Inequalities) Solve the polynomial inequality.

91. $(x+1)(3x-1)>0$
92. $(x+1)(3x-1)^2 \le 0$
93. $(x+1)^2(-3x+5) \ge 0$
94. $(2x-5)(3x+2)(-x)<0$
95. $(x-4)^{40}(x+3)^{51} \le 0$
96. $x^2(3x-5)(-x+2)^3 \le 0$
97. $-2x(3-x)(4+3x)(2-x)>0$
98. $2x(1-x)(x+2)(3-x) \le 0$
99. $3x^3-14x^2-5x \ge 0$
100. $3x^3+16x^2>-5x$

Exercises 101-106. For $f(x) = x^2-x+2$, determine:

101. $f(-2)$
102. $f(a)$
103. $f(x+1)$
104. $f(x+h)$
105. $f(x+h)-f(x)$
106. $\frac{f(x+h)-f(x)}{h}$, for $h \ne 0$

Exercises 107-112. For $f(x) = -2x^2-3x+1$, determine:

107. $f(-2)$
108. $f(a)$
109. $f(x+1)$
110. $f(x+h)$
111. $f(x+h)-f(x)$
112. $\frac{f(x+h)-f(x)}{h}$, for $h \ne 0$

Exercises 113-116. (Piecewise-Defined Functions) One may wish to consider a function such as the function h which acts like $f(x) = x^2$ for $x<0$ and like $g(x) = x+1$ for $x \ge 0$. Such a function is said to be a **piecewise-defined** function and is represented in the following manner:

$$h(x) = \begin{cases} x^2 & \text{if } x<0 \\ x+1 & \text{if } x \ge 0 \end{cases}$$

To evaluate h at a particular x you must first determine which of the two rules applies. For example:

$$h(-1) \underset{\uparrow}{=} (-1)^2 = 1, \quad \text{and} \quad h(9) \underset{\uparrow}{=} 9+1 = 10$$

top rule since $-1<0$ bottom rule since $9 \ge 0$

Evaluate the given function at $-5, -1, 0, 2,$ and 10

113. $f(x) = \begin{cases} x-3 & \text{if } x<0 \\ 2x & \text{if } x \ge 0 \end{cases}$

114. $f(x) = \begin{cases} x^2+1 & \text{if } x<2 \\ x-5 & \text{if } x \ge 2 \end{cases}$

115. $f(x) = \begin{cases} 2x & \text{if } x<0 \\ -x & \text{if } 0 \le x<5 \\ x^2 & \text{if } x \ge 5 \end{cases}$

116. $f(x) = \begin{cases} -4x+1 & \text{if } x<1 \\ x^2 & \text{if } 1 \le x<7 \\ -2x & \text{if } x \ge 7 \end{cases}$

§2. LIMITS

At the very heart of the calculus is the concept of a limit. Here is one of them:

$$\lim_{x \to 2} (3x + 5)$$

It is read: *The limit as x approaches 2 of the function* $3x + 5$.

It represents: The number $3x + 5$ approaches, as x approaches 2.

Clearly, as x gets closer and closer to 2, $3x$ will get closer and closer to 6, and $3x + 5$ will consequently approach 11. We therefore write:

$$\lim_{x \to 2} (3x + 5) = 11$$

By the same token,

$$\lim_{x \to 3} \frac{x}{x^2 + 5} = \frac{3}{14}$$

(as x approaches 3, the numerator approaches 3, and the denominator approaches $3^2 + 5 = 14$.)

CHECK YOUR UNDERSTANDING 2.11

Determine the given limit.

(a) $\lim_{x \to -1} (4x^2 + x)$ (b) $\lim_{x \to 2} \frac{x + 3}{x + 2}$ (c) $\lim_{x \to 3} [x(3x^2 + 1)]$

Answers: (a) 3 (b) $\frac{5}{4}$ (c) 84

At this point, you might be wondering what all the fuss is about. Up to now we could simply plug the relevant number into the expression to arrive at the limit, right? Yes, but:

$$\lim_{x \to 2} \frac{x^2 + x - 6}{x^2 - 4}$$

Attempting to substitute 2 for x in the above expression brings us to the **meaningless** form "$\frac{0}{0}$." However, if you let the value of x get closer and closer to 2; say $x = 1.99$, $x = 2.001$, $x = 1.9999$, $x = 2.00001$, and so on, you will find that $\frac{x^2 + x - 6}{x^2 - 4}$ will indeed approach a particular number. To find that number, we turn to a related algebra problem:

You can use your calculator to see what happens, but at some point, say for $x = 1.99999999$, you may receive an error message, since most calculators think that $1.99999999 = 2$. Poor things.

Simplify: $\frac{x^2 + x - 6}{x^2 - 4}$.

Solution: $\frac{x^2 + x - 6}{x^2 - 4} = \frac{(x + 3)(x - 2)}{(x + 2)(x - 2)} = \frac{x + 3}{x + 2}$, but:

the above is not totally correct, for one should really write:

We remind you that one cannot "cancel a 0."

$$\frac{x^2+x-6}{x^2-4} = \frac{(x+3)(x-2)}{(x+2)(x-2)} = \frac{x+3}{x+2} \text{ if } x \neq 2$$

conditional equality

In the limit process, however, the variable x **approaches 2** — it can get as close to 2 as you wish but it is **never equal to 2**. Thus:

$$\lim_{x\to 2}\frac{x^2+x-6}{x^2-4} = \lim_{x\to 2}\frac{(x+3)(x-2)}{(x+2)(x-2)} = \lim_{x\to 2}\frac{x+3}{x+2} = \frac{5}{4}$$

not conditional

We've encountered two types of limits:

DETERMINED FORM

Those like $\lim_{x\to 3}\frac{x}{x^2+5}$ and $\lim_{x\to 2}\frac{x+3}{x+2}$, which can be determined by simply plugging in the indicated x-value. (such limits are of **determined form**)

UNDETERMINED FORM

And a more interesting type, like $\lim_{x\to 2}\frac{x^2+x-6}{x^2-4}$, which cannot be simply evaluated at $x = 2$. (such limits are of **undetermined form**)

EXAMPLE 2.8 Evaluate:

(a) $\lim_{x\to 2}\frac{x^3-2x^2-3x}{x^2+2x-15}$ (b) $\lim_{x\to 3}\frac{x^3-2x^2-3x}{x^2+2x-15}$

(c) $\lim_{x\to -2}\frac{\frac{1}{2}+\frac{1}{x}}{x+2}$

SOLUTION: (a) Challenging the denominator at $x = 2$ we find that it does **not** turn out to be 0: $2^2+2\cdot 2-15 = -7$. So, being faced with a determined form, we simply plug in 2 in the expression to arrive at the answer:

A useful hint: The only way the denominator and numerator can be zero, is if each contains $(x-3)$ as a factor, and you can use this fact to help you factor the polynomials.

$$\lim_{x\to 2}\frac{x^3-2x^2-3x}{x^2+2x-15} = \frac{2^3-2\cdot 2^2-3\cdot 2}{2^2+2\cdot 2-15} = \frac{-6}{-7} = \frac{6}{7}$$

(b) We are confronted with an undetermined form, which we now transform into a determined form:

$$\lim_{x\to 3}\frac{x^3-2x^2-3x}{x^2+2x-15} = \lim_{x\to 3}\frac{x(x^2-2x-3)}{(x-3)(x+5)} = \lim_{x\to 3}\frac{x(x-3)(x+1)}{(x-3)(x+5)}$$

$$= \lim_{x\to 3}\frac{x(x+1)}{(x+5)} = \frac{3(3+1)}{3+5} = \frac{12}{8} = \frac{3}{2}$$

determined form

just plug in

Note that we carry the limit symbol until the limit is performed. An analogous situation:

$3 \cdot 2 + 4 = 6 + 4 = 10$

you write the "+" until the sum is performed

(c) We cannot simply substitute -2 for x (why not?). So, we do what has to be done:

$$\lim_{x \to -2} \frac{\frac{1}{2} + \frac{1}{x}}{x+2} = \lim_{x \to -2} \frac{\frac{x+2}{2x}}{x+2} = \lim_{x \to -2} \left(\frac{x+2}{2x} \cdot \frac{1}{x+2}\right)$$

$$= \lim_{x \to -2} \frac{1}{2x} = \frac{1}{2(-2)} = -\frac{1}{4}$$

see margin

CHECK YOUR UNDERSTANDING 2.12

Determine the given limit.

(a) $\lim_{x \to 1} \frac{x^2 + 3x - 4}{x^2 + 1}$ (b) $\lim_{x \to 1} \frac{x^2 + 3x - 4}{x^2 - 1}$ (c) $\lim_{x \to 1} \frac{x^2 - 1}{\frac{x}{2} - \frac{1}{2}}$

Answers: (a) 0 (b) $\frac{5}{2}$ (c) 4

A LIMIT NEED NOT EXIST

Does $\lim_{x \to 2} \frac{x+2}{x-2}$ exist? We certainly cannot substitute 2 for x in the expression, for that would yield a zero in the denominator. That, in and of itself, is not necessarily a problem (See Example 2.8). The problem is that, as x approaches 2, the denominator of $\frac{x+2}{x-2}$ gets closer and closer to 0 while its numerator tends to 4. The net result is that the quotient will just keep getting bigger and bigger: the limit does not exist (DNE).

Consider the function $f(x) = \begin{cases} 3x+1 & \text{if } x < 2 \\ x^2 & \text{if } x > 2 \end{cases}$. Does $\lim_{x \to 2} f(x)$ exist? No. Why not? Because:

As x tends to 2 from the left, the top rule is in effect, and $f(x)$ approaches $3 \cdot 2 + 1 = 7$. On the other hand, if x approaches 2 from the right, then the bottom rule is in effect, and $f(x)$ tends to $2^2 = 4$.

The notation $f(x) = \begin{cases} 3x+1 & \text{if } x < 2 \\ x^2 & \text{if } x > 2 \end{cases}$ is very transparent. For example: $f(1) = 3 \cdot 1 + 1 = 4$ (since $1 < 2$) while $f(3) = 3^2 = 9$ (since $3 > 2$) (See Exercise 113-116, page 130.)

One says that the **left-hand limit** of the above function equals 7 and that the **right-hand limit** equals 4; written:

$$\lim_{x \to 2^-} f(x) = 7 \quad \text{and} \quad \lim_{x \to 2^+} f(x) = 4$$

It follows that the "full-limit," $\lim_{x \to 2} f(x)$, does not exist (DNE).

In general, as you might anticipate:

$$\lim_{x \to c} f(x) = L \text{ if and only if } \lim_{x \to c^-} f(x) = L \text{ and } \lim_{x \to c^+} f(x) = L$$

We are assuming here that f is defined on either side of the number c.

CHECK YOUR UNDERSTANDING 2.13

Determine if the given limit exists, and if does, evaluate it.

(a) $\lim_{x \to 3} \frac{x^2 - 9}{x - 3}$ (b) $\lim_{x \to 3} \frac{x - 3}{x^2 - 6x + 9}$

(c) $\lim_{x \to 3} \begin{cases} 2x - 1 & \text{if } x < 3 \\ x + 5 & \text{if } x > 3 \end{cases}$ (d) $\lim_{x \to 3} \begin{cases} x - 1 & \text{if } x \le 3 \\ 2 & \text{if } x > 3 \end{cases}$

Answers: (a) 6 (b) DNE (c) DNE (d) 2

GEOMETRICAL INTERPRETATION OF THE LIMIT CONCEPT

Consider the functions, f, g, h, and k in Figure 2.3.

The solid dot above 3 in (a) depicts the value of the function at 3: $f(3) = 4$. Similarly: $g(3) = 7$ and $h(3) = 7$.

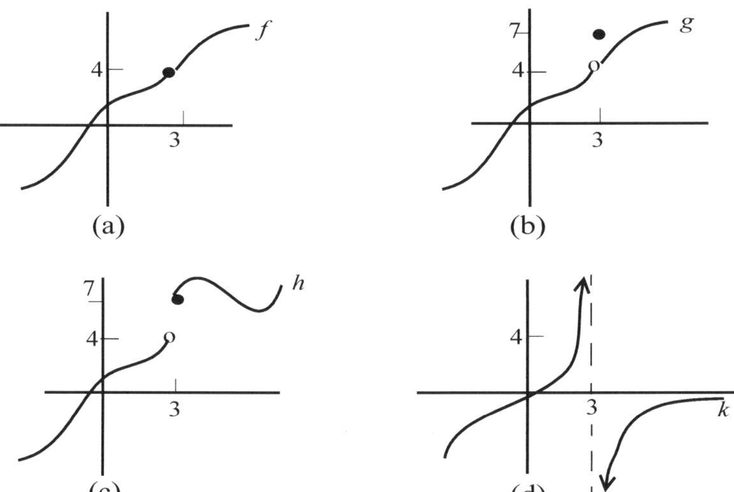

Figure 2.3

Looking at the function f in (a), we see that as x approaches 3, from **either the left or the right**, the function values (y-values) approach the number 4. Thus: $\lim_{x \to 3} f(x) = 4$. The function g in (b) differs from f only at $x = 3$, where it has a "hiccup." But that anomaly has absolutely no effect on the limit, since the limit does not care about what happens at 3 — it only cares about what happens as x **approaches** 3. The function h in (c) does **not** have a limit at $x = 3$, since $\lim_{x \to 3^-} h(x) = 4$ while $\lim_{x \to 3^+} h(x) = 7$. The function in (d) also does not have a limit at 3, as the function values get larger and larger, positively and negatively, as x approaches 3.

CHECK YOUR UNDERSTANDING 2.14

Referring to the graph of the function f below, determine if the given limit exists, and if it does, indicate its value.

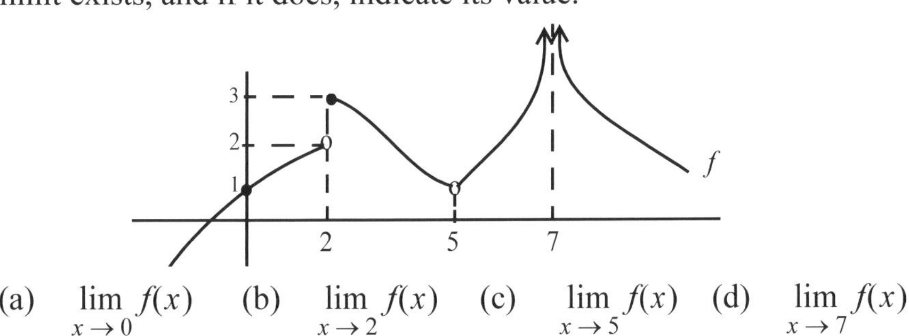

(a) $\lim_{x \to 0} f(x)$ (b) $\lim_{x \to 2} f(x)$ (c) $\lim_{x \to 5} f(x)$ (d) $\lim_{x \to 7} f(x)$

Answers: (a) 1 (b) DNE (c) 1 (d) DNE. Note: It is also acceptable to write: $\lim_{x \to 7} f(x) = \infty$ since the function values get larger and larger in the positive direction as x approaches 7 from either side.

PROPERTIES OF LIMITS

The following theorem formalizes results that you have been taking for granted all along.

It is certainly "believable," for example, that if $f(x)$ approaches L and if $g(x)$ approaching M as x tends to c, then $f(x) + g(x)$ must surely tend to $L + M$ as x approaches c.

Without a rigorous definition of the limit concept, however, we are not in a position to prove any of these statements.

THEOREM 2.4

LIMIT THEOREMS

If $\lim_{x \to c} f(x) = L$ and $\lim_{x \to c} g(x) = M$ then:

(i) $\lim_{x \to c} [f(x) + g(x)] = L + M$

(ii) $\lim_{x \to c} [f(x) - g(x)] = L - M$

(iii) $\lim_{x \to c} [f(x) \cdot g(x)] = L \cdot M$

(iv) $\lim_{x \to c} \dfrac{f(x)}{g(x)} = \dfrac{L}{M}$ if $M \neq 0$

(v) $\lim_{x \to c} [af(x)] = aL$

IN WORDS:
(i) The limit of a sum is the sum of the limits.
(ii) The limit of a difference is the difference of the limits.
(iii) The limit of a product is the product of the limits.
(iv) The limit of a quotient is the quotient of the limits (providing the limit of the denominator is not zero).
(v) The limit of a constant times a function is the constant times the limit of the function.

CONTINUITY

Let's reconsider the functions of Figure 2.3:

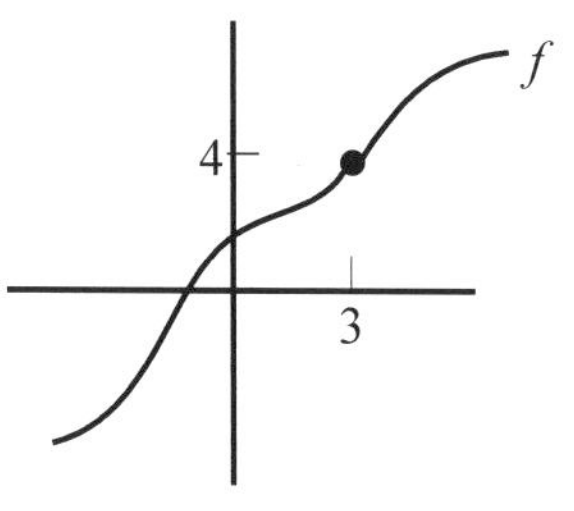

$\lim_{x \to 3} f(x) = 4$
(a)

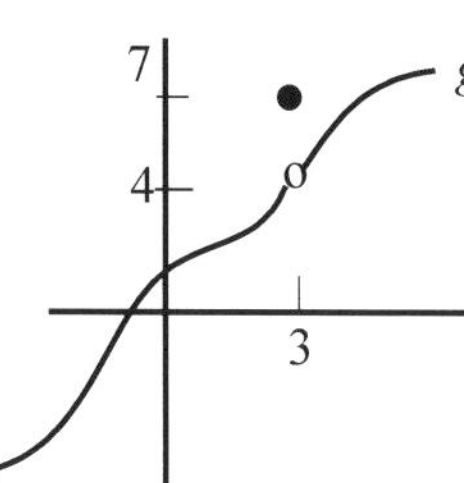

$\lim_{x \to 3} g(x) = 4$
(b)

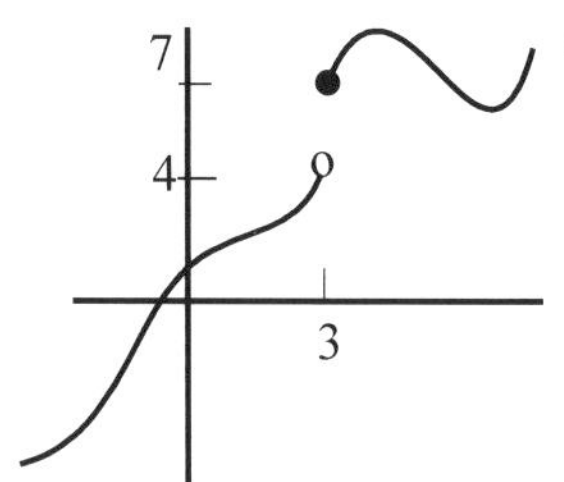

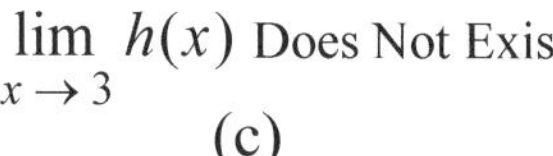

$\lim_{x \to 3} h(x)$ Does Not Exist
(c)

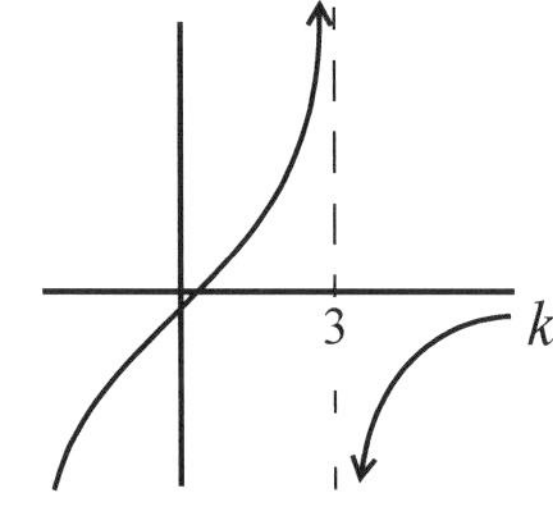

$\lim_{x \to 3} k(x)$ Does Not Exist
(d)

Figure 2.4

While the functions in (c) and (d) do not have limits as x approaches 3, both the functions in (a) and (b) do. The limit, being oblivious of what happens at 3, cannot tell you that the function g in (b) behaves in a somewhat peculiar fashion at $x = 3$. Another concept, one more sensitive than that of the limit, is called for:

A **continuous function** is a function that is continuous at every point in its domain.

DEFINITION 2.4
CONTINUITY

A function f is **continuous** at c if:

$$\lim_{x \to c} f(x) = f(c)$$

A function that is not continuous at c, is said to be **discontinuous** at that point.

In other words, for a function f to be continuous at c, three things must happen:

(1) f must be defined at c
(2) $\lim_{x \to c} f(x)$ must exist
(3) The limit must equal $f(c)$

Returning to Figure 2.4, we see that:

The function f in (a) is continuous at 3.
(limit equals 4, and $f(3) = 4$)
The function g in (b) is discontinuous at 3.
(limit equals 4, but $g(3) = 7$)
The functions in (c) and (d) are also discontinuous at 3
(limit does not even exist)

If a function f has a limit at c and that limit is not equal to $f(c)$ then f is said to have a **removable discontinuity** at c. [The function in Figure 2.4(b) has a removable discontinuity at 3].

If a function f fails even to have a limit at c, because the two one-sided limits exist but are not equal, then f is said to have a **jump discontinuity** at c. [The function in Figure 2.4(c) has a jump discontinuity at 3].

CHECK YOUR UNDERSTANDING 2.15

Indicate whether or not the given function f is continuous at $x = 3$.

(a) $f(x) = \dfrac{x^2 - 9}{x - 3}$ (b) $f(x) = \dfrac{x - 3}{x^2 - 6x + 9}$

(c) $f(x) = \begin{cases} 2x - 1 & \text{if } x < 3 \\ x + 3 & \text{if } x \geq 3 \end{cases}$ (d) $f(x) = \begin{cases} x - 1 & \text{if } x \leq 3 \\ 2 & \text{if } x > 3 \end{cases}$

Answers:
(a) Not continuous:
(b) Not continuous.
(c) Not continuous:
(d) Continuous.

EXERCISES

Exercises 1-27. Evaluate the given limit.

1. $\lim\limits_{x\to 3}(x^2-5)$
2. $\lim\limits_{x\to 0}(x^2-5)$
3. $\lim\limits_{x\to 3}\dfrac{x^2-5}{x+3}$
4. $\lim\limits_{x\to 3}\dfrac{x^2-9}{x+3}$
5. $\lim\limits_{x\to 5}\dfrac{x-5}{x+5}$
6. $\lim\limits_{x\to 5}\dfrac{x^2-5}{x-5}$
7. $\lim\limits_{x\to 5}\dfrac{x^3-25x}{x-5}$
8. $\lim\limits_{x\to 5}\dfrac{x^2-5}{x^2-3x-10}$
9. $\lim\limits_{x\to 5}\dfrac{x^2-25}{x^2-4x-5}$
10. $\lim\limits_{x\to 5}\dfrac{x^2-4x-5}{x^2-25}$
11. $\lim\limits_{x\to -2}\dfrac{x^2-4}{x+2}$
12. $\lim\limits_{x\to -2}\dfrac{x^2-4}{3x^2+6x+2}$
13. $\lim\limits_{x\to 3}\dfrac{x^2-2x-3}{x^2-x-6}$
14. $\lim\limits_{x\to 3}\dfrac{x^2-x-6}{x^2-2x-3}$
15. $\lim\limits_{x\to -3}\dfrac{x^2+4x+3}{x^2+3x}$
16. $\lim\limits_{x\to 4}\dfrac{x^2-4x-4}{x^2-x-12}$
17. $\lim\limits_{x\to 4}\dfrac{x^2-8x+16}{x^2-4x}$
18. $\lim\limits_{x\to -2}\dfrac{x^2+4x+4}{x^2+3x+2}$
19. $\lim\limits_{x\to -1}\dfrac{2x^3+5x^2+3x}{x^2-3x-4}$
20. $\lim\limits_{x\to 1}\dfrac{x^4-1}{x^2-1}$
21. $\lim\limits_{x\to -2}\dfrac{x^2+x-2}{x^4-16}$
22. $\lim\limits_{x\to -\sqrt{2}}\dfrac{x^2-2}{x+\sqrt{2}}$
23. $\lim\limits_{x\to 0}\dfrac{x^3+2x^2-3x}{x^3-2x^2-15x}$
24. $\lim\limits_{x\to -3}\dfrac{x^3+2x^2-3x}{x^3-2x^2-15x}$
25. $\lim\limits_{x\to -1}\dfrac{\frac{1}{x}+1}{x^2-1}$
26. $\lim\limits_{x\to 0}\left(\dfrac{1}{x}-\dfrac{1}{x^2+x}\right)$
27. $\lim\limits_{x\to 2}\dfrac{\frac{x^2}{x-1}-4}{\frac{1}{x+2}-\frac{1}{4}}$

Exercises 28-31. Referring to the graph of the function f, determine if $\lim\limits_{x\to 3}f(x)$ exists. If it does, indicate its value. Is the function continuous at 3?

28.

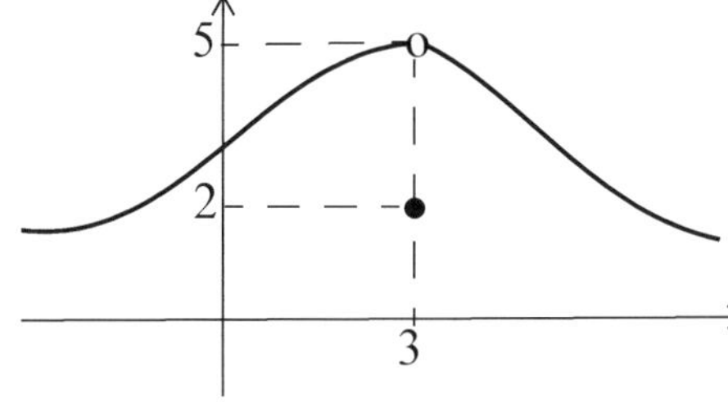

29.

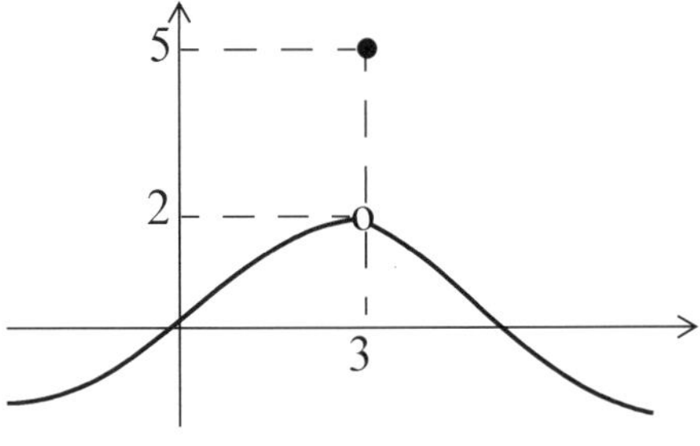

30.

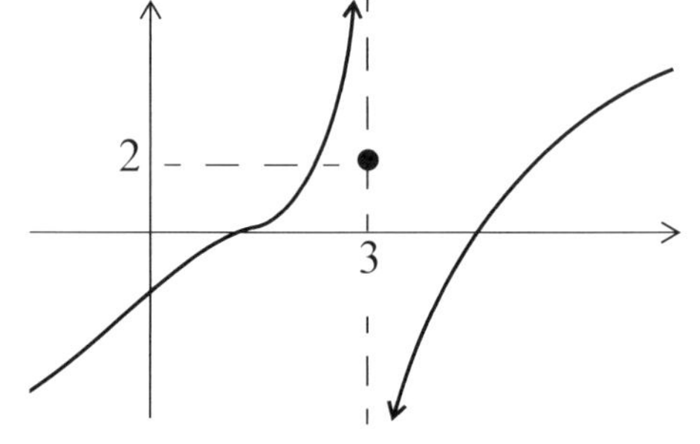

31.

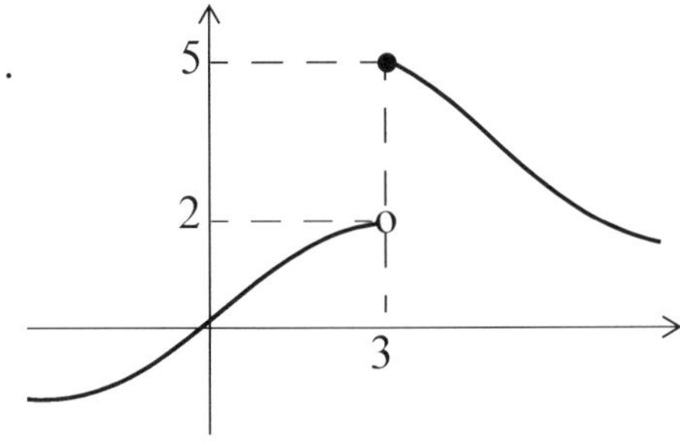

Exercises 32-35. (Piecewise-Defined Functions) Determine if $\lim_{x \to 2} f(x)$ exists. Is the function continuous at 2? (See Exercises 113-116, page 130.)

32. $f(x) = \begin{cases} x+2 & \text{if } x < 2 \\ x^2 & \text{if } x > 2 \end{cases}$

33. $f(x) = \begin{cases} x+2 & \text{if } x < 2 \\ x^2 & \text{if } x \geq 2 \end{cases}$

34. $f(x) = \begin{cases} x+1 & \text{if } x < 2 \\ x^2 & \text{if } x > 2 \end{cases}$

35. $f(x) = \begin{cases} x+2 & \text{if } x < 2 \\ 2 & \text{if } x = 2 \\ x^2 & \text{if } x > 2 \end{cases}$

Exercises 36-40. (Construction) Sketch the graph of a function f satisfying the given conditions.

36. $f(1) = 5$ and $\lim_{x \to 1} f(x) = 5$

37. $f(3) = 1$ and $\lim_{x \to 3} f(x) = -1$

38. $f(1) = 5$ and $\lim_{x \to 1} f(x) = 6$.

39. $f(1) = 5$ and $\lim_{x \to 1} f(x)$ does not exist.

40. f is not continuous at 1, 2, 3, but is defined at those points.

§3. TANGENT LINES AND THE DERIVATIVE

The tangent line to a point on the circle is that line which touches the circle only at that point:

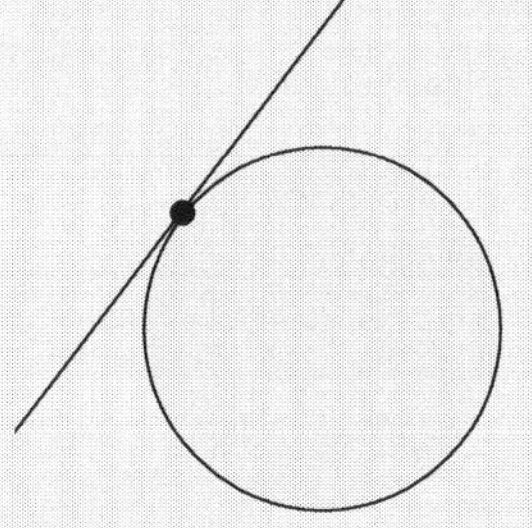

This will not do for more general curves. Line 2 in Figure 2.5, for example, touches the curve at more than one point.

Consider the two lines of Figure 2.5. Which do you feel better represents the tangent line to the curve at the point $(c, f(c))$? Chances are that you chose line 2, and might have based that decision on the concept of a tangent line to a circle (see margin). Our goal in this section is to define the "tangent line" of Figure 2.5 in such a way that it conforms with our predisposed notion of tangency.

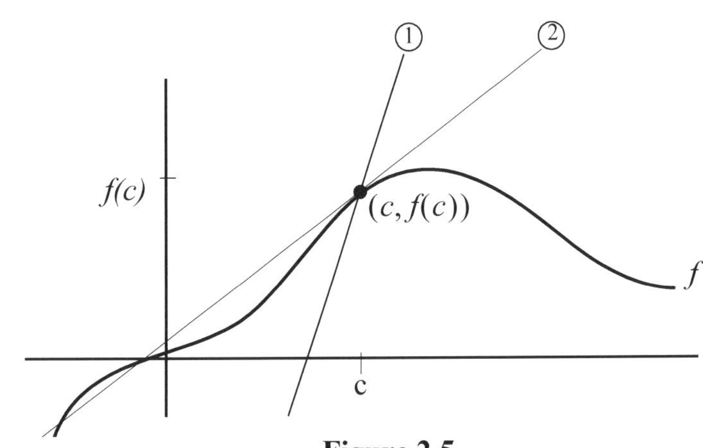

Figure 2.5

But why bother? What's so special about tangent lines? For one thing, near the point of interest, a tangent line offers a nice approximation for the given function [see Figure 2.6(a)]. For another, tangent lines can be used to find where maxima and minima occur [see Figure 2.6(b)] and this enables one to solve a host of practical problems.

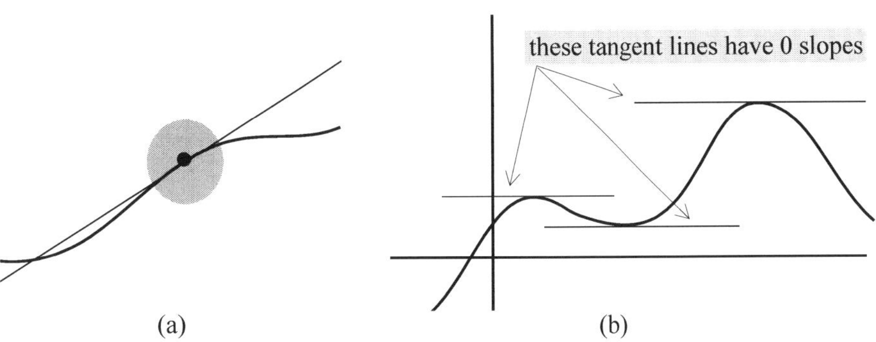

Figure 2.6

Returning to the situation depicted in Figure 2.5, we demand that the tangent line at $(c, f(c))$ passes through that point (a reasonable demand). Consequently, since a (non-vertical) line is determined by its slope and any point on the line, it suffices to define the slope of the tangent line we seek.

We call the line W_h, to remind us that we got it by moving h units from c along the x-axis. (If h were negative, then $c+h$ would lie to the left of c.)

But there is a problem. We need 2 points to find the slope of a line, and here we have but the one: $(c, f(c))$. So, consider Figure 2.7 where the would-be tangent line T is represented in dotted form (it really doesn't exist, until we define it). A solid line W_h passing through the two points on the curve $(c, f(c))$ and $(c+h, f(c+h))$ also appears.

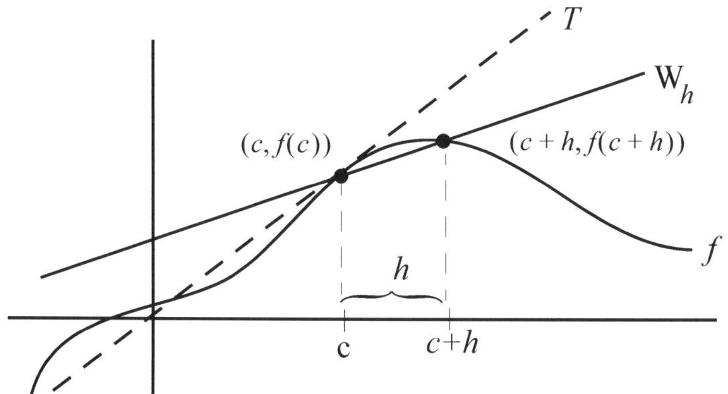

Figure 2.7

The line W_h is wrong — it is not the line we seek. But we can do something with W_h which we were not able to do with our phantom line T; we can calculate its slope:

the Greek letter Δ, pronounced "delta," is often used to denote "***a change in***.

$$m = \frac{f(c+h)-f(c)}{(c+h)-c} = \frac{f(c+h)-f(c)}{h} \quad \left(\frac{\text{change in y}}{\text{change in x}} : \frac{\Delta y}{\Delta x}\right)$$

see margin

Yes, the line W_h is wrong, but it is easy to see that the lines W_h will pivot closer and closer to T as h gets smaller and smaller! It is therefore totally natural to define:

$$\text{slope of } T = \lim_{h \to 0} \frac{f(c+h)-f(c)}{h}$$

The limit (the slope of T) is called the derivative of f at c, and is denoted by $f'(c)$:

DEFINITION 2.5
DERIVATIVE OF A FUNCTION AT A POINT

The **derivative of a function f at c** is the number $f'(c)$ given by:

$$f'(c) = \lim_{h \to 0} \frac{f(c+h)-f(c)}{h}$$

(Providing, of course, that the above limit exists.)

The graph of:

$f(x) = -3x^2 + 6x - 1$ appears below.

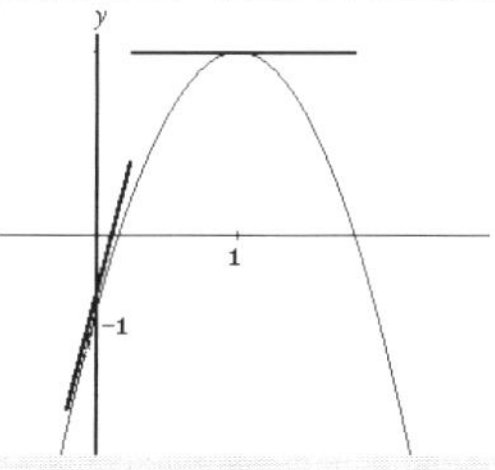

From the graph, we can anticipate that $f'(0)$ will be a positive number (tangent line climbs, and rather rapidly), and that $f'(1) = 0$ (why?).

EXAMPLE 2.9 Determine $f'(0)$ and $f'(1)$ for the function:

$$f(x) = -3x^2 + 6x - 1$$

SOLUTION: Using Definition 2.5 with $c = 0$, we have:

$$f'(0) = \lim_{h \to 0} \frac{f(0+h) - f(0)}{h} = \lim_{h \to 0} \frac{f(h) - f(0)}{h}$$

($c = 0$) — undetermined form

$$= \lim_{h \to 0} \frac{(-3h^2 + 6h - 1) - (-1)}{h} = \lim_{h \to 0} \frac{-3h^2 + 6h}{h}$$

$$= \lim_{h \to 0} \frac{h(-3h+6)}{h} = \lim_{h \to 0} (-3h + 6) = 6$$

determined form, just plug in 0 for h

It is important that you are able to deal with $f(1+h)$. Once you do that correctly the rest is rather easy: just manipulate the numerator into a form which enables you to cancel that bothersome h in the denominator of the expression [go from an undetermined form to a determined form].

Repeating the process with $c = 1$, we have:

$$f'(1) = \lim_{h \to 0} \frac{f(1+h) - f(1)}{h} = \lim_{h \to 0} \frac{[-3(1+h)^2 + 6(1+h) - 1] - 2}{h}$$

$$= \lim_{h \to 0} \frac{[-3(1 + 2h + h^2) + 6 + 6h - 1] - 2}{h}$$

$$= \lim_{h \to 0} \frac{-3 - 6h - 3h^2 + 6 + 6h - 1 - 2}{h} = \lim_{h \to 0} \frac{-3h^2}{h}$$

$$= \lim_{h \to 0} -3h = 0$$

CHECK YOUR UNDERSTANDING 2.16

Determine $f'(2)$ for the function $f(x) = -3x^2 + 6x - 1$ of Example 2.9.

Answer: -6

We did some work in Example 2.9 to find $f'(0)$, and repeated the same process to find $f'(1)$. We could have saved some time by finding the derivative function, $f'(x)$, and then evaluating it at 0 and at 1; where:

Note that the $f'(c)$ of Definition 2.5 is a **number**: the slope of the tangent line at $(c, f(c))$. On the other hand, $f'(x)$ is a **function** whose value at x is the slope of the tangent line at the point $(x, f(x))$.

DEFINITION 2.6
DERIVATIVE FUNCTION

The **derivative of a function f** is the function $f'(x)$ given by:

$$f'(x) = \lim_{h \to 0} \frac{f(x+h) - f(x)}{h}$$

(providing of course, that the above limit exists)

EXAMPLE 2.10 Find the derivative, $f'(x)$, of the function $f(x) = -3x^2 + 6x - 1$, and then use it to determine $f'(0)$, $f'(1)$, and $f'(-1)$.

SOLUTION: Turning to Definition 2.7, we have:

Note that we are just duplicating the process of Example 2.9, but with "x" instead of "0" or "1."

$$f'(x) = \lim_{h \to 0} \frac{f(x+h) - f(x)}{h} = \lim_{h \to 0} \frac{[-3(x+h)^2 + 6(x+h) - 1] - (-3x^2 + 6x - 1)}{h}$$

$$= \lim_{h \to 0} \frac{[-3(x^2 + 2xh + h^2) + 6x + 6h - 1] + 3x^2 - 6x + 1}{h}$$

Derivatives are limits of undetermined form. If the derivative exists, then the h in the denominator **has** to eventually cancel with an h-factor in the numerator. For this to happen, all terms in the numerator that do not "contain an h" must drop out.

$$= \lim_{h \to 0} \frac{-3x^2 - \mathbf{6xh} - \mathbf{3h^2} + 6x + \mathbf{6h} - 1 + 3x^2 - 6x + 1}{h}$$

$$= \lim_{h \to 0} \frac{-6xh - 3h^2 + 6h}{h} = \lim_{h \to 0} \frac{h(-6x - 3h + 6)}{h}$$

$$= \lim_{h \to 0} (-6x - 3h + 6) = -6x + 6$$

We have shown that the derivative of $f(x) = -3x^2 + 6x - 1$ is:

$$f'(x) = -6x + 6$$

In particular:
$$f'(0) = -6 \cdot 0 + 6 = 6,$$
$$f'(1) = -6 \cdot 1 + 6 = 0,$$
$$f'(-1) = -6(-1) + 6 = 12.$$

Some of the more powerful graphing calculators, like the TI-89 and Voyage, can give you the derivative of a function:

```
F1  F2      F3   F4    F5     F6
    Algebra Calc Other PrgmIO Clean Up

▪ d/dx(-3·x^2 + 6·x - 1)          6 - 6·x
```

The "double-d" notation is also used to denote the derivative of a function.

GRAPHING CALCULATOR GLIMPSE 2.1

Graphing calculators can give the derivative of a function at a given point:

```
nDeriv(-3X²+6X-1
,X,0)
                6
nDeriv(-3X²+6X-1
,X,1)
                0
```

```
nDeriv(-3X²+6X-1
,X,-1)
               12
```

Answer: $f'(x) = 2x + 1$

$f'(0) = 1$, $f'(1) = 3$, $f'(-1) = -1$

CHECK YOUR UNDERSTANDING 2.17

Find the derivative, $f'(x)$, of the function $f(x) = x^2 + x + 1$, and use it to determine $f'(0)$, $f'(1)$, and $f'(-1)$.

EXAMPLE 2.11 Determine the tangent line to the graph of the function $f(x) = -3x^2 + 6x - 1$, at $x = 2$.

SOLUTION: Whenever you see the word "line" you should think of:

$$y = mx + b$$

slope ↑ (under m), ↑ y-intercept (under b)

As it was in Example 2.3, page 120, the first step is to find the slope, m. In that example, we had a couple of points to work with. Here, we have something just as good:

$$y = mx + b$$

↑ $f'(2)$ (slope of tangent line at $x = 2$)

In Example 2.10, we showed that the derivative of the given function is $f'(x) = -6x + 6$. Consequently, the slope of our tangent line is $f'(2) = -6 \cdot 2 + 6 = \mathbf{-6}$; bringing us to:

$$y = -\mathbf{6}x + b \qquad (*)$$

The procedure for finding b has not changed. We need to know a point on the line, and we do:

The tangent line must pass through the point on the curve whose x-coordinate is 2, namely, the point:

$$(2, f(\mathbf{2})) = (2, -\mathbf{1})$$

$$f(\mathbf{2}) = -3 \cdot 2^2 + 6 \cdot 2 - 1 = -\mathbf{1}$$

Since the point $(2, -1)$ satisfies (*):

$$y = -6x + b$$
$$-1 = -6 \cdot 2 + b$$
$$b = 11$$

Tangent line: $\boxed{y = -6x + 11}$

GRAPHING CALCULATOR GLIMPSE 2.2

You can instruct most graphing calculators to draw the tangent line at a specified point on the graph of a function:

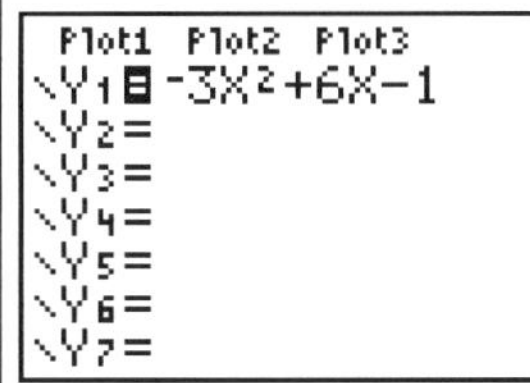

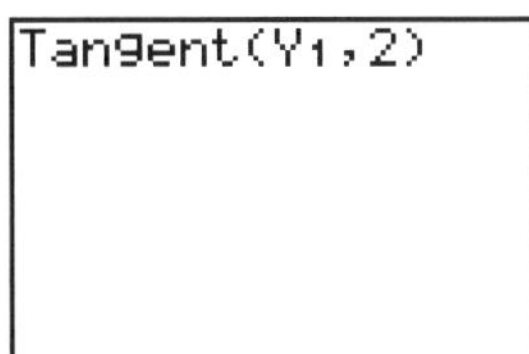

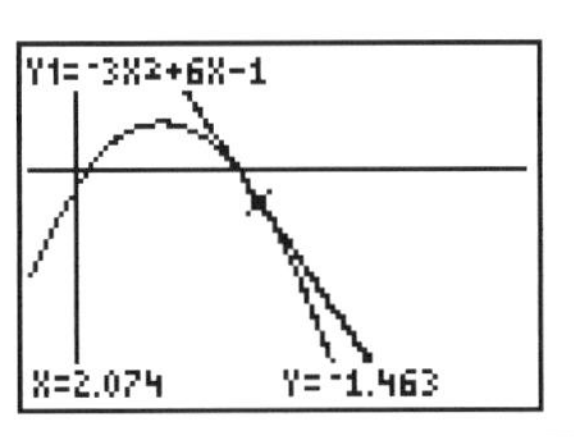

CHECK YOUR UNDERSTANDING 2.18

Find the tangent line to the graph of the function $f(x) = x^2 + x + 1$ at $x = 2$.

Answer: $y = 5x - 3$

GEOMETRICAL INSIGHTS INTO THE DERIVATIVE

Consider the two graphs of Figure 2.8. Geometrically speaking, you can see that the function f in (a) has a (positive) derivative at $x = c$ (tangent line exists and has positive slope). In contrast, the function g in (b) is not differentiable at $x = c$. Do you see why not?

If a function is differentiable at c, then the graph has to be "smooth" at that point [as in Figure 2.8(a)]. If the graph "changes direction abruptly" at that point [as in Figure 2.8(b)], then the function will not be differentiable at c.

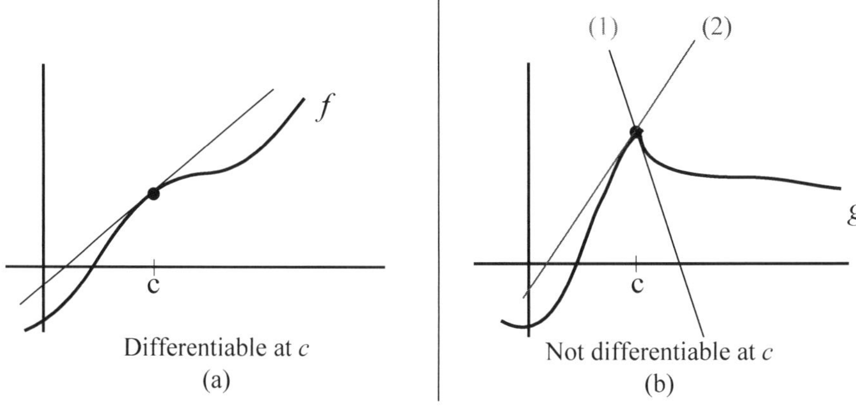

Figure 2.8

Because: Lines are notoriously straight, and the graph of g has a sharp bend at c [the "would-be tangent line (1)" is of negative slope, while the "would-be tangent line **(2)**" is of positive slope]. Since no line can approximate the graph of g at c, there cannot be a tangent line at c [in other words: $f'(c)$ does not exist]. To be more specific, we call your attention to the graph of the absolute value function in Figure 2.9.

The absolute value function, $|x|$, returns the magnitude of x. For example: $|5| = 5$ and $|-5| = 5$

$$\text{Abs}(x) = |x| = \begin{cases} x & \text{if } x \geq 0 \\ -x & \text{if } x < 0 \end{cases}$$

The absolute value function, and its graph

Figure 2.9

From our previous discussion, we can "see" that the absolute value function is not differentiable at $x = 0$; a fact which we now formally establish:

EXAMPLE 2.12 Show that the absolute value function $f(x) = |x|$ is not differentiable at 0.

SOLUTION: We go directly to Definition 2.5:

$$f'(c) = \lim_{h \to 0} \frac{f(c+h) - f(c)}{h}$$

For $f(x) = |x|$, and $c = 0$, the above limit takes the form:

$$\lim_{h \to 0} \frac{|0+h| - |0|}{h} = \lim_{h \to 0} \frac{|h|}{h}$$

The derivative is a "special limit," but a limit nonetheless. Just as $\lim_{x \to 2}(4x-5)$ approaches the number 3 as the variable x approaches 2 from either side of 2, $\frac{|h|}{h}$ would have to approach some number as the variable h approaches 0 from **either side** of 0. This is **not** the case here.

Recalling that the symbol $\lim_{h \to 0^-} f(x)$ indicates the limit as h approaches 0 from the left, and the symbol $\lim_{h \to 0^+} f(x)$ indicates the limit as h approaches 0 from the right, we have:.

since h is negative

$$\lim_{h \to 0^-} f(x) = \lim_{h \to 0^-} \frac{|h|}{h} = \frac{-h}{h} = -1$$

and:

since h is positive

$$\lim_{h \to 0^+} f(x) = \lim_{h \to 0^+} \frac{|h|}{h} = \frac{h}{h} = 1$$

Since the left-hand limit is different than the right-hand limit, the limit does not exist, and consequently the absolute value function is not differentiable at 0.

CONTINUITY AND THE DERIVATIVE

The following result, which you are invited to establish in the exercises, says that differentiability is a stronger condition than continuity:

THEOREM 2.5 If a function f is differentiable at c, then f is continuous at c.

To put it another way: if a function is not continuous at c, then it is not differentiable at c.

Here is the pecking order of the concepts of limit, continuity, and differentiability:

$$\text{Differentiability} \Rightarrow \text{Continuity} \Rightarrow \text{Limit Exists}$$

read: *implies*

Figure 2.10 demonstrates that neither of the above two implications is reversible.

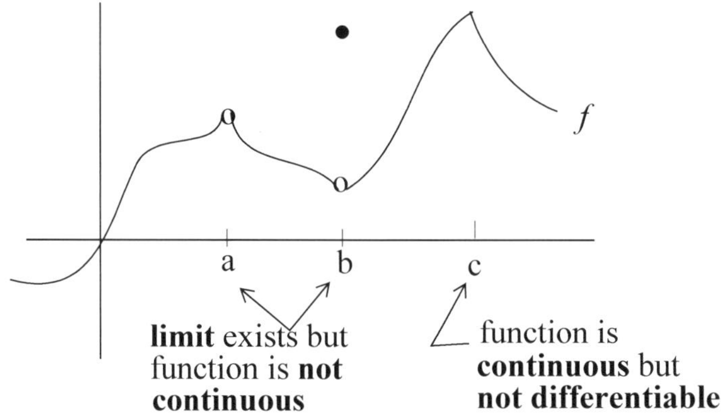

Figure 2.10

EXERCISES

Exercises 1-2. Sketch a tangent line to the graph of the function *f*, at the indicated point and then estimate the value of $f'(2)$, $f'(4)$, and $f'(7)$.

1. 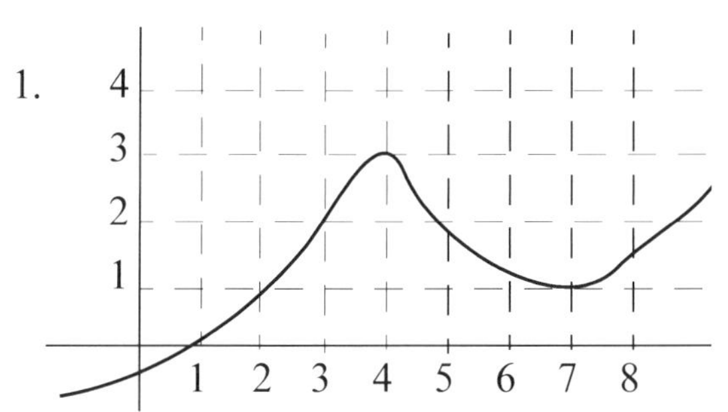

2. 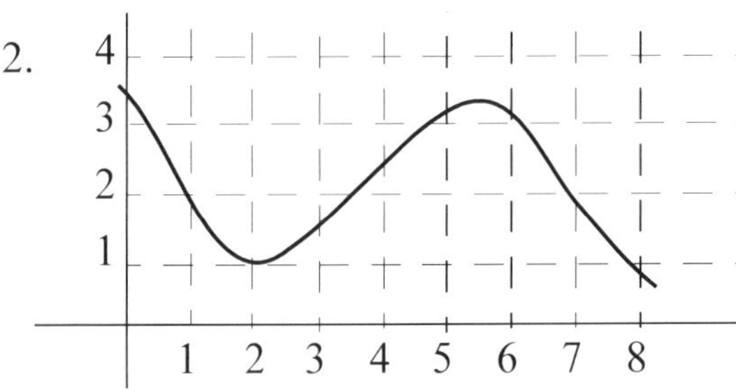

Exercises 3-11. Use Definition 2.5 to determine $f'(2)$ for the given function.

3. $f(x) = 5x + 1$
4. $f(x) = -x + 4$
5. $f(x) = 4x^2$
6. $f(x) = 3x^2 + x$
7. $f(x) = -x^2 + 3x - 1$
8. $f(x) = -3x^2 + 3x - 500$
9. $f(x) = 55$
10. $f(x) = x^3$
11. $f(x) = x^3 + x + 1$

Exercises 12-20. Use Definition 2.6 to determine $f'(x)$ for the given function.

12. $f(x) = x$
13. $f(x) = -5x - 4000$
14. $f(x) = 7x + 500$
15. $f(x) = 3x^2$
16. $f(x) = x^2 + 2x$
17. $f(x) = -2x^2 + x - 2$
18. $f(x) = 5$
19. $f(x) = 101$
20. $f(x) = -x^3 - 3$

Exercises 21-28. Find the equation of the tangent line to the graph of the given function at the indicated point.

21. $f(x) = -3x$ at $x = 5$
22. $f(x) = -3x + 5$ at $x = 500$
23. $f(x) = x^2 + 2x$ at $x = 0$
24. $f(x) = x^2 + 2x$ at $x = 1$
25. $f(x) = x^2 + 2x$ at $x = 2$
26. $f(x) = 2x^2 + x + 1$ at $x = 1$
27. $f(x) = 11$ at $x = -9$
28. $f(x) = 11$ at $x = 99$

29. Give a geometrical argument involving tangent lines to suggest that the derivative of the function $f(x) = x$ equals 1, and then establish the result via Definition 2.6.

30. Give a geometrical argument involving tangent lines to suggest that the derivative of the function $f(x) = c$, where c is a constant, equals 0; and then establish the result via Definition 2.6.

31. Give a geometrical argument involving tangent lines to suggest that the derivative of the linear function $f(x) = mx + b$ equals m, and then establish the result via Definition 2.6.

32. Pair off each function [A] through [F] with its corresponding derivative function [1] through [6]. Suggestion: Think "slope of tangent lines." Where the tangent line is horizontal, for example, the derivative at that point will be zero; where the tangent line has a positive slope, the derivative will be positive (the greater the slope, the greater the value of the derivative), and so on

[A]
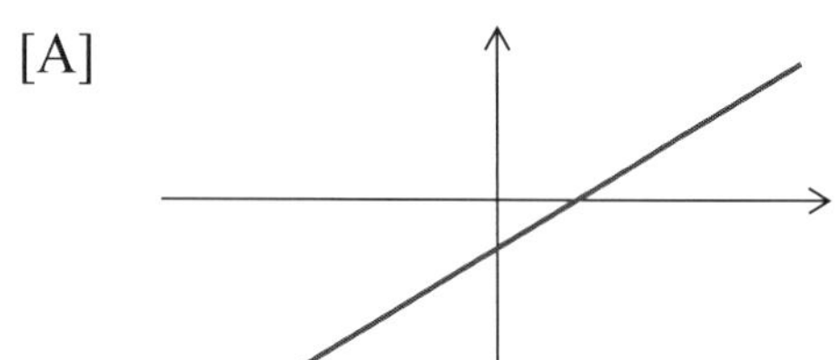

[1]
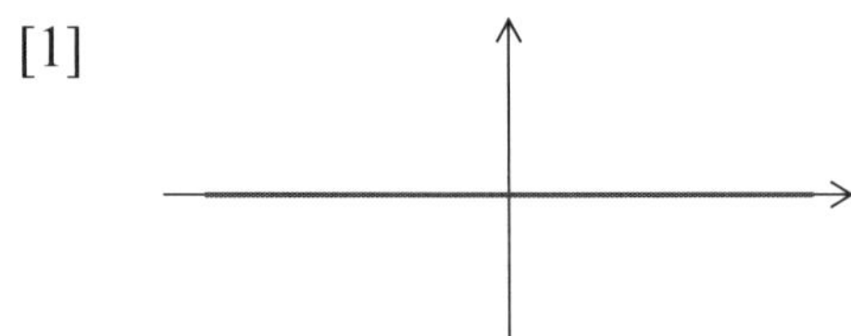

[B]
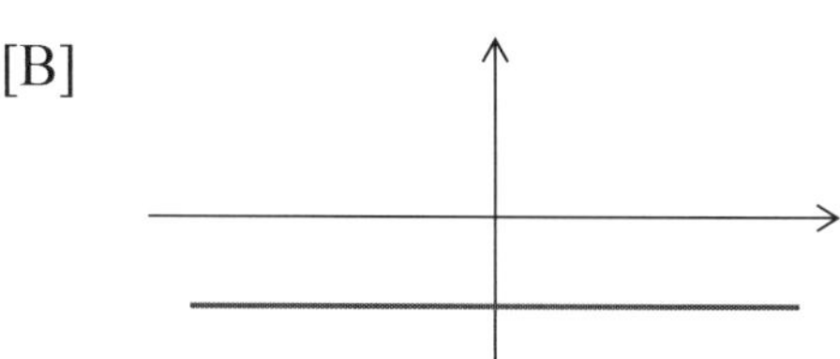

[2]
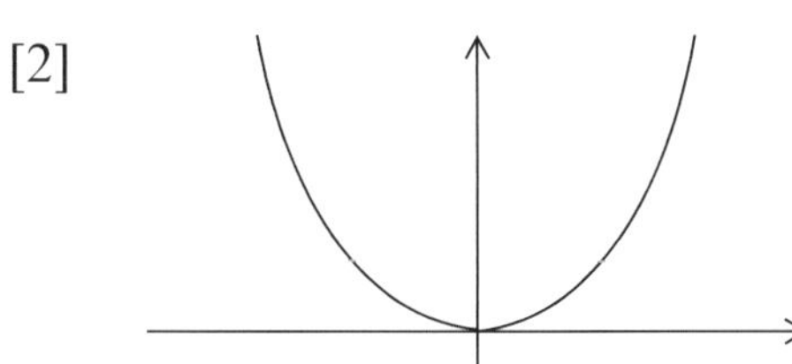

[C]
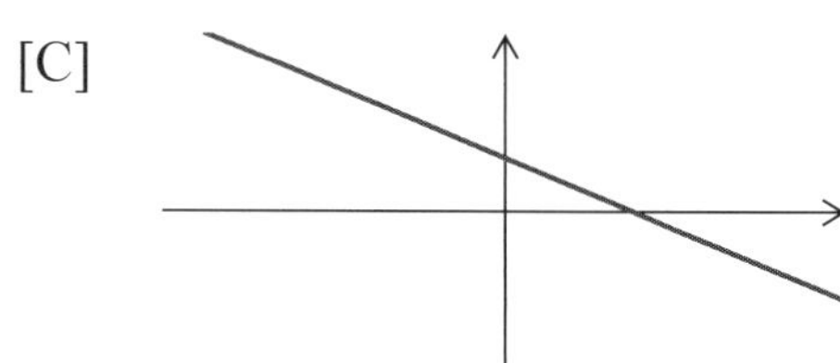

[3]
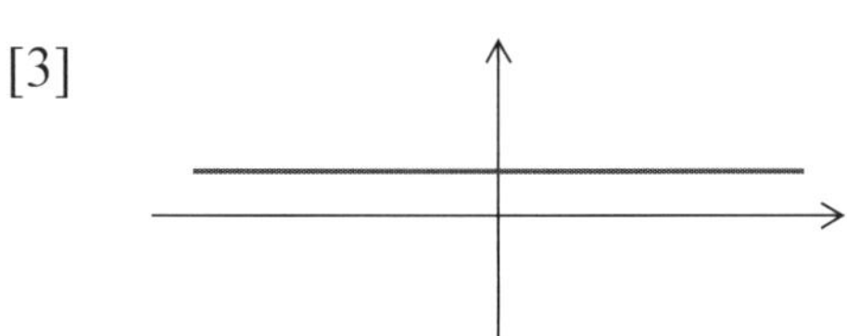

[D]
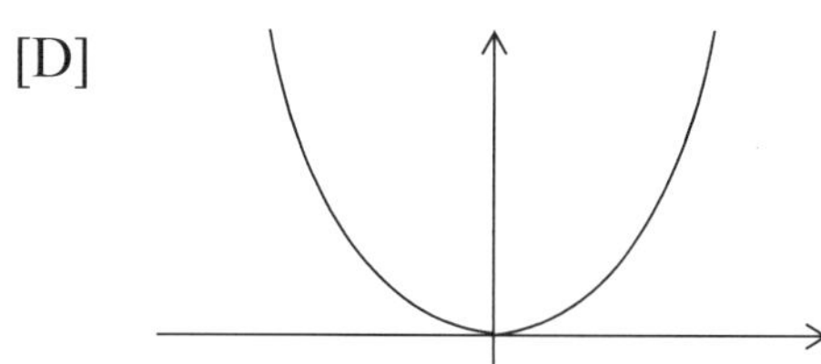

[4]
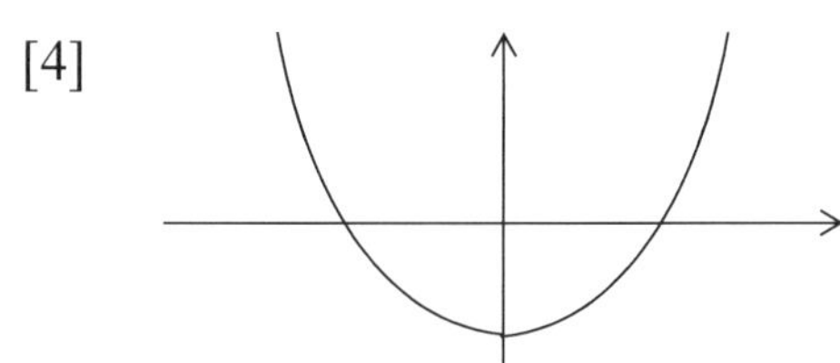

[E]
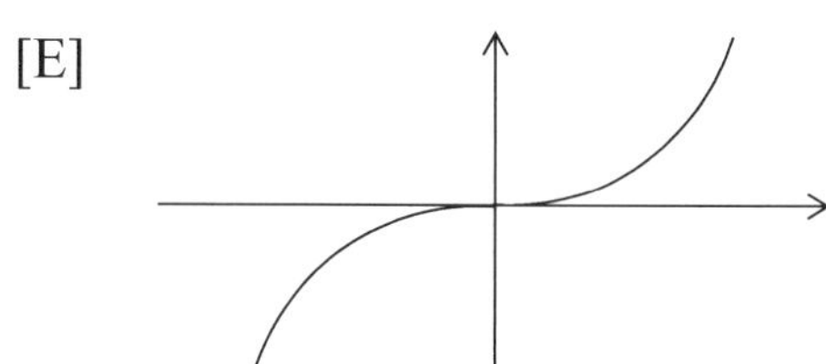

[5]
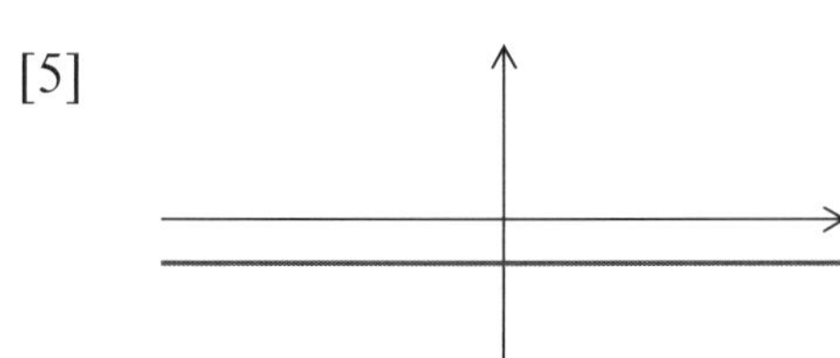

[F]
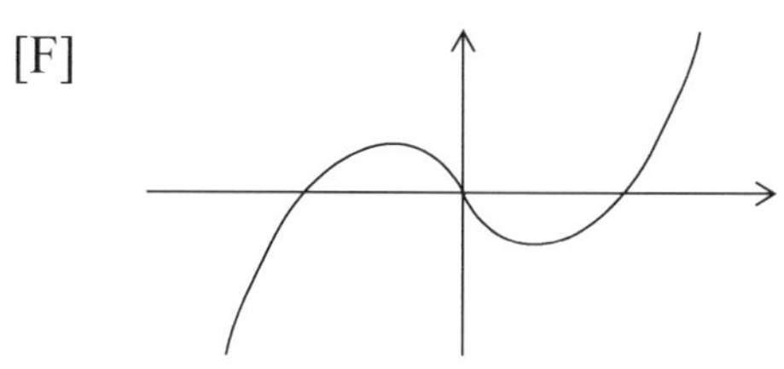

[6]
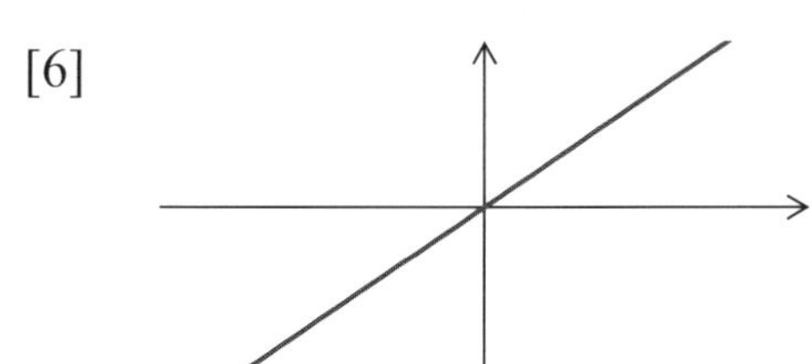

Exercises 33-34. Referring to the given graph of the function f, indicate where the function: fails to have a limit, is discontinuous (not continuous), and where it is not differentiable.

33.

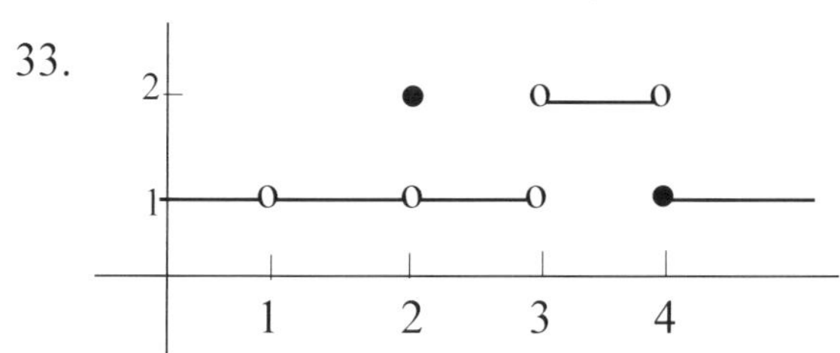

34.

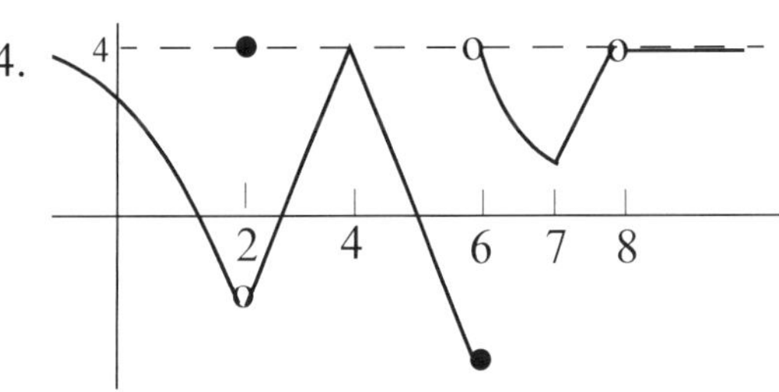

Exercises 35-36. Show that the function is not differentiable at $x = 2$. (See Exercises 113-116, page 130.)

35. $f(x) = \begin{cases} 2x+2 & \text{if } x \le 2 \\ 3x & \text{if } x > 2 \end{cases}$

36. $f(x) = \begin{cases} x & \text{if } x < 2 \\ x^2-2 & \text{if } x \ge 2 \end{cases}$

Exercises 37-38. Sketch the graph of $y = f'(x)$ from the given graph of the function $y = f(x)$. Suggestion: Think "slope of tangent lines." Where the tangent line is horizontal, for example, the derivative at that point will be zero; where the tangent line has a positive slope, the derivative will be positive (the greater the slope, the greater the value of the derivative), and so on.

37.

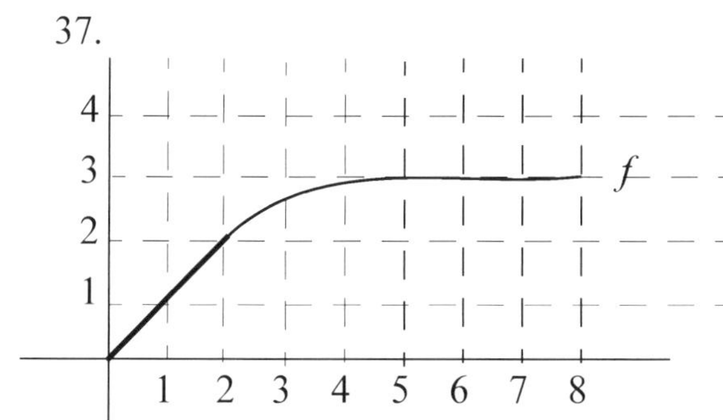

38.

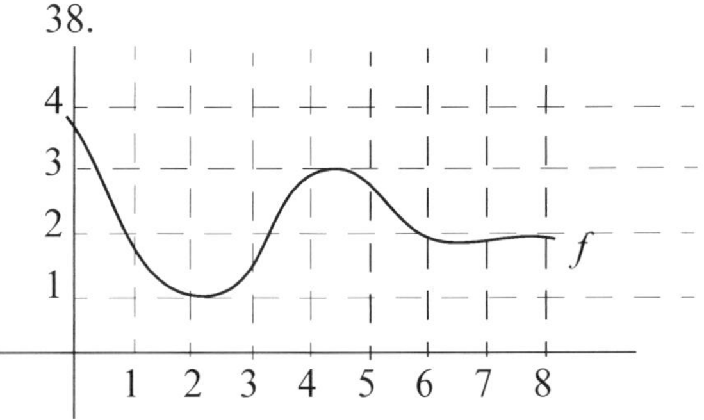

39. **(Theory)** Prove that if a function f is differentiable at c, then $f'(c) = \lim_{x \to c} \frac{f(x)-f(c)}{x-c}$.

Suggestion: let $h = x - c$

40. **(Theory)** Prove that if a function f is differentiable at c, then f is continuous at c.

Suggestion: Begin with $\lim_{x \to c} [f(x)-f(c)] = \lim_{x \to c} \left[\frac{f(x)-f(c)}{x-c} \cdot (x-c)\right]$.

§4. DIFFERENTIATION FORMULAS

In this section we introduce formulas that will enable you to differentiate a function such as

$$f(x) = 4x^5 - 3x^2 - 5x + 2$$

in a matter of seconds. Our main goal is to render you proficient in the use of those formulas. You are, however, invited to establish some of them in the exercises.

For the sake of brevity, we use the expression $(x^n)'$ to represent the derivative of the function $f(x) = x^n$.

Here is the most indispensable of the differentiation formulas:

THEOREM 2.6
POWER RULE

For any number n:

$$(x^n)' = nx^{n-1}$$

For example:

$$(x^5)' = 5x^4 \leftarrow \text{one power less}$$

$$(x^{-5})' = (-5)x^{-6} \leftarrow \text{one power less: } -5-1 = -6$$

A particularly important case:

$$x' = (x^1)' = 1x^{1-1} = x^0 = 1$$

You were invited to establish the above derivative formula:

$$x' = 1$$

in Exercise 29, page 147. Here is a geometrical argument:

The graph of the function $f(x) = x$ is a line of slope **1**. Since the tangent line to any line at any point must be the line itself, $x' = \mathbf{1}$.

While in a geometrical mode, let us also point out that the derivative of any constant function must be 0:

The graph of the function $f(x) = c$ is a horizontal line of slope zero. Since the tangent line to that line at any point must be the line itself:

$$c' = 0$$

You were invited to establish this result in Exercise 30, page 147.

Before turning to other differentiation formulas we note that for given functions f and g, and any constant c, we define the functions $f+g$, $f-g$, fg, $\frac{f}{g}$, and cf in a natural fashion:

$$(f+g)(x) = f(x)+g(x) \qquad (f-g)(x) = f(x)-g(x)$$

$$(fg)(x) = f(x)g(x) \qquad \left(\frac{f}{g}\right)(x) = \frac{f(x)}{g(x)} \qquad (cf)(x) = cf(x)$$

providing $g(x) \neq 0$

You are invited to establish these results, which can be generalized to any finite sum or difference of differentiable functions, in the exercises.

THEOREM 2.7 For any two differentiable functions f and g:

$$(f+g)'(x) = [f(x)+g(x)]' = f'(x)+g'(x)$$

and

$$(f-g)'(x) = [f(x)-g(x)]' = f'(x)-g'(x)$$

IN WORDS: The derivative of a sum (difference) is the sum (difference) of the derivatives.

For example:

$$(x^5+x^3)' = (x^5)'+(x^3)' = 5x^4+3x^2$$

$$(x^4-x^3+x)' = 4x^3-3x^2+1$$

$$(x^7+x^3+x-x^{-2})' = 7x^6+3x^2+1+2x^{-3}$$

$$(x^9+x^2-x^{-4}+325)' = 9x^8+2x+4x^{-5}$$

(what happened to the 325?)

You are invited to establish this result in the exercises.

THEOREM 2.8 For any differentiable function f and any constant c:

$$(cf)'(x) = [cf(x)]' = cf'(x)$$

IN WORDS: The derivative of a constant times a function is the constant times the derivative of the function.

For example:

Generally, one omits that middle step and simply multiplies the power **5** of x with the coefficient **4** of x^5, lowering the power by 1: $4(x^5)' = 20x^4$

$$(4x^5)' = \mathbf{4}(x^{\mathbf{5}})' = \boxed{4(5x^4)} = \mathbf{20}x^4$$

($4\cdot 5$; see comment in margin)

$$(3x^7+6x^2+4x-3)' = 21x^6+12x+4$$

With the above theorems at hand, you are now in a position to quickly differentiate any polynomial function, and beyond.

EXAMPLE 2.13 Differentiate:

(a) $f(x) = 3x^4-2x^3+x-2$

(b) $f(x) = 5x^3+2x^{-3}+7x-1$

(c) $f(x) = (x^2+3)(x^3+x^2)$

(d) $f(x) = \dfrac{5x^6+4x^2-6}{2x^2}$

Since the derivative of a constant is zero, it really doesn't matter if the constant term in the expression for f is -2 or 100. Replacing the -2 with 100 would simply raise the graph of f by 102 units. That would not change the shape of the graph, and cannot therefore alter the derivative of the function.

SOLUTION: The dominant player is the "power rule" of Theorem 2.6:

$$(x^n)' = nx^{n-1}:$$

(a)

$$\text{If: } f(x) = 3x^4 - 2x^3 + x - 2$$

derivative of a constant is 0; one power less; derivative of x is 1

$$\text{Then: } f'(x) = 12x^3 - 6x^2 + 1$$

(b)

$$\text{If: } f(x) = 5x^3 + 2x^{-3} + 7x - 1$$

one power less

$$\text{Then: } f'(x) = 15x^2 - 6x^{-4} + 7 = 15x^2 - \frac{6}{x^4} + 7$$

(c) To differentiate the function $f(x) = (x^2+3)(x^3+x^2)$, we first perform the product to arrive at a sum of **powers-of-x**:

$$f(x) = (x^2+3)(x^3+x^2) = x^2(x^3+x^2) + 3(x^3+x^2)$$
$$= x^5 + x^4 + 3x^3 + 3x^2$$

and then differentiate:

$$f'(x) = (x^5 + x^4 + 3x^3 + 3x^2)' = 5x^4 + 4x^3 + 9x^2 + 6x$$

(d) To differentiate the function $f(x) = \dfrac{5x^6 + 4x^2 - 6}{2x^2}$, we first express it as a sum of **powers-of-x**:

$$f(x) = \frac{5x^6 + 4x^2 - 6}{2x^2} = \frac{5x^6}{2x^2} + \frac{4x^2}{2x^2} - \frac{6}{2x^2} = \frac{5}{2}x^4 + 2 - 3x^{-2}$$

and then differentiate:

$$f'(x) = \left(\frac{5}{2}x^4 + 2 - 3x^{-2}\right)' = \frac{5}{2}(4x^3) + 0 - 3(-2x^{-3})$$
$$= 10x^3 + 6x^{-3} = 10x^3 + \frac{6}{x^3}$$

CHECK YOUR UNDERSTANDING 2.19

Differentiate the given function.

(a) $-4x^3 + 2x^2 - 3x + 5$ (b) $f(x) = x^2(3x^3 + 2x - 5)$

(c) $f(x) = \dfrac{5x^3 + 2x - 7}{2x^2}$

Answers:
(a) $-12x^2 + 4x - 3$
(b) $15x^4 + 6x^2 - 10x$
(c) $\frac{5}{2} - \frac{1}{x^2} + \frac{7}{x^3}$

EXAMPLE 2.14 Find the tangent line to the graph of the function $f(x) = 2x^3 + x^2 - 5x + 1$ at $x = 1$.

SOLUTION: The word "line," translates to:

$$y = mx + b$$

To find the slope of the tangent line we differentiate:

$$f'(x) = (2x^3 + x^2 - 5x + 1)' = 6x^2 + 2x - 5$$

and evaluate the derivative at $x = 1$:

$$y = \mathbf{m}x + b$$
$$\uparrow$$
$$f'(1)$$

$$m = f'(1) = 6(1)^2 + 2(1) - 5 = \mathbf{3}$$

At this point, we know that the tangent line is of the form:

$$y = \mathbf{3}x + b$$

To find b, we make use of the fact that the tangent line passes through the point:

$$(1, f(\mathbf{1})) = (1, -\mathbf{1})$$

$$2(1)^3 + (1)^2 - 5(1) + 1 = 2 + 1 - 5 + 1 = -\mathbf{1}$$

Finding b:

Since the line passes through (1, -1)

$$y = 3x + b$$
$$-1 = 3(1) + b$$
$$b = -4$$

We now have the tangent line: $y = 3x - 4$.

CHECK YOUR UNDERSTANDING 2.20

Find the equation of the tangent line to the graph of the function $f(x) = \frac{x^3 - x}{x^2}$ at $x = 2$.

Answer: $y = \frac{5}{4}x - 1$

PRODUCT AND QUOTIENT RULES

It would be nice if the derivative of a product were the product of the derivatives. Alas, however, this cannot be; for:

$$\text{While } x^5 = (x^3)(x^2),\ \ (\boldsymbol{x^5})' \neq (\boldsymbol{x^3})'(\boldsymbol{x^2})'$$

$$\text{Since: } (x^5)' = 5x^4 \text{ while } (x^3)'(x^2)' = (3x^2)(2x) = 6x^3$$

Niceties aside, here is the truth of the matter:

Note:
$$\begin{aligned}(x^5)' &= (x^3 \cdot x^2)' \\ &= x^3(x^2)' + x^2(x^3)' \\ &= x^3(2x) + x^2(3x^2) \\ &= 2x^4 + 3x^4 \\ &= 5x^4\end{aligned}$$

THEOREM 2.9 If f and g are differentiable, then so is fg, and:
$$(fg)'(x) = [f(x)g(x)]' = f(x)g'(x) + g(x)f'(x)$$

IN WORDS: The derivative of a product of two functions is one function times the derivative of the other, plus the other times the derivative of the one.

EXAMPLE 2.15 Differentiate:
$$f(x) = (x^3 + x^2)(x^2 + 3)$$
(a) Using the product rule.
(b) Without using the product rule.

SOLUTION:

(a) $$[f(x)g(x)]' = f(x)g'(x) + g(x)f'(x)$$

$$\begin{aligned}[(x^3 + x^2)(x^2 + 3)]' &= (x^3 + x^2)(x^2 + 3)' + (x^2 + 3)(x^3 + x^2)' \\ &= (x^3 + x^2)(2x) + (x^2 + 3)(3x^2 + 2x) \\ &= 2x^4 + 2x^3 + 3x^4 + 2x^3 + 9x^2 + 6x \\ &= 5x^4 + 4x^3 + 9x^2 + 6x\end{aligned}$$

The expanding approach is certainly the easier way to go.

(b) $$\begin{aligned}[(x^3 + x^2)(x^2 + 3)]' &= (x^5 + x^4 + 3x^3 + 3x^2)' \\ &= 5x^4 + 4x^3 + 9x^2 + 6x\end{aligned}$$

CHECK YOUR UNDERSTANDING 2.21

Determine the derivative of $f(x) = (x^3 + x)(3x^2 + x - 1)$ with and without using the product rule.

Answer:
$15x^4 + 4x^3 + 6x^2 + 2x - 1$

Just as the derivative of a product is **not** the product of the derivatives, the derivative of a quotient is **not** the quotient of the derivatives:

THEOREM 2.10 If f and g are differentiable, then so is $\frac{f}{g}$ and:
$$\left(\frac{f}{g}\right)'(x) = \left[\frac{f(x)}{g(x)}\right]' = \frac{g(x)f'(x) - f(x)g'(x)}{[g(x)]^2}$$
providing $g(x) \neq 0$

IN WORDS: The derivative of a quotient is the denominator times the derivative of the numerator, minus the numerator times the derivative of the denominator — all over the denominator squared.

EXAMPLE 2.16 Differentiate:

$$f(x) = \frac{5x^4 + 4x^2 - 3}{2x^2}$$

(a) Using the quotient rule.
(b) Without using the quotient rule.

SOLUTION:

(a) $$\left[\frac{f(x)}{g(x)}\right]' = \frac{g(x)f'(x) - f(x)g'(x)}{[g(x)]^2}$$

$$\left[\frac{5x^4 + 4x^2 - 3}{2x^2}\right]' = \frac{(2x^2)(5x^4 + 4x^2 - 3)' - (5x^4 + 4x^2 - 3)(2x^2)'}{(2x^2)^2}$$

$$= \frac{(2x^2)(20x^3 + 8x) - (5x^4 + 4x^2 - 3)4x}{4x^4}$$

$$= \frac{40x^5 + 16x^3 - 20x^5 - 16x^3 + 12x}{4x^4}$$

$$= \frac{20x^5 + 12x}{4x^4} = 5x + \frac{3}{x^3}$$

(b) $$\left[\frac{5x^4 + 4x^2 - 3}{2x^2}\right]' = \left[\frac{5x^4}{2x^2} + \frac{4x^2}{2x^2} - \frac{3}{2x^2}\right]' = \left[\frac{5}{2}x^2 + 2 - \frac{3}{2}x^{-2}\right]' = 5x + 0 + 3x^{-3} = 5x + \frac{3}{x^3}$$

Option (b) is clearly the better choice in this example, but it is not available in the next example:

EXAMPLE 2.17 Determine the derivative of:

$$f(x) = \frac{5x^4 + 4x^2 - 3}{2x^2 + 1}$$

SOLUTION:

$$\left[\frac{f(x)}{g(x)}\right]' = \frac{g(x)f'(x) - f(x)g'(x)}{[g(x)]^2}$$

$$f'(x) = \frac{(2x^2 + 1)(5x^4 + 4x^2 - 3)' - (5x^4 + 4x^2 - 3)(2x^2 + 1)'}{(2x^2 + 1)^2}$$

$$= \frac{(2x^2 + 1)(20x^3 + 8x) - (5x^4 + 4x^2 - 3)(4x)}{(2x^2 + 1)^2}$$

$$= \frac{(40x^5 + 36x^3 + 8x) - (20x^5 + 16x^3 - 12x)}{(2x^2 + 1)^2}$$

$$= \frac{20x^5 + 20x^3 + 20x}{(2x^2 + 1)^2}$$

CHECK YOUR UNDERSTANDING 2.22

Differentiate:

$$f(x) = \frac{2x^2 + 3x}{2x^2 + 3}$$

Answer:
$$\frac{-6x^2 + 12x + 9}{(2x^2 + 3)^2}$$

Higher Order Derivatives

As you might expect, the second derivative of a function f is the derivative of its first derivative $f''(x) = [f'(x)]'$.

For example, if $f(x) = 2x^5 + x^2 - 5x + 2$, then:

$$f'(x) = 10x^4 + 2x - 5$$
$$f''(x) = 40x^3 + 2$$

Answer:

$$f''(x) = 30x + \frac{2}{x^3}$$
$$f'''(x) = 30 - \frac{6}{x^4}$$

CHECK YOUR UNDERSTANDING 2.23

Determine the second and third derivatives of the function.

$$f(x) = 5x^3 - x + \frac{1}{x}$$

Let's finish up this section by listing all the derivative formulas:

We are assuming, here, that the functions f and g are differentiable.

THEOREM 2.11

(a) $(x^n)' = nx^{n-1}$

(b) $x' = 1$ and $c' = 0$ for any constant c.

(c) $[cf(x)]' = cf'(x)$ for any constant c.

(d) $[f(x) \pm g(x)]' = f'(x) \pm g'(x)$

(e) $[f(x)g(x)]' = f(x)g'(x) + g(x)f'(x)$

(f) $\left[\frac{f(x)}{g(x)}\right]' = \frac{g(x) \cdot f'(x) - f(x) \cdot g'(x)}{[g(x)]^2}$

EXERCISES

Exercises 1-24. Differentiate the given function.

1. $f(x) = 2x+7$
2. $f(x) = -3x^2-2x+5$
3. $f(x) = 3x^5+4x^3-7$
4. $f(x) = 4x^4+7x^3-3x-2$
5. $g(x) = 2x^{-3}+4$
6. $f(x) = x^2-1+3x^{-3}$
7. $g(x) = 7x^3+5x^2-4x+x^{-4}+1$
8. $f(x) = \frac{1}{3}x^3+\frac{1}{5}x^2-x-1$
9. $g(x) = x+\frac{2}{x}+\frac{3}{x^2}$
10. $g(x) = 5-\frac{3}{x^3}$
11. $k(x) = \frac{x^5+3x-5}{x^3}$
12. $k(x) = \frac{4x^4+x^3-2x^2+3x+7}{x^2}$
13. $K(x) = (x^3+2x)(3x^3+2x+3)$
14. $K(x) = (4x^4+2x^3+x^2)(x^3+x+1)$
15. $F(x) = \frac{3x^2+2x-5}{x}$
16. $F(x) = \frac{-x^5+3x-4}{x^2}$
17. $F(x) = \frac{3x^2+2x-5}{x+4}$
18. $F(x) = \frac{-x^5+3x-4}{x^2+2x}$
19. $f(x) = \frac{5}{3x^2+1}$
20. $g(x) = \frac{1}{(x-2)^2}$
21. $H(x) = \frac{x}{2x+1}+\frac{x}{3x-1}$
22. $H(x) = \frac{2x+2}{2x-1}-3x^2+x+5$
23. $f(x) = \left(\frac{x}{3x+1}\right)(x^2+2x)$
24. $f(x) = \left(\frac{x}{3x+1}\right)\left(\frac{x^2+2x}{x+3}\right)$

Exercises 25-28. (Higher Order Derivatives) Determine the second derivatives of the given function.

25. $f(x) = x^3-6x^2+12x$
26. $f(x) = \frac{2}{3}x^3-\frac{1}{2}x^2-x+1$
27. $f(x) = x^3+2x^2-\frac{1}{x}$
28. $f(x) = \frac{x^2+1}{3x}$

Exercises 29-42. Evaluate the given expression at the indicated point, if:

$$f(0) = 1, f(1) = 3, f(2) = 6, f'(0) = 2, f'(1) = 6, f'(2) = 0$$
$$g(0) = 3, g(1) = 2, g(2) = 5, g'(0) = 1, g'(1) = 2, g'(2) = 2$$
$$h(0) = 0, h(1) = 6, h(2) = 2, h'(0) = 3, h'(1) = 1, h'(2) = 1$$

29. $[f(x)+g(x)]'$ at $x = 1$
30. $[f(x)g(x)]'$ at $x = 1$

31. $\left[\dfrac{f(x)}{g(x)}\right]'$ at $x = 1$

32. $\left[\dfrac{g(x)}{f(x)}\right]'$ at $x = 1$

33. $[f(x)+g(x)+h(x)]'$ at $x = 2$

34. $[f(x)+g(x)h(x)]'$ at $x = 2$

35. $[f(x)g(x)+h(x)]'$ at $x = 2$

36. $\left[\dfrac{f(x)+g(x)}{h(x)}\right]'$ at $x = 1$

37. $\left[\dfrac{f(x)g(x)}{h(x)}\right]'$ at $x = 1$

38. $\left[\dfrac{f(x)-g(x)}{h(x)+1}\right]'$ at $x = 0$

39. $\left[\dfrac{g(t)}{h(t)}+g(t)\right]'$ at $t = g(1)$

40. $\dfrac{f(s)}{g'(s)}+g(2)h'(s)$ at $s = g'(1)$

41. $\left[\dfrac{f(x)+g(x)+h(x)}{f(x)-g(x)-h(x)}\right]'$ at $x = g[f(0)]$

42. $h'(x)+h(1)\left[\dfrac{g(x)f(x)}{h(x)}\right]'$ at $x = g'[f(0)]$

Exercises 43-48. (Tangent Line) Determine the tangent line to the graph of the given function at the indicated point.

43. $f(x) = 3x^2-x-1$ at $x = 1$

44. $f(x) = -x^3-2x+2$ at $x = 0$

45. $f(x) = \dfrac{x^5+2x}{x^4}$ at $x = -1$

46. $f(x) = \dfrac{3x^3+x^2-2x-1}{x^2}$ at $x = 1$

47. $f(x) = \dfrac{x^2+2x}{x-1}$ at $x = -1$

48. $f(x) = \dfrac{3x^3+x^2-2x-1}{x^2+1}$ at $x = 1$

Exercises 49-52. (Horizontal Tangent Line) Determine where the graph of the given function has a horizontal tangent line.

49. $f(x) = x^3-6x^2+12x$

50. $f(x) = \dfrac{2}{3}x^3-\dfrac{1}{2}x^2-x+1$

51. $f(x) = \dfrac{2}{3}x^3-\dfrac{1}{2}x^2-x+2$

52. $f(x) = \dfrac{x^2+1}{3x}$

Exercises 53-57. (Theory)

53. Use Definition 2.6, page 141 to establish the validity of Theorem 2.7.

54. Use Definition 2.6, page 141 to establish the validity of Theorem 2.8.

55. Use Definition 2.6, page 141, to verify that $(x^2)' = 2x$, then use the product theorem (Theorem 2.9) to verify that $(x^4)' = 4x^3$, and that $(x^5)' = 5x^4$.

56. Assume that for a positive integer n $(x^n)' = nx^n$. Use the product theorem (Theorem 2.9) to show that $(x^{n+1})' = (n+1)x^n$.

§5. OTHER INTERPRETATIONS OF THE DERIVATIVE

Up until now we have taken a geometrical point of view of the derivative: SLOPE OF TANGENT LINES. In this section we will take a more dynamic interpretation; namely:

The rate of change of one quantity with respect to another.

The above interpretation is the preferred choice in science and business where one is often concerned with the effect that a change in a quantity will have on others:

If the temperature is decreased, what happens to the pressure?

If production level is increased, what happens to the profit?

We begin by noting that the Greek letter Δ, pronounced "delta," is often used to denote "***a change in***." In particular:

change in y divided by change in x → $\frac{\Delta y}{\Delta x}$

DEFINITION 2.7
AVERAGE RATE OF CHANGE

The **average rate of change** of a function $y = f(x)$ from $x = a$ to $x = b$ is given by:

$$\frac{\Delta y}{\Delta x} = \frac{f(b) - f(a)}{b - a}$$

EXAMPLE 2.18 On April first, the price for a gallon of gasoline was \$3.55. By the end of the month the cost went up to \$3.67 per gallon.

(a) What was the average cost for a gallon of gasoline during the month of April?

(b) What was the average daily rate of change of the cost of a gallon of gasoline during April?

SOLUTION:

(a) Two values are given: \$3.55 and \$3.67. The average of those two values is simply:

$$\frac{\$3.55 + \$3.67}{2} = \frac{\$7.22}{2} = \$3.61$$

(b) To find the average rate of change of cost per day, we turn to Definition 2.7:

$$\frac{\Delta \text{ cost}}{\Delta \text{ days}} = \frac{\$3.67 - \$3.55}{(30-0)\text{days}} = \frac{\$0.12}{30 \text{ days}} = \$0.004 \text{ per day}$$

CHECK YOUR UNDERSTANDING 2.24

At noon, the temperature in Central Park was $80°$, and by 6 pm it dropped to $68°$. Determine:

(a) Average temperature

(b) Average hourly rate of change of the temperature, during that time period.

Answers: (a) $74°F$
(b) $-2°F$ per hour

Letting Δx approach zero in the expression for the **average rate of change** $\frac{\Delta y}{\Delta x}$ of a function $y = f(x)$ brings us to the following definition:

DEFINITION 2.8
INSTANTANEOUS RATE OF CHANGE

The **instantaneous rate of change** of the function $y = f(x)$ with respect to the variable x is denoted by $\frac{dy}{dx}$, and is given by:

$$\frac{dy}{dx} = \lim_{\Delta x \to 0} \frac{\Delta y}{\Delta x}$$

(providing the limit exists)

But the above limit can also be written in the form:

$$\lim_{\Delta x \to 0} \frac{\Delta y}{\Delta x} = \lim_{\Delta x \to 0} \frac{\overbrace{f(x+\Delta x)-f(x)}^{\text{change in } y\text{-value}}}{\Delta x}$$

which is simply the derivative function of Definition 2.6, page 141, with Δx replacing h:

$$f'(x) = \lim_{\underset{\Delta x}{h \to 0}} \frac{f(x+\overset{\Delta x}{h})-f(x)}{\underset{\Delta x}{h}}$$

Bringing us to:

DIFFERENT NOTATION AND INTERPRETATION OF <u>THE</u> DERIVATIVE:

$$f'(x) = \lim_{h \to 0} \frac{f(x+h)-f(x)}{h}$$

$$\frac{dy}{dx} = \lim_{\Delta x \to 0} \frac{\Delta y}{\Delta x}$$

slope of tangent line

Instantaneous rate of change

It is important to note that notation and interpretation not withstanding, we are talking about **one and the same** mathematical object: The **derivative function**.

EXAMPLE 2.19 For the function:

$$y = f(x) = 2x^3 + x - 1$$

determine:

(a) The average rate of change of the function from $x = 1$ to $x = 3$.

(b) The slope of the line passing through the points $(1, f(1))$ and $(3, f(3))$.

(c) The instantaneous rate of change of the function at $x = 1$.

(d) The slope of the tangent line to the graph of the function at the point $(1, f(1))$.

SOLUTION:

One and the same (a) and (b)

(a) Average rate of change: $\frac{\Delta y}{\Delta x} = \frac{f(3) - f(1)}{3 - 1} = \frac{56 - 2}{2} = 27$

(b) Slope of line passing through the two points:

$$\frac{\Delta y}{\Delta x} = \frac{f(3) - f(1)}{3 - 1} = \frac{56 - 2}{2} = 27$$

One and the same (c) and (d)

(c) $\frac{d}{dx}(2x^3 + x - 1) = 6x^2 + 1$

Instantaneous rate of change at $x = 1: 6(1)^2 + 1 = 7$

(d) $f'(x) = (2x^3 + x - 1)' = 6x^2 + 1$

Slope of tangent line at $x = 1: 6(1)^2 + 1 = 7$.

We point out that for $y = f(x)$, you can use $\left.\frac{dy}{dx}\right|_{x = c}$ in lieu of $f'(c)$.

For example, if $y = f(x) = 2x^3 + x - 1$, then:

$$f'(x) = 6x^2 + 1, \quad \text{and} \quad f'(5) = 6 \cdot 5^2 + 1 = 151$$

$$\text{OR:} \quad \frac{dy}{dx} = 6x^2 + 1, \quad \text{and} \quad \left.\frac{dy}{dx}\right|_{x = 5} = 6 \cdot 5^2 + 1 = 151$$

CHECK YOUR UNDERSTANDING 2.25

For $y = f(x) = 3x^4 + 2x^3 - x + 1$ determine:

(a) $f'(x)$ (b) $\frac{dy}{dx}$ (c) $f'(2)$ (d) $\left.\frac{dy}{dx}\right|_{x = 2}$

Answers: (a) and (b): $12x^3 + 6x^2 - 1$
(c) and (d): 119

VELOCITY AND ACCELERATION

If a particle moves along a straight line, then its position at any instant of time is defined to be its "directed distance" from a fixed reference point.

We note that the instantaneous rate of change of position with respect to time is called **velocity** and the instantaneous rate of change of velocity with respect to time is called **acceleration**. We also note that the magnitude of the velocity of an object is said to be its **speed**.

By convention, when a particle is moving along a vertical line, a positive velocity denotes movement in an upward direction, while negative velocity denotes movement in a downward direction.

EXAMPLE 2.20 If a ball is thrown straight upward from ground level at a speed of 96 feet per second, then (while in flight) its position (in feet) from the ground t seconds later is given by:

$$s(t) = -16t^2 + 96t$$

(a) Determine the velocity of the ball at $t = 1$, and at $t = 4$.
(b) Find the speed of the ball when it hits the ground.
(c) How high will the ball go?
(d) Find the acceleration of the ball.

SOLUTION: Since velocity, $v(t)$, is the rate of change of position with respect to time, we have:

$$v(t) = \frac{ds}{dt} = \frac{d}{dt}(-16t^2 + 96t) = -32t + 96 \text{ feet per second}$$

(ds: feet; dt: seconds)

(a) In particular,

$$v(1) = -32 \cdot 1 + 96 = 64\frac{\text{ft}}{\text{sec}}$$

Being positive, we conclude that, at time $t = 1$, the ball is rising at a speed of $64\frac{\text{ft}}{\text{sec}}$.

From:

$$v(4) = -32 \cdot 4 + 96 = -32\ \frac{\text{ft}}{\text{sec}}$$

we conclude that at time $t = 4$ the ball is falling at a speed of 32 feet per second:

Since both position and velocity are functions of time, the key to answering (b), and (c) is to first focus on finding the particular t of interest. We'll point them out to you along the way.

(b) Setting the position function to zero (ground is reference point), we find how long it takes for the ball to hit the ground:

$$s(t) = -16t^2 + 96t = 0$$
$$-16t(t-6) = 0$$

ball is at ground level (leaves from ground level) $\rightarrow t = 0$ or $t = 6 \leftarrow$ ball is at ground level (hits the ground)

↑ the t of interest

Knowing that it takes 6 seconds for the ball to hit the ground, the impact velocity is determined by evaluating the velocity function at that time:

$$v(6) = -32 \cdot 6 + 96 = -96\,\frac{\text{ft}}{\text{sec}}$$

$$\text{Impace speed} = |v(6)| = 96\frac{\text{ft}}{\text{sec}}$$

(c) We know that the instant at which the ball is at its highest level, it is standing still. Consequently, we set velocity to 0, and solve for the time it takes for the ball to reach its zenith:

$$v(t) = -32t + 96 = 0$$
$$-32t = -96$$
$$t = 3 \text{ seconds}$$

↑ the t of interest

Knowing it reaches its highest level in 3 seconds, we find the maximum height of the ball by evaluating the position function at that time:

$$s(3) = -16 \cdot 3^2 + 96 \cdot 3 = 144 \text{ feet}$$

(d) Differentiating the velocity function, we come to the acceleration function:

$$a(t) = \frac{dv}{dt} = \frac{d}{dt}(-32t + 96) = -32$$

(dv: feet per second; dt: second; −32: feet per second squared)

Note on the units:

$$\frac{\frac{\text{ft}}{\text{sec}}}{\text{sec}} = \frac{\text{ft}}{\text{sec}} \cdot \frac{1}{\text{sec}} = \frac{\text{ft}}{\text{sec}^2}$$

That constant acceleration is due to the force of gravity (the negative sign indicates that the force is in the downward direction).

CHECK YOUR UNDERSTANDING 2.26

If a stone is dropped from a height of 144 feet, then its position (above the ground) t seconds later is given by:

$$s(t) = -16t^2 + 144$$

Determine the velocity of the stone when it hits the ground.

Answer: −96 ft/sec.

AN ECONOMIST'S NAME FOR THE DERIVATIVE

In business and economics, the word ***marginal*** is used synonymously with rate of change, or derivative. The derivative of a cost function with respect to the number of units produced, for example, is said to be the marginal cost function and is denoted by $\overline{MC}(x)$ (instead of $C'(x)$ or $\frac{dC}{dx}$). Forgetting names and nomenclature for a moment, suppose that you know that for a function $y = f(x)$, $f'(27) = 3$. This tells you that if x goes from 27 to 28, then the function value will increase by approximately 3 units (over 1— up about 3). By the same token, suppose that $C(x)$ gives total cost (in dollars) for producing x units, and that $\overline{MC}(27) = C'(27) = 3$. This tells you that if production goes from 27 units to 28 units, then the total cost will increase by approximately 3 dollars (over 1— up about 3). In general:

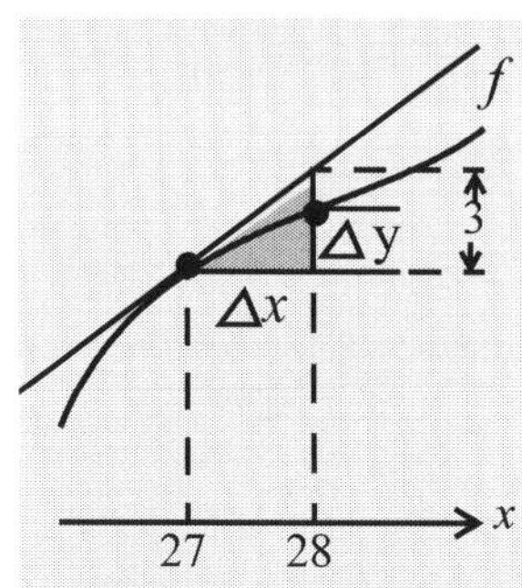

One produces and sells, say 27 cars, or 28 cars; but not 27.3 cars. In economics, the smallest "physical" Δx is 1 ("the **next** one").

> If $C(x)$ denotes the total cost of producing x units, then the **marginal cost**, $\overline{MC} = C'(x)$, represents the (approximate) additional cost for the production of the **next** unit.
>
> If $R(x)$ denotes the total revenue from the sale of x units, then the **marginal revenue**, $\overline{MR} = R'(x)$, represents the (approximate) additional revenue resulting from the sale of the **next** unit.
>
> If $P(x)$ denotes the total profit from the sale of x units, then the **marginal profit**, $\overline{MP} = P'(x)$, represents the (approximate) additional profit (perhaps negative) resulting from the sale of the **next** unit.

EXAMPLE 2.21 Weekly profits, in dollars, for the production of x units is approximated by the function:

$$P(x) = -2500 + 20x - \frac{x^2}{300} \quad (x < 5000)$$

(a) Determine the marginal profit when 1500 units are produced and sold.
(b) Indicate what the answer in (a) predicts.

Analyzing the Given Function:
The "$x < 5000$" tells us that this function can only be used to approximate the company's profit up to a production level of 5000 units.

We observe that, initially, the company will make a profit of **about** \$20 for the sale of each unit (the $20x$ term). We also find that profit **diminishes somewhat** as production increases (the $-\frac{x^2}{300}$ term).

SOLUTION:

(a) Noting that *marginal* is but another name for *derivative*, we have:

$$\overline{MP}(x) = P'(x) = \left(-2500 + 20x - \frac{x^2}{300}\right)' = 20 - \frac{x}{150}$$

Evaluating, we find the rate of change when $x = 1500$:

$$\overline{MP}(1500) = 20 - \frac{1500}{150} = 10$$

Since the rate of change is positive, the slope of the tangent line to the graph of the profit function at that point is positive. This means that the graph of the profit function is increasing, indicating that it pays to produce more than1500. (At this point, we cannot say exactly how much should be produced in order to maximize profit; that issue is addressed in Section 2.7).

(b) An approximate profit of \$10 will be realized from the sale of the 1501th unit.

Answers: (a) -10
(b) Profit will decrease by approximately \$10 from the sale of the 4501th unit.

CHECK YOUR UNDERSTANDING 2.27

Referring to Example 2.21:

(a) Determine marginal profit at $x = 4500$.

(b) Indicate what the answer in (a) predicts.

Analyzing the Given Function:

The constant term, 19.7 thousand dollars, represents the monthly **fixed cost** of the company (rent, tax, insurance, etc.): Cost to produce nothing.

The non-constant part of the function, $4.8x - 0.03x^2$, is called the **variable cost**.

Note that initially each boat costs about 4.8 thousand to produce (the $4.8x$ factor), with that cost diminishing somewhat as production increases (the $-0.03x^2$ factor).

EXAMPLE 2.22

Suppose that the total monthly cost, in thousands of dollars, for manufacturing x sailboats per month is given by:

$$C(x) = 19.7 + 4.8x - 0.03x^2 \qquad (x \le 35)$$

(a) Find the marginal cost function.

(b) Find the marginal cost at a production level of 25 boats per month, and interpret the result.

(c) Calculate $C(26) - C(25)$ and compare the result with that of (b).

SOLUTION:

(a) Noting that *marginal* is but another name for *derivative*, we have:

$$\overline{MC}(x) = C'(x) = (19.7 + 4.8x - 0.03x^2)' = 4.8 - 0.06x$$

(b) At $x = 25$: $\overline{MC}(25) = C'(25) = 4.8 - 0.06(25) = 3.3$

The most important part of this problem, is to realize what the above number predicts:

It will cost the company **approximately** \$3,300 to produce the 26th boat (the **next** boat).

(c) The following calculation gives the **exact** additional cost for the production of the 26th unit:

cost of producing 26 units → cost of producing 25 units

$$C(26) - C(25) =$$

$$[19.7 + 4.8(26) - 0.03(26)^2] - [19.7 + 4.8(25) - 0.03(25)^2] = 3.27$$

It will cost the company \$3,270 to produce the 26th unit, demonstrating that the marginal cost is a good approximation for the actual cost of production.

You may be wondering why one should bother to find the approximate additional cost for manufacturing the 26th unit in (b), when the exact additional cost can so easily be determined. The point, as you will see in Section 2.7, is that it is not the local information of marginal cost, say at $x = 25$, that is of primary interest, but the global information stemming from the marginal cost function itself.

> Answers:
> (a) $\overline{MR}(25) = \$4{,}900$
> (b) Approximately \$4,900 will be realized from the sale of the 26th boat.

CHECK YOUR UNDERSTANDING 2.28

Assume that the revenue function (in thousands of dollars) for the sailboat company of Example 2.22 is given by:

$$R(x) = 5.9x - 0.02x^2$$

(a) Determine marginal revenue from the sale of 25 boats.

(b) Indicate what the answer in (a) predicts.

EXAMPLE 2.23 Suppose that the monthly cost, in thousands of dollars, for manufacturing x sailboats per month is given by:

$$C(x) = 19.7 + 4.8x - 0.03x^2 \qquad (x \le 35)$$

And that the revenue function is given by

$$R(x) = 5.9x - 0.05x^2$$

Determine the profit function and the marginal profit function from the production and sale of 25 boats.

SOLUTION: By definition: Profit = Revenue – Cost Hence:

$$P(x) = R(x) - C(x) = [5.9x - 0.05x^2] - [19.7 + 4.8x - 0.03x^2]$$
$$= -19.7 + 1.1x - 0.02x^2$$

> You can also use:
> $P'(x) = R'(x) - C'(x)$

Marginal Profit: $\overline{MP}(x) = P'(x) = 1.1 - 0.04x$, and

$$\overline{MP}(25) = 1.1 - 0.04(25) = 0.1$$

From the above we can conclude that an additional profit of approximately \$100 will be realized from the sale of the 26th boat.

> Answer: The company will realize a profit of approximately \$76 from the sale of the 101th unit.

CHECK YOUR UNDERSTANDING 2.29

Given the total monthly cost function (in dollars):

$$C(x) = 2500 + 325x - 0.005x^2$$

And the total revenue function:

$$R(x) = 400x$$

Determine the marginal profit from the production and sale of 100 units, and interpret the result.

EXERCISES

1. **World Population (billions)**

Year	1992	1993	1994	1995	1996	1997	1998	1999	2000
Population	5.45	5.53	5.61	5.69	5.77	5.85	5.92	6.00	6.08

Determine the average rate of change of the world population from:

(a) 1992 to1995 (b) 1995 to 2000 (c) 1992 to 2000

Exercises 2-5. For given f, a, and b, determine:

(a) The slope of the line passing through the points $(a, f(a))$ and $(b, f(b))$.

(b) The average rate of change of the function from $x = a$ to $x = b$.

(c) The slope of the tangent line at $(a, f(a))$.

(d) The (instantaneous) rate of change of the function at $x = a$.

2. $f(x) = 2x^3 + 3x - 1,\ a = 1, b = 2$

3. $f(x) = x^3 + 2x - 3,\ a = 0, b = \frac{1}{2}$

4. $f(x) = x^2(2x + 3),\ a = -1, b = 3$

5. $f(x) = \frac{x^2 + 1}{x},\ a = 1, b = 3$

6. **(Volume)** The volume of a cube is given by $V = x^3$, where x is the length of a side. Determine the (instantaneous) rate of change of volume with respect to the length of the side when $x = 1$, and when $x = 3$.

7. **(Area)** The area of a circle of radius r is given by $A = \pi r^2$. Find the (instantaneous) rate of change of the area of a circle with respect to the radius when $r = 2$.

8. **(Free falling object)** A stone is dropped from a bridge h feet above the water, and hits the water below in three seconds. Its position (in feet) t seconds later is given by the formula $s(t) = -16t^2 + h$.

   (a) What is the height of the bridge?

   (b) Find the impact speed of the stone.

9. **(Free falling object)** A ball is tossed directly upward from the ground at a speed of 128 feet per second, and its position (in feet) t seconds later is given by the formula $s(t) = -16t^2 + 128t$.

   (a) Determine the velocity at $t = 2$.

   (b) Find the maximum height reached by the object.

   (c) What is the speed of the object when it hits the ground?

10. **(Free falling object)** An object is thrown straight upward and its position (in feet) t seconds later is given by $s(t) = -16t^2 + 96t + 100$.

    (a) What does $s(0)$ represent?

    (b) What does the "96" represent? Suggestion: Consider the velocity function.

    (c) Determine the velocity at $t = 2$.

    (d) Find the maximum height reached by the object.

    (e) What is the speed of the object when it hits the ground?

11. **(Marginal Cost)** Suppose that the total cost (in dollars) for manufacturing x compact disk players per month is given by:

    $$C(x) = 1000 + 23x - \frac{x^2}{25} \quad \text{for} \quad x \leq 250$$

    (a) What does the "1000" represent?

    (b) Find the marginal cost function.

    (c) Find the marginal cost at a production level of 100 units.

    (d) What does the answer in (c) predict?

    (e) Calculate $C(101) - C(100)$ and compare the result with that of (c).

12. **(Marginal Cost)** Repeat Exercise 11 for:

    $$C(x) = 5000 + 55x - 0.02x^2 \quad \text{for} \quad x \leq 250$$

13. **(Marginal Cost)** Suppose that the total cost, in dollars, for manufacturing x automobiles per week is given by.

    $$C(x) = 37{,}000 + 22.5x - 0.01x^2 \quad \text{for} \quad x \leq 300$$

    (a) What does the "37,000" represent?

    (b) Find the marginal cost function.

    (c) Find the marginal cost at a production level of 50 units.

    (d) What does the answer in (c) predict?

    (e) Calculate $C(51) - C(50)$ and compare the result with that of (c).

14. **(Marginal Revenue)** The total revenue, in dollars, from the sale of x compact disk players is given by:

$$R(x) = 37x - \frac{x^2}{10000}$$

(a) Find the marginal revenue function.

(b) Find the marginal revenue at a production level of 100 units.

(c) What does the answer in (b) predict?

(d) Calculate $R(101) - R(100)$ and compare the result with that of (b).

15. **(Marginal Revenue)** Repeat Exercise 14 for $R(x) = 50x + 0.02x^2$

16. **(Marginal Profit)** The total monthly cost in dollars for manufacturing x compact disk players per month is given by:

$$C(x) = 1000 + 23x \qquad (x \le 1000)$$

The revenue realized, in dollars, from the sale of x disk players is given by:

$$R(x) = 37x - \frac{x^2}{100}$$

(a) Find the profit function and the marginal profit function.

(b) Find the marginal profit at $x = 100$.

(c) What does the answer in (b) predict?

(d) Calculate $P(101) - P(100)$ and compare the result with that of (b).

17. **(Marginal Profit)** Repeat Exercise 16 for:

$$C(x) = 2500 + 50x \text{ and } R(x) = 65x + 0.12x^2$$

Exercises 18-19. (Marginal Cost) Referring to the graph of the total cost (in thousands of dollars) as a function of the number of units produced, approximate the marginal cost at $x = 1$, and at $x = 7$.

18.

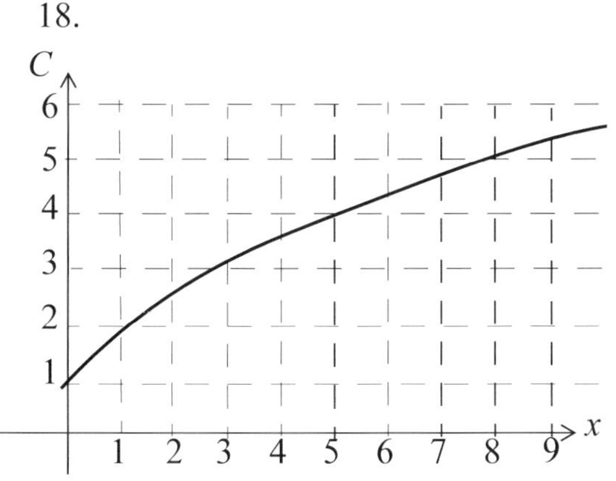

19.

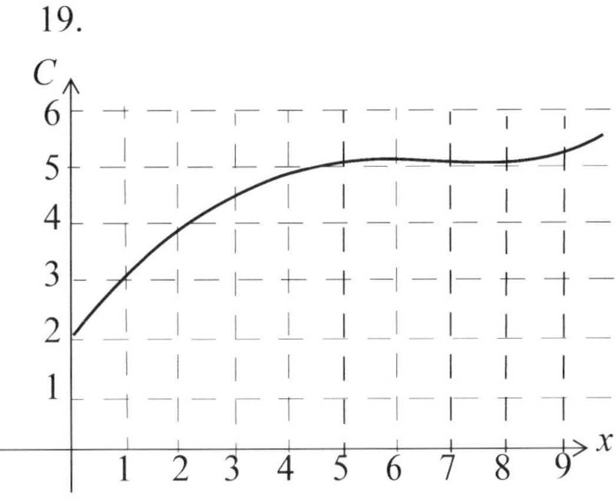

§6. Graphing Polynomial Functions

There is no question that graphing utilities can graph most functions better and faster than any of us, but this does not diminish the importance of this discussion, as it will serve to reinforce important concepts.

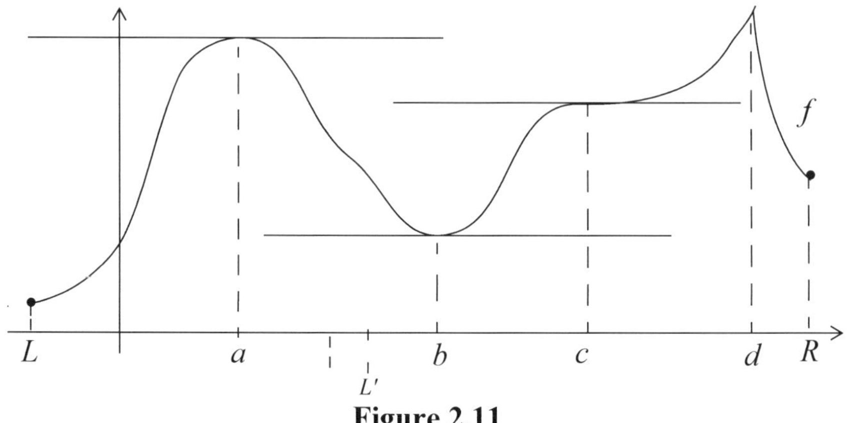

Figure 2.11

The above figure depicts the graph of a function f over the interval $[L, R]$ (left end point L, and right end point R). Forgetting those endpoints momentarily, we cite three observations that can be drawn from the figure, in order of their importance:

A **local maximum (local minimum)** point on the graph of a function is a point on the graph which is as high or higher (low or lower) than any other point in its immediate vicinity.
When one says that a maximum (minimum) occurs at $x = a$, one generally means that a **local** maximum (**local** minimum) occurs at that point.
The terms **absolute** maximum (**absolute** minimum) value of a function denotes the largest (smallest) value the function assumes within a specified domain.

1. Where a (local) maximum or (local) minimum occurs, if the function is differentiable, then its derivative must be zero (see a and b in Figure 2.11 and their associated **horizontal** tangent lines).
2. The derivative of a function may be zero, without either a maximum or minimum occurring at that point (see c).
3. A maximum or minimum can occur without the derivative being zero at that point, but only because the derivative does not exist there (see d, where the function fails to be differentiable).

The story concerning endpoints is easily told:

4. An endpoint maximum or minimum occurs at the end points. For example, a minimum occurs at L in Figure 2.11, and also at R: the graph climbs as you move to the right of L, and it falls as you approach R.

Note if the left end point is moved to L', then a maximum will occur at that new endpoint.

Endpoints aside, we have observed that a maximum or minimum **can only** occur where the derivative of the function is zero or where the derivative does not exist. Since one is often concerned with locating optimal values, any x where $f'(x) = 0$, or where $f'(x)$ does not exist is said to be a **critical point** of the function.

EXAMPLE 2.24 Find the critical points and associated values of the function:

$$f(x) = 3x^5 - 5x^3$$

SOLUTION: To find the critical points, we differentiate, set equal to zero, and solve. Evaluating the function at those critical points yields their corresponding values:

$$f'(x) = 15x^4 - 15x^2 = 0$$
$$15x^2(x^2 - 1) = 0$$
$$15x^2(x+1)(x-1) = 0$$

$$f(-1) = 3(-1)^5 - 5(-1)^3 = -3 + 5 = 2$$
$$f(1) = 3(1)^5 - 5(1)^3 = 3 - 5 = -2$$

Critical Points:	$x = 0$	$x = -1$	$x = 1$
Values:	$f(0) = 0$	$f(-1) = 2$	$f(1) = -2$

CHECK YOUR UNDERSTANDING 2.30

Find the critical points and associated values of the function:

$$f(x) = \frac{x^4}{4} + \frac{x^3}{3} - x^2 + 1$$

Answer: Critical points: $x = -2$, $x = 0$, and $x = 1$.
Values: $f(-2) = -\frac{5}{3}$, $f(0) = 1$, and $f(1) = \frac{7}{12}$.

FIRST DERIVATIVE TEST

Finding the critical points of a function $y = f(x)$ reveals those x's at which a (local) maximum or (local) minimum might occur. In Example 2.24 we found that the function $f(x) = 3x^5 - 5x^3$ has a horizontal tangent line at $x = -1$, $x = 0$, and at $x = 1$. But what exactly occurs at those points — a maximum, a minimum, or neither? This question will be resolved in Example 2.26.

A function is said to be increasing where its graph is climbing, and decreasing where it is falling. To be more precise: A function f is **increasing** over an interval I, if for every two points, a and b in I, with $a < b$, $f(a) < f(b)$.
A function f is **decreasing** over an interval I, if for every two points, a and b in I, with $a < b$, $f(a) > f(b)$.

For now, recalling Figure 2.11, we point out that the graph of a function increases to the left of a maximum point, and decreases to its right [Figure 2.12(a)]; and that the graph decreases to the left of a minimum point, and then increases to its right [Figure 2.12(a)]:

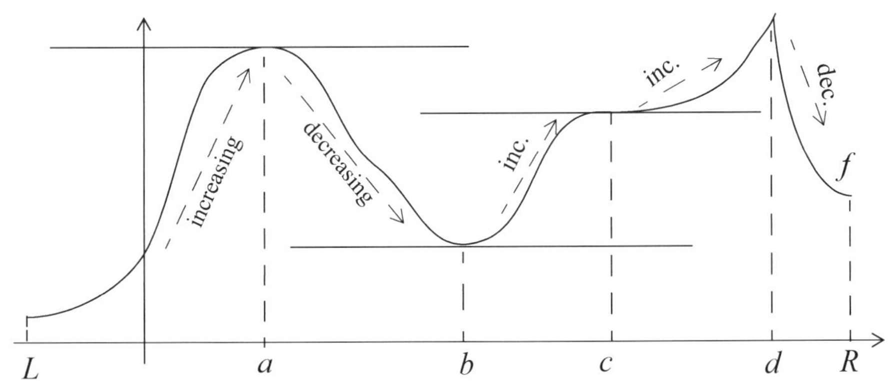

Figure 2.12

And here is another geometrical observation:

> Where the derivative of a function is positive, the tangent line has a positive slope, and the function is increasing.
> Where the derivative is negative, the tangent line has a negative slope, and the function is decreasing.

Recall that the derivative gives the slope of a tangent line.

Putting this together we have:

THEOREM 2.12
THE FIRST DERIVATIVE TEST

Let $f'(c) = 0$. If the derivative of f is positive just to the left of c, and negative just to the right of c, then the function has a maximum at c. If the derivative is negative just to the left of c, and positive just to the right of c, then the function has a minimum at c.

EXAMPLE 2.25 Assume that a function f has the following derivative:

$$f'(x) = (x-5)(2x-5)(-x+1)$$

Find where the function assumes maximum and minimum values.

SOLUTION: We employ the first three steps of Example 2.6, page 123, to determine the sign of the given derivative function:

Step 1. Indicate the zeros of $f'(x)$ on the number line, placing the letter ***c*** above each of the three odd zeros, to remind us that its sign will **c**hange about those zeros:

c c c

1 $\frac{5}{2}$ 5

Step 2. Noting that $f'(x) = (x-5)(2x-5)(-x+1)$ is negative to the right of 5 (see margin), we placed a negative sign over the interval $(5, \infty)$:

To the right of 5, the first two factors, $(x-5)$ and $(2x-5)$, are positive while the last factor, $(-x+1)$, is negative. The product of the three factors is therefore negative.

c c c $-$

1 $\frac{5}{2}$ 5

Step 3. Since each of the three zeros is odd, the sign will change as you move across those zeros, bringing us to:

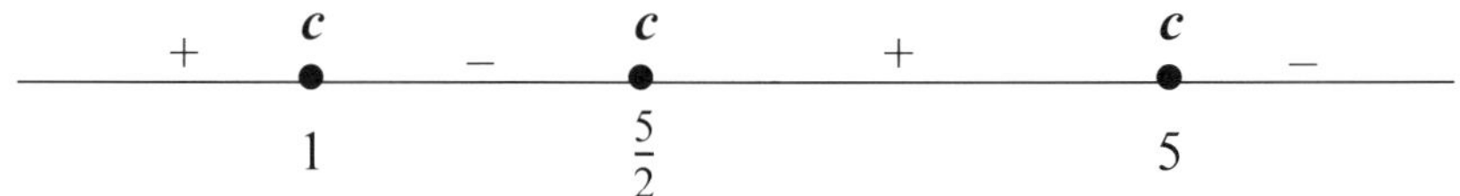

Knowing that the function is increasing where the derivative is positive and decreasing where the derivative is negative enables us to conclude that a maximum occurs at $x = 1$ and at $x = 5$, and a minimum occurs at $x = \frac{5}{2}$:

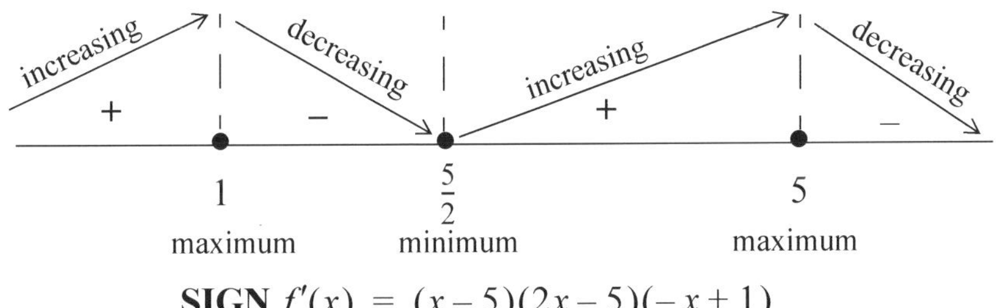

SIGN $f'(x) = (x-5)(2x-5)(-x+1)$

CHECK YOUR UNDERSTANDING 2.31

Assume that a function f has the following derivative:

$$f'(x) = (x-5)(2x-5)^2(-x+1)$$

Find where the function assumes maximum and minimums values.

Answer: minimum at 1 and a maximum at 5.

POLYNOMIAL GRAPHS

We begin by noting that the graphs of polynomial functions of the form $f(x) = x^n$ basically fall into two categories: n—even and n—odd:

$f(x) = x^n$

***n* even**	***n* odd**
Shapes similar to those of the functions $y = x^2$, and $y = x^4$, below. Each such graph passes through the origin, and the points (−1,1) and (1,1). The larger the exponent, the flatter is the graph over $(-1, 1)$ and the steeper outside of $(-1, 1)$.	Shapes similar to those of the functions $y = x^3$, and $y = x^5$, below. Each such graph passes through the origin, and the points (−1,−1) and (1,1). The larger the exponent, the flatter is the graph over $(-1, 1)$ and the steeper outside of $(-1, 1)$
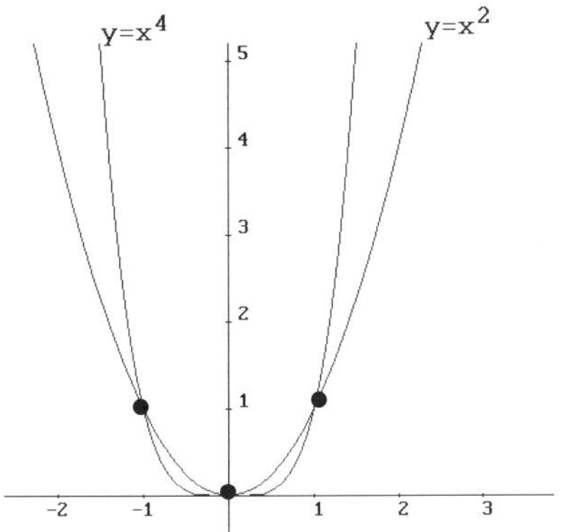	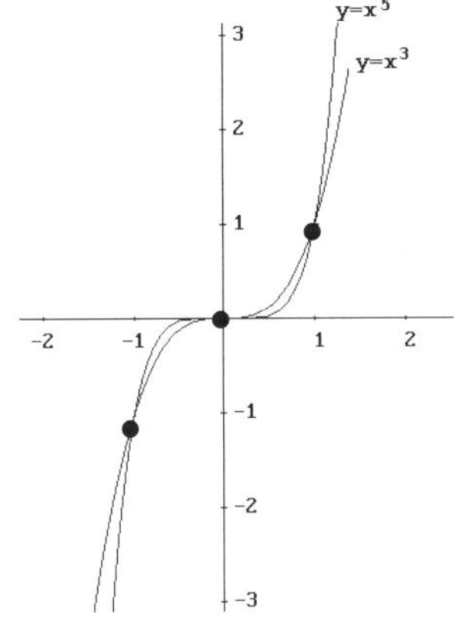

Even with a graphing utility, you can only view a finite portion of the graph of a function. It is important that you are able to determine how a graph behaves far away from the origin (as x tends to $\pm\infty$, written: $x \to \pm\infty$). The following fact reveals the nature of the graph of a polynomial function as $x \to \pm\infty$:

> Far away from the origin the graph of the polynomial function:
> $$p(x) = a_n x^n + a_{n-1} x^{n-1} + \cdots + a_1 x + a_0$$
> resembles, in shape, that of its leading term $g(x) = a_n x^n$

For example, as $x \to \pm\infty$, the graph of $p(x) = \mathbf{6x^4} - 3x^3 - 15x^2 - 10$ resembles that of $g(x) = \mathbf{6x^4}$. This makes sense, since as x gets larger and larger in magnitude, the term $6x^4$ becomes more and more dominant (see margin).

$\mathbf{6x^4} - 3x^3 - 15x^2 - 10$
$= \mathbf{6x^4}\left(1 - \frac{3}{6x} - \frac{15}{6x^2} - \frac{10}{6x^4}\right)$
and $\left(1 - \frac{3}{6x} - \frac{15}{6x^2} - \frac{10}{6x^4}\right) \to 1$
as $x \to \pm\infty$.

CHECK YOUR UNDERSTANDING 2.32

Determine a function $g(x) = ax^n$, whose graph resembles that of the given polynomial function f as $x \to \pm\infty$.

(a) $f(x) = x^3 - x$ (b) $f(x) = (2x^3 + x)(x^2 - 5x + 1)(x - 1)$

Answers:
(a) $g(x) = x^3$
(b) $g(x) = 2x^6$

EXAMPLE 2.26 Determine where the function:

$$f(x) = 3x^5 - 5x^3$$

assumes maximum/minimum values. Sketch the graph of the function.

This is the function of Example 2.24.

SOLUTION: Basically, two steps are needed to find where maxima and minima occur:

Step 1: Differentiate and factor:

$$f'(x) = 15x^4 - 15x^2 = 15x^2(x+1)(x-1)$$

Step 2: SIGN $f'(x)$:

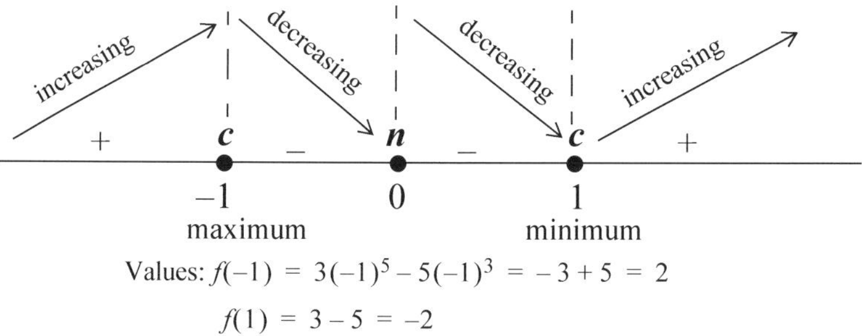

Values: $f(-1) = 3(-1)^5 - 5(-1)^3 = -3 + 5 = 2$

$f(1) = 3 - 5 = -2$

Note that unlike the odd zeros at ± 1, the sign of the function does not change as you cross over the even zero at 0. See discussion on page 123.

Step 3: Plot maximum point $(-1, 2)$ and minimum point $(1, -2)$ [Figure 2.13(a)]. We also sketched a bit of the graph about the origin in Figure 2.13(a) to reflect the fact that the graph has a horizontal tangent line at that point (derivative is zero at $x = 0$).

Step 4: Far from the origin: as $x \to \pm\infty$:

Knowing that the graph of the function $f(x) = 3x^5 - 5x^3$ resembles that of its leading term $g(x) = 3x^5$ lead us to the graph-parts labeled A and B in Figure 2.13(a) (see margin).

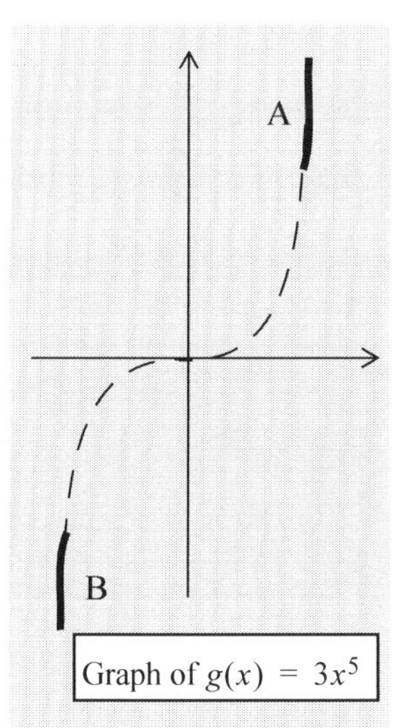

Graph of $g(x) = 3x^5$

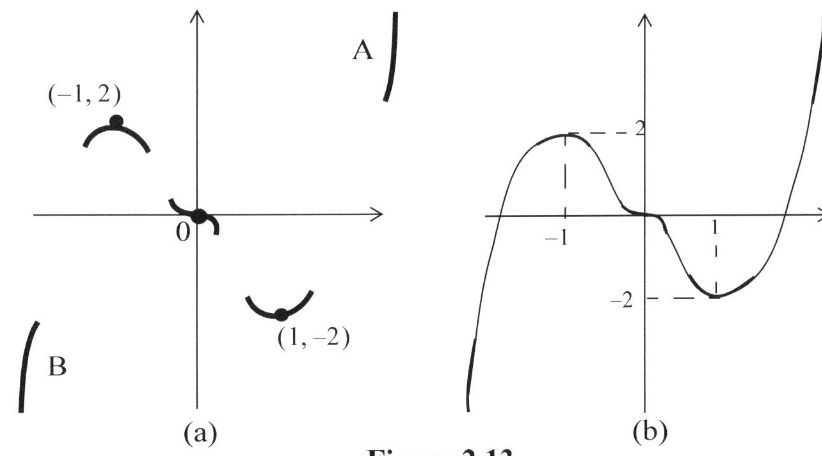

(a) (b)

Figure 2.13

Step 5: Figure 2.13(a) along with the increasing/decreasing information in Step 2, lead us to the graph of the function in Figure 2.13(b).

GRAPHING CALCULATOR GLIMPSE 2.3

Note how the graph of $f(x) = 3x^5 - 5x^3$ reflects the fact that the tangent line at $x = 0$ (the x-axis is the tangent line) is horizontal, without either a maximum or minimum occurring at that point.

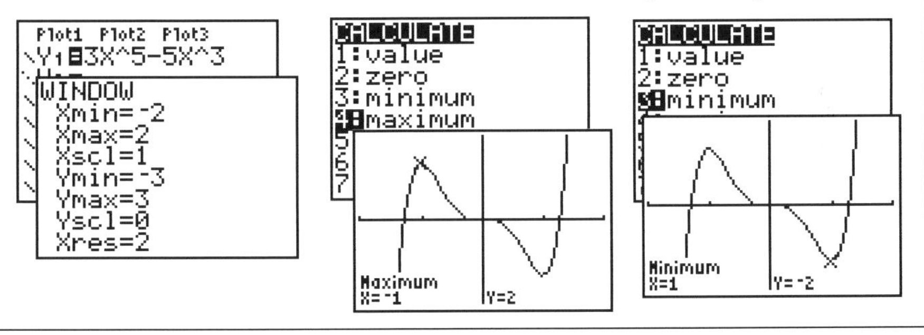

Yes, but how much does one learn!

This is the function of CYU 2.30.

EXAMPLE 2.27 Sketch the graph of the function:

$$f(x) = \frac{x^4}{4} + \frac{x^3}{3} - x^2 + 1$$

SOLUTION:

Step 1: Differentiate and factor:

$$f'(x) = \frac{4x^3}{4} + \frac{3x^2}{3} - 2x = x^3 + x^2 - 2x$$
$$= x(x^2 + x - 2) = x(x+2)(x-1)$$

Step 2: SIGN $f'(x)$:

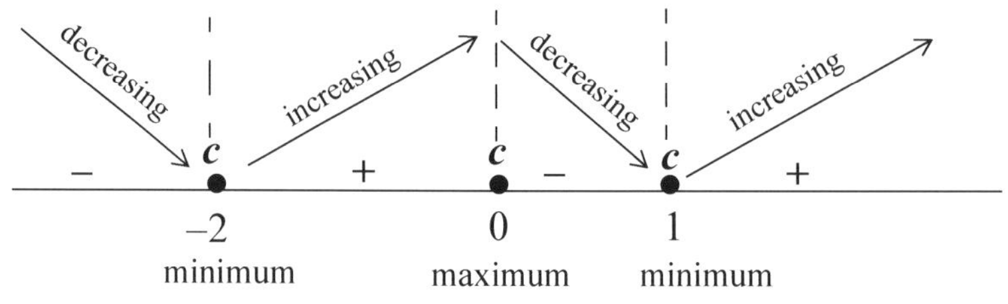

SIGN $f'(x) = x(x+2)(x-1)$

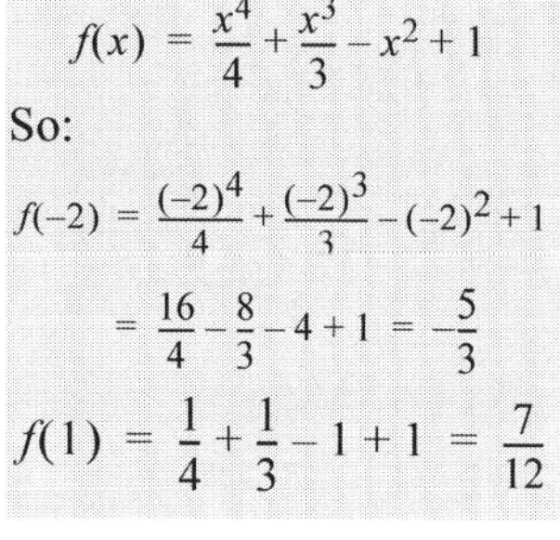

$f(x) = \frac{x^4}{4} + \frac{x^3}{3} - x^2 + 1$

So:

$f(-2) = \frac{(-2)^4}{4} + \frac{(-2)^3}{3} - (-2)^2 + 1$

$= \frac{16}{4} - \frac{8}{3} - 4 + 1 = -\frac{5}{3}$

$f(1) = \frac{1}{4} + \frac{1}{3} - 1 + 1 = \frac{7}{12}$

Step 3: Plot maximum and minimum points.

Minimum points: $(-2, f(-2)) = (-2, -\frac{5}{3})$ and $(1, f(1)) = (1, \frac{7}{12})$ (margin); maximum point: $(0, f(0)) = (0, 1)$ [see Figure 2.14(a)].

Step 4: As $x \to \pm\infty$. The graph of $f(x) = \frac{x^4}{4} + \frac{x^3}{3} - x^2 + 1$ resembles that of $y = \frac{x^4}{4}$ [see A and B of Figure 2.14(a)].

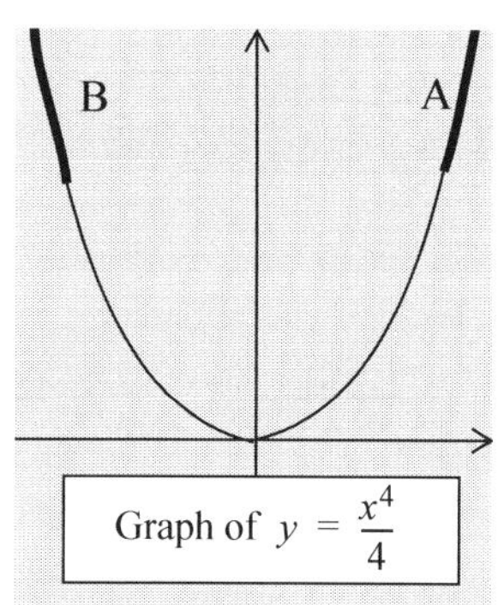

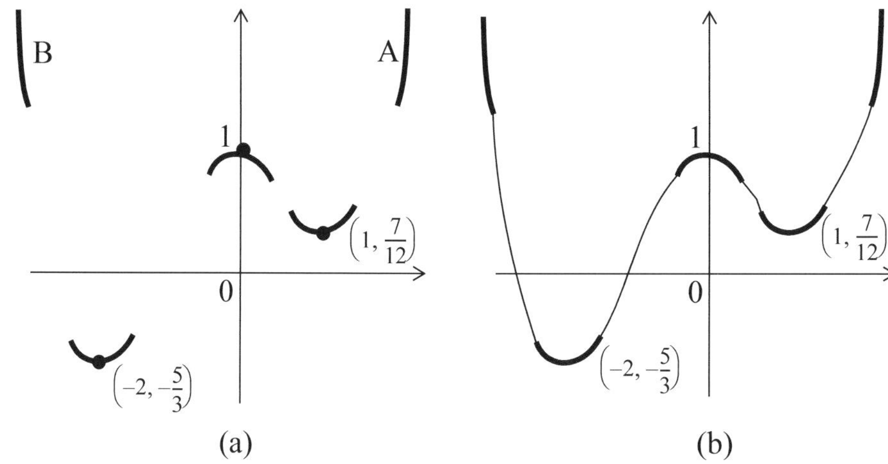

Figure 2.14

Step 5: Figure 2.14(a) along with the increasing/decreasing information in Step 2, lead us to the graph of the function in Figure 2.14(b).

CHECK YOUR UNDERSTANDING 2.33

Sketch the graph of the function $f(x) = \frac{x^5}{5} - \frac{x^3}{3} + 1$.

Answer: See A-16.

CONCAVITY AND POINTS OF INFLECTION

Throughout this discussion, we are assuming that the second derivative of the function under consideration exists.

The graph of the function f of Figure 2.15(a) is said to be **concave up** (it bends up) while that in (b) is said to be **concave down** (it bends down).

The slopes of the tangent lines to the concave up curve are increasing as you move from left to right along the curve and the first derivative is therefore increasing. It follows that the first derivative's derivative (the second derivative) must be positive: $f''(x) > 0$.

The slopes of the tangent lines to the concave down curve of Figure 2.15(b) are decreasing; which is to say: the first derivative is decreasing. Consequently: $f''(x) < 0$.

CONCAVE UP AND CONCAVE DOWN

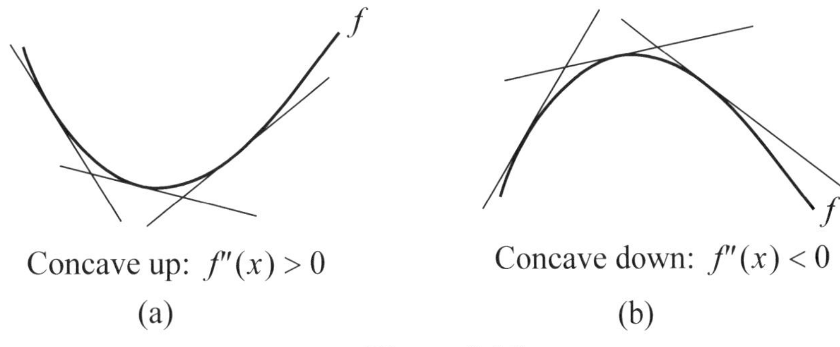

Figure 2.15

A point on the curve about which concavity changes is said to be a **point of inflection**:

The point of inflection of the decreasing function in Figure 2.16(a) might represent where the losses of a company are starting to diminish. The point of inflection of the increasing function in Figure 2.16(b) might represent the point at which the rate of growth of the national debt is beginning to decrease.

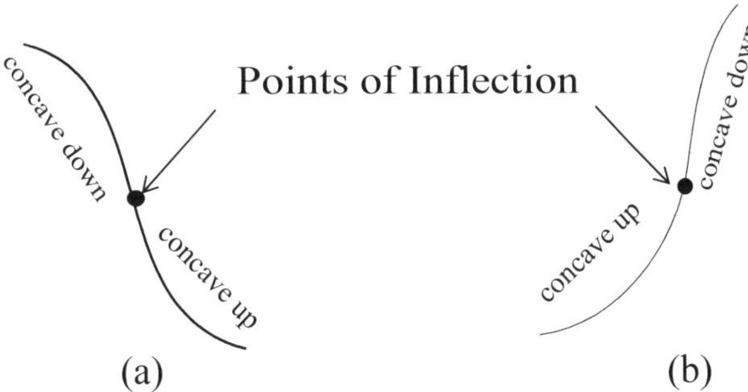

Figure 2.16

Summarizing:

$f''(x) > 0$: Concave up

$f''(x) < 0$: Concave down

A point $(c, f(c))$ about which concavity changes (from up to down, or vice versa) is said to be a **point of inflection.**

EXAMPLE 2.28 Sketch the graph of:

$$f(x) = x^4 - 4x^3 + 15$$

Label maximum, minimum, and inflection points.

SOLUTION: We proceed as in the previous examples but will now be concerned with both the sign of the first derivative and that of the second derivative.

$$f'(x) = (x^4 - 4x^3 + 15)' = 4x^3 - 12x^2 = 4x^2(x-3)$$

SIGN $f'(x)$ tells us where the function is increasing and where it is decreasing, and therefore where maxima and minima occur.

SIGN $f'(x)$: dec $-$ | n at 0 (horizontal tangent line) | dec $-$ | c at 3 (min) | inc $+$

Values: $f(0) = 15$, $f(3) = 3^4 - 4 \cdot 3^3 + 15 = -12$

$$f''(x) = (4x^3 - 12x^2)' = 12x^2 - 24x = 12x(x-2)$$

SIGN $f''(x)$ tells us where the function is concave up and where it is concave down, and therefore where inflection points occur.

SIGN $f''(x)$: concave up $+$ | c at 0 | concave down $-$ | c at 2 | concave up $+$

inflection points at 0 and 2

Values: $f(0) = 15$, $f(2) = 2^4 - 4 \cdot 2^3 + 15 = -1$

As $x \to \pm\infty$. The graph of $f(x) = x^4 - 4x^3 + 15$ resembles that of $y = x^4$ (margin).

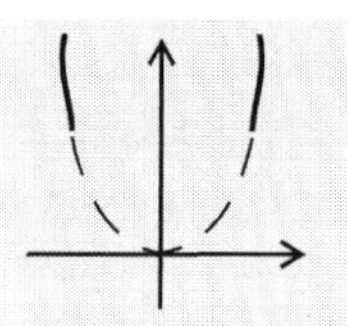

Figure 2.17 reveals the fruit of our labor.

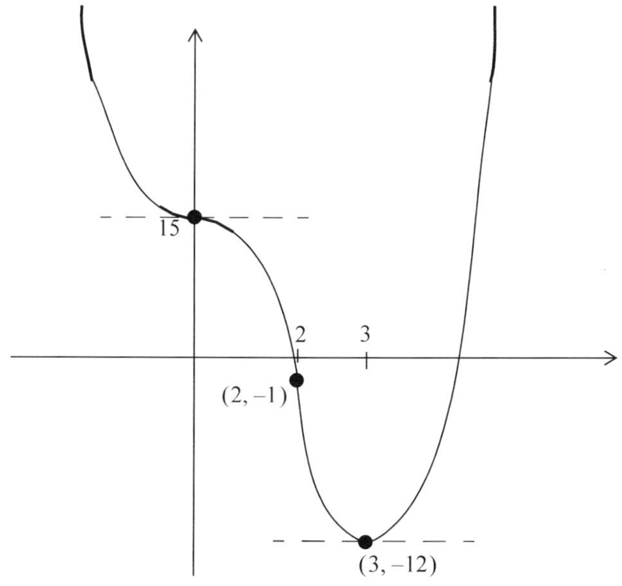

Figure 2.17

CHECK YOUR UNDERSTANDING 2.34

Find the values of x at which an inflection point occurs on the graph of the function $f(x) = 3x^5 - 5x^3$ of Example 2.26.

Answer: $0, \pm\sqrt{2}$,

END POINTS AND ABSOLUTE MAXIMA AND MINIMA

You can use SIGN $f'(x)$ to analyze end-point situations. Consider the following example.

We remind you that the absolute maximum (minimum) value of the function, is the largest (smallest) value the function assumes over the indicated region (in this case, the interval $[-4, 5]$.

EXAMPLE 2.29 Determine the absolute maximum value and the absolute minimum value of the function:

$$f(x) = \frac{x^3}{3} - x^2 - 3x + 1$$

over the interval $[-4, 5]$.

SOLUTION: Differentiate and Factor:

$$f'(x) = x^2 - 2x - 3 = (x+1)(x-3)$$

SIGN and Interpret:

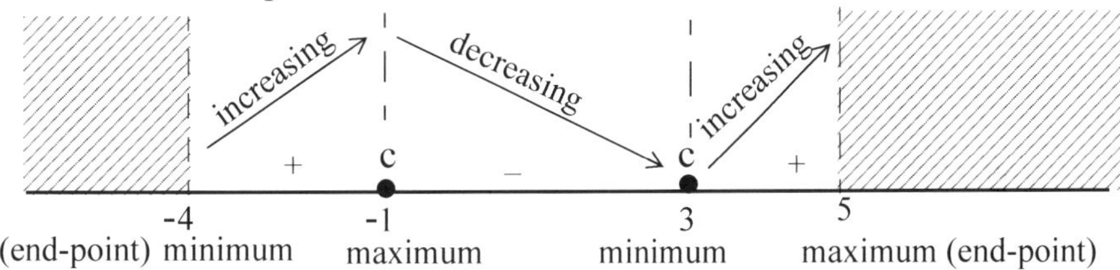

SIGN $f'(x) = (x+1)(x-3)$

Figure 2.18

You can see that there are two contenders for the absolute maximum value: at $x = -1$ and at the right endpoint $x = 5$, as well as two contenders for the absolute minimum value: at the left end-point $x = -4$, and at $x = 3$. To see who wins out, we calculate:

Absolute Maximum Contenders	Absolute Minimum Contenders
A draw: → $f(-1) = \frac{8}{3}$, $f(5) = \frac{8}{3}$	The winner: → $f(-4) = -\frac{73}{3}$ $f(3) = -8$

Conclusion. The absolute maximum value of $\frac{8}{3}$ occurs at $x = -1$ and at $x = 5$. The absolute minimum value of $-\frac{73}{3}$ occurs at $x = -4$.

CHECK YOUR UNDERSTANDING 2.35

Determine the absolute maximum value and the absolute minimum value of the function $f(x) = \frac{x^4}{4} + \frac{x^3}{3} - x^2 + 1$ of Example 2.27 over the interval $[-2, 2]$.

Answer: Absolute minimum of $-\frac{5}{3}$ at $x = -2$. Absolute maximum of $\frac{11}{3}$ at $x = 2$.

EXERCISES

Exercises 1-8. Find the critical points and associated values of the given function.

1. $f(x) = \frac{3x^4}{2} - 4x^3 - 9x^2$
2. $f(x) = \frac{3x^4}{2} + 4x^3 - 9x^2$
3. $g(x) = 3x^5 - 5x^3 + 1$
4. $g(x) = \frac{x^5}{5} - \frac{5x^3}{3} + 4x$
5. $h(x) = \frac{x-1}{x^2}$
6. $f(x) = x + \frac{1}{x}$
7. $h(x) = \frac{x^2}{x-1}$
8. $f(t) = -\frac{x^3}{x+1}$

Exercises 9-16. Sketch the graph of the given function. Label maximum, minimum, and inflection points.

9. $f(x) = x^3 - 9x$
10. $g(x) = x^4 - 4x^2 + 1$
11. $g(x) = x^4 - 2x^2 + 3$
12. $f(x) = x^3 - x^2$
13. $g(x) = 3x^5 - 5x^3 + 10$
14. $g(x) = \frac{x^5}{5} - \frac{5x^3}{3} + 4x$
15. $h(x) = x^4 - \frac{4}{3}x^3 + 1$
16. $s(t) = \frac{x^4}{4} - x^3 + 1$

17. Verify that the graph of the function $f(x) = x^2$ is concave up everywhere. Generalize your proof to accommodate the graph of any function of the form $f(x) = x^{2n}$, where n is a positive integer.

18. Verify that the graph of the function $f(x) = x^3$ is concave up to the right of the origin and concave down to the left of the origin. Generalize your proof to accommodate the graph of any function of the form $f(x) = x^{2n+1}$, where n is a positive integer.

19. Determine where the absolute maximum value and the absolute minimum value of the function $f(x) = x^4 - 2x^2 + 1$ occur over the specified interval.

    (a) $[-2, 1]$ (b) $[-1, 2]$ (c) $[0, 2]$ (d) $[2, 3]$

20. Determine where the absolute maximum value and the absolute minimum value of the function $f(x) = x^2 - 5x + 3$ occur over the specified interval.

    (a) $[-2, 4]$ (b) $[-2, 10]$ (c) $[-2, 1]$ (d) $[3, 4]$ (e) $\left[\frac{3}{2}, \frac{7}{2}\right]$

21. Show that the graph of the quadratic function $f(x) = ax^2 + bx + c$ is concave up everywhere if $a > 0$ and concave down everywhere if $a < 0$. The graph is said to be a parabola and its minimum or maximum point is said to be its vertex. Prove that the vertex occurs at $x = -\dfrac{b}{2a}$.

22. (a) Prove that the graph of the cubic function $f(x) = ax^3 + bx^2 + cx + d$ has only one inflection point, and that it occurs at $x = -\dfrac{b}{3a}$.

(b) Show that the graph of $f(x) = ax^3 + bx^2 + cx + d$ bends down to the left of the inflection point if $a > 0$ and bends up to the left of the inflection point if $a < 0$.

(c) Find a and b such that $(1, 2)$ is a maximum point on the graph of $f(x) = ax^3 + bx^2$.

(d) Find a, b, and c such that $(1, 2)$ is a maximum point on the graph of $f(x) = ax^3 + bx^2 + cx$ and that a minimum occurs at $x = -1$.

(c) Find a and b such that the graph of the $f(x) = ax^3 + bx^2$ has a point of inflection at $(1, 2)$.

(d) Find a, b, and c such that the graph of the $f(x) = ax^3 + bx^2 + cx$ has a point of inflection at $(1, 2)$ and a maximum at $x = -1$.

Exercises 23-26 (Second Derivative Test) If $f'(c) = 0$ and $f''(c) < 0$, then $f(c)$ is a (local) maximum value of the function. If $f'(c) = 0$ and $f''(c) > 0$, then $f(c)$ is a (local) minimum value of the function. The second derivative test is inconclusive if $f''(c) = 0$.

A geometrical argument:

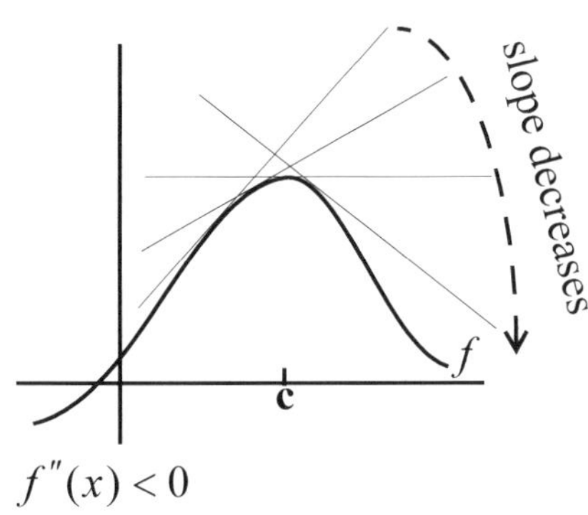

$f''(x) < 0$

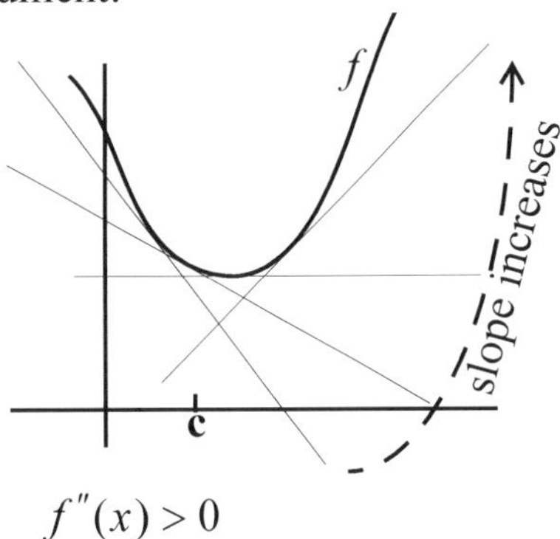

$f''(x) > 0$

Use the second derivative test to locate where the given function assumes maximum values, and where it assumes minimum values:

23. $f(x) = x^3 - x^2$

24. $g(x) = x^4 - x^2 + 3$

25. $g(x) = 3x^5 - 5x^3 + 10$

26. $g(x) = \dfrac{x^5}{5} - \dfrac{5x^3}{3} + 4x$

27. **(Inconclusive Second Derivative Test)** Show that $f''(0) = 0$ for each of the following functions, and then go on to show that one of the functions has a maximum at 0, another has a minimum at 0, and the remaining one has neither a maximum nor a minimum at 0.

(i) $f(x) = x^3$ (ii) $f(x) = x^4$ (iii) $f(x) = -x^4$

An optimization problem is one where a maximum or minimum value of a function is to be determined.

§7. Optimization Problems

Let's start things off with an example:

EXAMPLE 2.30

MAXIMUM PROFIT

A company can produce up to 500 units per month. Its profit, in terms of number of units produced is given by:

$$P(x) = -\frac{x^3}{30} + 9x^2 + 400x - 7500$$

How many units should the company produce to maximize profit?

SOLUTION: Differentiating the profit function we have:

We remind you that the derivative of the profit function is called the marginal profit function.

$$P'(x) = -\frac{x^2}{10} + 18x + 400 = -\frac{1}{10}(x^2 - 180x - 4000)$$

$$= -\frac{1}{10}(x - 200)(x + 20)$$

(or $\overline{MP}(x)$)

The critical points occur at $x = 200$, and at $x = -20$. We have no interest in the latter point, since production cannot be negative. Applying the first derivative test we see that the maximum does occur at a production level of $x = 200$:

The first derivative test clearly shows that profit continues to decrease after 200.

SIGN $P'(x) = -\frac{1}{10}(x-200)(x+20)$: $-$ c at -20, 0, $+$, c at 200 (maximum), $-$

GRAPHING CALCULATOR GLIMPSE 2.4

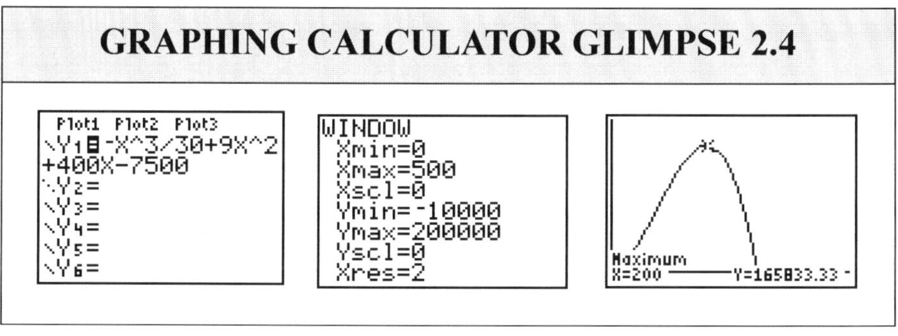

CHECK YOUR UNDERSTANDING 2.36

The total monthly profit (in thousands of dollars) from the sale of x sailboats is given by:

$$P(x) = 5.7x - 0.025x^2 \qquad (x \le 150)$$

Find the maximum profit.

Answer: $324,900

Example 2.30 was kind of easy in that we were conveniently given the function to be optimized. The real challenge in most optimization problems, however, is to extricate the function to be maximized or minimized from a given word problem. The first three steps in the following procedure can help you meet that challenge:

Step 1. See the problem.

Step 2. Express the quantity to be optimized in terms of any convenient number of variables.

Step 3. In the event that the expression in Step 3 involves more than one variable, use the given information to arrive at an expression involving but one variable.

Step 4. Differentiate, set equal to zero, and solve (find the critical points). If necessary, analyze the nature of the critical points, and any endpoint of the domain.

EXAMPLE 2.31 A 1,250 ft^2 rectangular region is to be enclosed. The region is adjacent to an already existing fence and only three of its sides will require new fencing. Determine the dimensions of the region requiring the least amount of new fencing.

SOLUTION:

Step 1: SEE THE PROBLEM

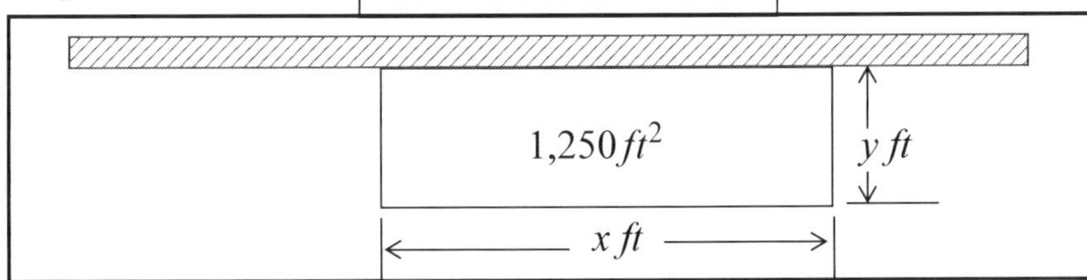

Step 2: You want to minimize **F**encing. To better focus your attention, we suggest that you immediately write down:

$$\boldsymbol{F} =$$

and then fill in the right side of the equation:

$$\boldsymbol{F} = x + 2y$$

Step 3: We need to get rid of one of the two variables on the right side. In order to do so, we look for a relation between those variables and find it in the given area information:

$$xy = 1250$$

$$y = \frac{\mathbf{1250}}{\boldsymbol{x}} \qquad (*)$$

We can now express $\boldsymbol{F}$ as a function of one variable:

$$\boldsymbol{F}(x) = x + 2y = x + 2\left(\frac{\mathbf{1250}}{x}\right) = x + \frac{2500}{x} = x + 2500x^{-1}$$

Step 4: Differentiate, set equal to 0, and solve:

$$F'(x) = 1 - 2500x^{-2} = 0$$

$$1 = \frac{2500}{x^2}$$

$$x^2 = 2500$$

$$x = \pm 50$$

If you doubt that a minimum occurs at 50, you can consider the SIGN of

$$F'(x) = 1 - \frac{2500}{x2} = \frac{x^2 - 2500}{x^2} = \frac{(x+50)(x-50)}{x^2}$$

which, for $x > 0$, is positive to the right of 50 and negative to the left of 50:

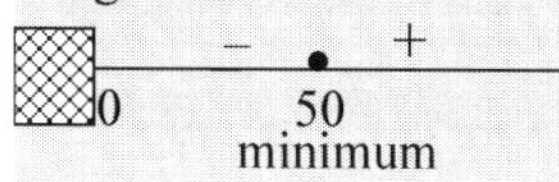

At this point, we know that the graph of the volume function has a horizontal tangent line at $x = 50$ and at $x = -50$. You can forget about the -50, for x represents a length. Since a minimum amount of fencing surely exists, it must occur at $x = 50$ (margin).

Conclusion: The dimensions of the region requiring the least amount of fencing are:

$$x = 50 \text{ feet, and } y = \frac{1250}{x} = \frac{1250}{50} = 25 \text{ feet}$$

↑ side opposite the existing fence ↑ from (*)

CHECK YOUR UNDERSTANDING 2.37

Determine the maximum rectangular area that can be enclosed using 1200 feet of fencing.

Answer: 90,000 ft^2.

EXAMPLE 2.32
MAXIMUM VOLUME

Best Box Company is to manufacture open-top boxes from 12 in. by 12 in. pieces of cardboard. The construction process consists of two steps: (1) cutting the same size squares from each corner of the cardboard, and (2) folding the resulting cross-like configuration into a box. What size square should be cut out, if the resulting box is to have the largest possible volume?

SOLUTION:

Step 1: SEE THE PROBLEM

12
x
x
12
x
12-2x
12-2x

In general:

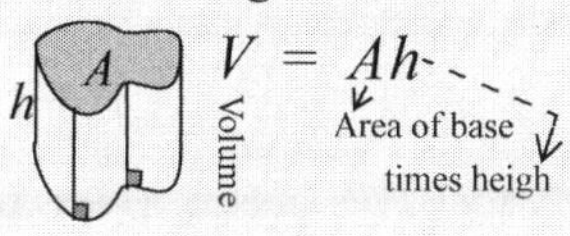

Step 2: Volume is to be maximized, and from the above, we see that:

$$V(x) = (12-2x)(12-2x)x = 4x^3 - 48x^2 + 144x$$

Step 3: Not applicable, since volume is already expressed as a function of **one** variable.

Step 4: Differentiate, set equal to 0, and solve:

$$V'(x) = 12x^2 - 96x + 144 = 0$$
$$12(x-6)(x-2) = 0$$
$$x = 6 \text{ or } x = 2$$

If you want, you can use the first derivative test to analyze the nature of the two critical values:

+ c − c +
2 max. 6. min.
SIGN $V'(x)$

At this point we know that there are two critical points: $x = 6$ and $x = 2$. We can forget about the 6, for if you cut a square of length 6 from the piece of cardboard, nothing will remain. Since a box of maximum-volume certainly exists, and since we are down to one horizontal tangent line, $x = 2$ must be the answer.

Conclusion: For maximum volume, cut out a square if size 2 in.

CHECK YOUR UNDERSTANDING 2.38

An eight cubic foot closed box with square base is to be constructed. Determine the minimum amount of cardboard required to manufacture the box.

Answer: 24 square feet.

EXAMPLE 2.33
MAXIMUM YIELD

When 20 peach trees are planted per acre, each tree will yield 200 peaches. For every additional tree planted per acre, the yield of each tree diminishes by 5 peaches. How many trees per acre should be planted to maximize yield?

SOLUTION:

Step 1:

SEE THE PROBLEM

20 trees/acre ⟶	200 per tree
$20+x$ trees/acre ⟶	$200-5x$ per tree

Step 2: Letting x denote the number of trees above 20 to be planted per acre, we express the yield per acre as a function of x:

number of trees per acre — number of peaches per tree

$$Y(x) = \underbrace{(20+x)}\underbrace{(200-5x)} = -5x^2 + 100x + 4000$$

Step 3: Not applicable [The function to be maximized, $Y(x)$, is already expressed in terms of one variable].

SIGN $Y'(x) = 10(-x+10)$:

+ c –

0 maximum

Step 4: $Y'(x) = -10x + 100 = 0$

$$x = \frac{100}{10} = 10$$

We conclude that maximum yield will be attained with the planting of 30 trees.

CHECK YOUR UNDERSTANDING 2.39

A car-rental agency can rent 150 cars per day at a rate of \$24 per day. For each price increase of \$1 per day, 5 less cars are rented. What rate should be charged to maximize the revenue of the agency?

Answer: \$27.00

EXAMPLE 2.34

MINIMIZE COST

The Big-Box manufacturing firm has received an order for 1,500 shipping boxes. The firm has 25 machines that can be used to manufacture the boxes, each of which can produce 30 boxes per hour. It will cost the firm \$55 to set up each machine. Once set up, the machines are fully automated, and can all be supervised by a single worker, earning \$14 per hour. How many machines should be used to minimize the cost of production?

SOLUTION:

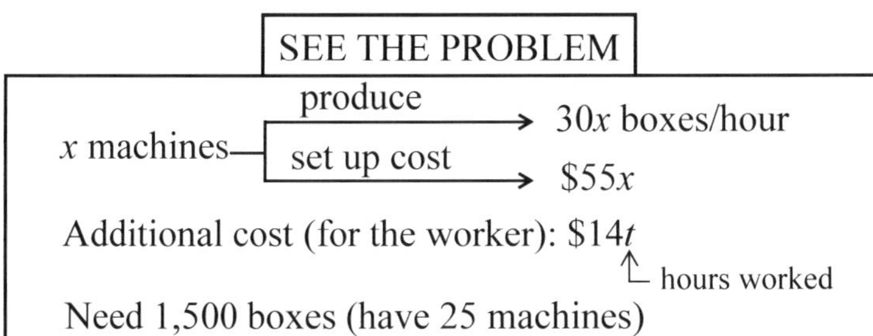

Letting C denote the cost of production when using x machines, and t denote the time required to produce the 1500 units when x machines are operating we have:

$$C = 55x + 14t \text{ (dollars)}$$

We express the variable t in terms of x, and find it helpful to consider units along the way:

$$\text{Time, t, to produce 1,500 boxes using x machines} = \frac{1500 \text{ boxes}}{30x\,\frac{\text{boxes}}{\text{hour}}}$$

$$= \frac{50 \text{ boxes}}{x} \cdot \frac{\text{hour}}{\text{boxes}} = \frac{50}{x} \text{ hours}$$

We can now express cost as a function of one variable:

$$C(x) = 55x + 14 \cdot \frac{50}{x} = 55x + 700x^{-1}$$

Bringing us to the routine part of the solution process:

$$C'(x) = 55 - 700x^{-2} = 0$$

$$55 = \frac{700}{x^2}$$

$$x^2 = \frac{700}{55}$$

$$x = \sqrt{\frac{700}{55}} \approx 3.6$$

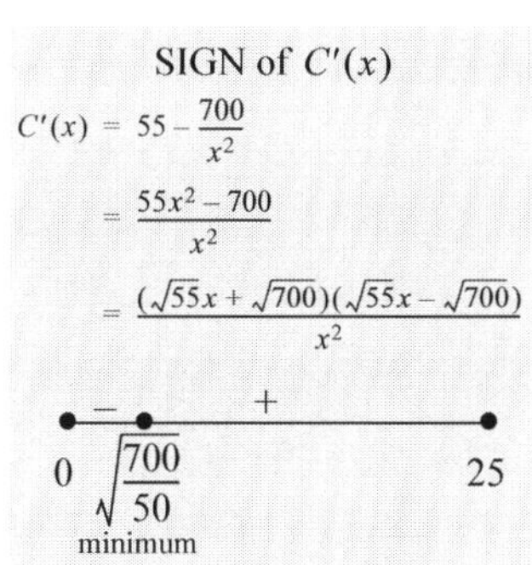

Since 3.6 lies nearly half-way between 3 and 4, one must calculate both $C(3)$ and $C(4)$ directly prior to making a final decision. If you do, you will find that $C(3) \approx 398$ and that $C(4) = 395$.

Conclusion: To minimize cost of production, the company should utilize four machines.

CHECK YOUR UNDERSTANDING 2.40

A sailboat company can manufacture up to 500 boats per year. The number of boats, n, that the company can sell per year can be approximated by the function $n = 500 - \frac{p}{100}$, where p is the price (in dollars) for a single boat. What yearly production level will maximize profit, if it costs the company $50{,}000 + 10{,}000n$ to produce n boats?

Answer: 200 boats

USING A GRAPHING CALCULATOR TO FIND APPROXIMATE SOLUTIONS TO OPTIMIZATION PROBLEMS.

EXAMPLE 2.35
MINIMAL POLLUTION

Two chemical plants are located 12 miles apart. The pollution count from plant A in parts per million, at a distance of x miles from plant A, is given by $(4K)/x^2$ for some constant K. The pollution count from the cleaner plant B, at a distance of x miles from plant B, is one quarter that of A. Determine the point on the line joining the two plants where the pollution count is minimal.

SOLUTION:

SEE THE PROBLEM

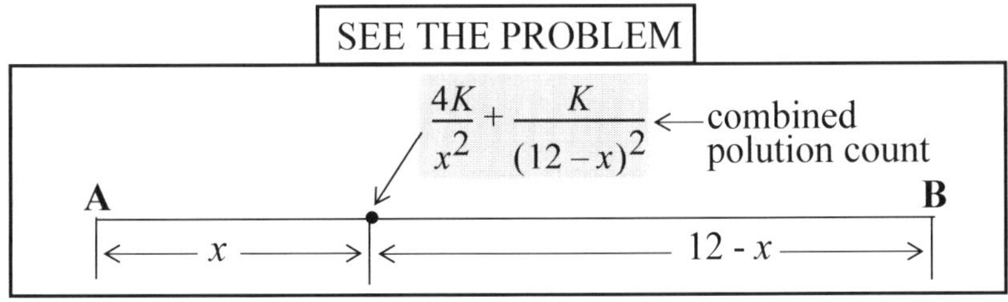

We see that $\frac{4K}{x^2}+\frac{K}{(12-x)^2}$ is to be minimized, but what are we going to do with the K? We factor it out:

$$\frac{4K}{x^2}+\frac{K}{(12-x)^2}=K\left(\frac{4}{x^2}+\frac{1}{(12-x)^2}\right)$$

Conclusion: Minimum pollution count will occur where the function:

$$\frac{4}{x^2}+\frac{1}{(12-x)^2}$$

is minimum (what about the "K?"). Taking the easy way out:

Actually, it's not so much an "*easy way out*," for when it comes to optimization problems the real challenge is to express the quantity to be optimized as a function of one variable. The rest involves a routine process which can be relegated to machines.

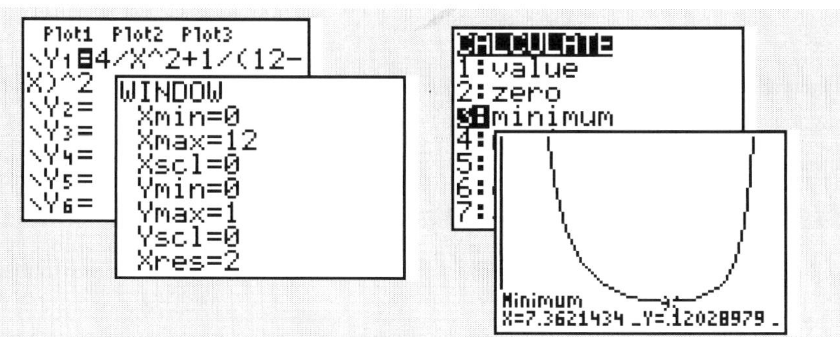

We determined that the minimum pollution count between the two plants occurs at a distance of approximately 7.36 miles from plant A.

CHECK YOUR UNDERSTANDING 2.41

A cylindrical drum is to hold 65 cubic feet of chemical waste. Metal for the top of the drum costs \$2 per square foot, and \$3 per square foot for the bottom. Metal for the side of the drum costs \$2.50 per square foot. Use a graphing utility to approximate, to three decimal places, the dimensions of the drum which will minimize cost.

Note: The area of a circle of radius r is $A = \pi r^2$, and its circumference is $C = 2\pi r$.

Answer: $r \approx 2.179$ feet and $h \approx 4.358$ feet.

EXERCISES

Exercises 1-2. (**Maximum Profit**) Determine the production level that will maximize profit, for the given cost and revenue functions. Verify that maximum profit occurs when marginal revenue equals marginal cost.

1. $C(x) = 5000 + 100x$, $R(x) = 200x - \dfrac{x^2}{10}$, for $0 \le x \le 2000$.

2. $C(x) = 1200 + 300x + \dfrac{250x^2}{100 + x}$, $R(x) = 500x$, for $0 \le x \le 2000$.

3. (**Maximize Yield**) The cabbage yield, Y, in bushels per acre, is given by $Y(x) = 60 + 16x - 4x^2$, where x denotes the number of quarts of fertilizer used per acre. How many quarts of fertilizer should be used to maximize yield?

4. (**Minimize Cost**) Determine the minimum monthly cost of a company if its monthly costs for producing x units is given by $C(x) = x^2 - 150x + 500$.

5. (**Air Velocity**) When a person coughs, the radius r of the trachea decreases. The velocity of air in the trachea during a cough can be approximated by the function $v(r) = ar^2(r_0 - r)$, where a is a constant, and r_0 is the radius of the trachea in a relaxed state. Determine the radius at which the velocity is greatest.

6. (**Maximum and Minimum Bacteria Concentration**) The bacteria concentration (in parts per million) in a swimming pool, t days after a chlorine treatment, is given by $C(t) = 10^{-6}(2t^3 - 15t^2 + 24t + 86)$, $(t < 6)$. How many days after the treatment is the concentration the highest? The lowest? Note: Do you have to worry about the 10^{-6}?

7. (**Minimum Area**) A poster is to surround 1200 in^2 of printing material with a top and bottom margin of 4 in. and side margins of 3 in. Find the outside dimensions of the minimum poster area.

8. (**Smallest Perimeter**) A rectangle has an area of 900 in^2. What are the dimensions of the rectangle of smallest perimeter?

9. (**Minimum Fencing**) A fenced-in rectangular garden is divided into 2 areas by a fence running parallel to one side of the rectangle. Find the dimensions of the garden that minimizes the amount of fencing needed, if the garden is to have an area of 15,000 square feet.

10. (**Minimum Cost**) A fenced-in rectangular garden is divided into 3 areas by two fences running parallel to one side of the rectangle. The two fences cost \$6 per running foot, and the outside fencing costs \$4 per running foot. Find the dimensions of the garden that minimizes the total cost of fencing, if the garden is to have an area of 8,000 square feet.

11. (**Maximum Revenue**) A chemical company charges \$900 per ton of a certain chemical, for orders up to and including 10 tons. For every ton in access of 10 ordered by a customer, the cost of each ordered ton will be reduced by \$30. Find the maximum revenue for the company.

12. (**Maximum Revenue**) A car-rental agency can rent 100 cars per day at a rate of $40 per day. For each price increase of $1 per day, 2 less cars are rented. What rate should be charged to maximize the revenue of the agency?

13. (**Maximum Revenue**) A car-rental agency can rent 150 cars per day at a rate of $24 per day. For each price increase of $1 per day, 5 less cars are rented. What should the agency charge in order to maximize revenue?

14. (**Maximum Profit**) It costs the college bookstore $7 for a student supplement to a text. The bookstore is currently selling 300 copies of the supplement at $12 each. It is estimated that it will be able to sell 10 additional copies for each 25¢ reduction in price, and 5 copies less for each 25¢ increase in price. At what price should the bookstore sell the supplements in order to maximize profit?

15. (**Maximum Volume**) A box with a square base is to be constructed. For mailing purposes, the perimeter of the base (the girth of the box), plus the length of the box, cannot exceed 108 in. Find the dimensions of the box of greatest volume.

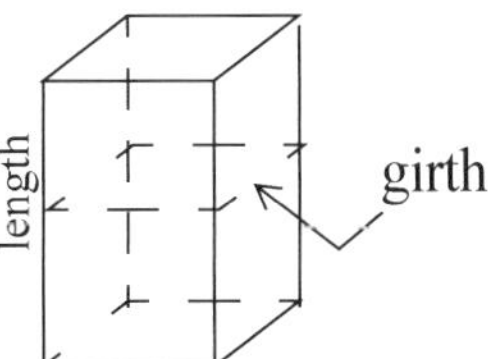

16. (**Maximum Volume**) A shipping crate with base twice as long as it is wide is to be shipped by freighter. The shipping company requires that the sum of the three dimensions of the crate cannot exceed 288 inches. What are the dimensions of the crate of maximum volume?

17. (**Minimum Surface Area**) Find the dimensions of a 1000 cubic foot closed box, with square base, whose surface area is a minimum.

18. (**Minimum Material Cost**) A closed crate with square base is to have a volume of 250 cubic feet. The material for the top and bottom of the crate costs $2 per square foot, and the material for the sides costs $1 per square foot. Find the minimum cost of material.

19. (**Maximum Light Emission**) A Norman window is a window in the shape of a rectangle surmounted by a semicircle. Find the dimensions of the window that admits the most light if the perimeter of the window (total outside length) is 15 feet. (Assume that the same type of glass is used for both parts of the window.)

    Area of a circle: πr^2; circumference of a circle: $2\pi r$

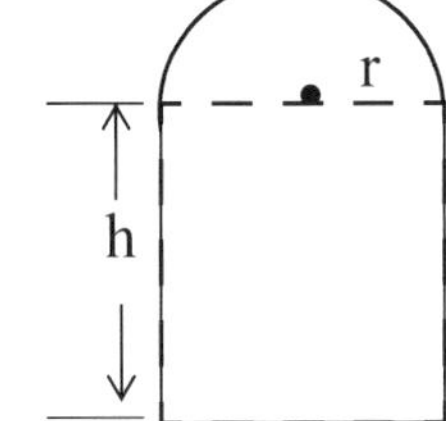

20. (**Maximum Light Emission**) Repeat Exercise 19, if an opaque semi-circular glass is to be used which emits half the light of the rectangular pane of glass.

21. (**Minimum Cost**) A cylindrical drum is to hold 65 cubic feet of chemical waste. Metal for the top of the drum costs $2 per square foot, and $3 per square foot for the bottom. Metal for the side of the drum costs $2.50 per square foot. Find the dimensions for minimal material cost.

USE A GRAPHING CALCULATOR TO APPROXIMATE, TO TWO DECIMAL PLACES, THE SOLUTION OF THE OPTIMIZATION PROBLEMS IN EXERCISES 22 THROUGH 27.

22. (**Maximum Drug Concentration**) The concentration (in milligrams per cubic centimeter) of a particular drug in a patient's bloodstream, t hours after the drug has been administered has been modeled by $C(t) = \frac{0.2t}{0.9t^2 + 5t + 3}$. How many hours after the drug is administered will the concentration be at its maximum? What is the maximum concentration?

23. **(Maximum Profit)** Given the cost and revenue functions:

$$C(x) = 5000 + 100x + \frac{x^2}{\sqrt{x + 500}} \quad \text{and} \quad R(x) = 130x - \sqrt{\frac{x^2}{x + 100}}$$

Determine the number of units, x, which should be manufactured in order to maximize profit.

You will need to use the Pythagorean Theorem for Exercises 24 through 27.

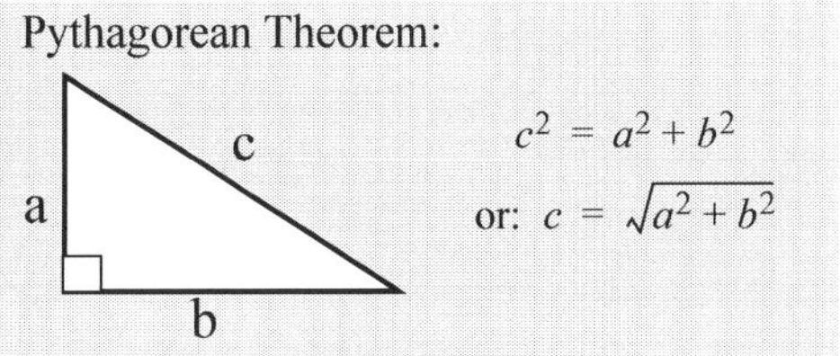

24. (**Minimal Distance Between Two Cars**) At noon, car A is 41 miles due west of car B, and traveling east at a constant speed of 55 miles per hour. Meanwhile, car B is traveling north at 40 miles per hour. At what time will the two cars be closest to each other? Suggestion: minimize the square of the distance, rather than the distance itself.

25. (**Minimal Construction Cost**) A cable is to be run from a power plant, on one side of a river that is 600 feet wide, to a tower on the other side, which is 2000 feet downstream. The cost of laying the cable in the water is \$6 per foot, while the cost on land is \$4 per foot. Find the most economical route for laying the cable.

26. (**Minimum Commuting Time**) A lighthouse lies 2 miles offshore directly across from point A of a straight coastline. The lighthouse keeper lives 5 miles down the coast from point A. What is the minimum time it will take the lighthouse keeper to commute to work, rowing his boat at 3 miles per hour, and walking at 5 miles per hour?

27. (**Minimum Cable Length**) A power line runs north-south. Town A is 3 miles due east from a point a on the power line, and town B is 5 miles due west from a point b on the power line that is 9 miles north of a. A transformer, on the power line, is to accommodate both towns. Where should it be located so as to minimize the combined cable lengths to A and B?

28. **(Minimal Distance Between Three Points)** A straight road runs from North to South. Point A is 5 miles due West of point E on the road. If you walk 10 miles south of A and then go 30 miles due East, you will reach point B. If you walk 10 miles due South of B and then go 15 miles due west, then you will reach point C. Use a graphing utility to determine, to 2 decimal places, the point P on the road whose combined distance from the three points A, B, and C is minimal.

§8. THE INDEFINITE INTEGRAL

A question for you:

$$(?)' = 3x^2$$

A similar question:
$(?)^2 = 49$
Answer 7 and -7.

One answer: x^3, since: $(x^3)' = 3x^2$

Two more answers: $x^3 + 9$, and $x^3 - 173$.

Are there antiderivatives of $f(x) = 3x^2$ that are not of the type $x^3 + C$ for some constant C? No:

A proof of this result lies outside the scope of this text.

THEOREM 2.13 If $f'(x) = g'(x)$ then $f(x) = g(x) + C$ for some constant C.

DEFINITION 2.9 ANTIDERIVATIVE An **antiderivative** of a function f is a function whose derivative is f.

For example, $\frac{x^4}{4}$ and $\frac{x^4}{4} + 1$ are antiderivatives of x^3, since:

$$\left(\frac{x^4}{4}\right)' = x^3, \text{ and } \left(\frac{x^4}{4} + 1\right)' = x^3$$

CHECK YOUR UNDERSTANDING 2.42

Find two different antiderivatives of the function $f(x) = 8x^7$.

Answer: See A-20

The fact that all antiderivatives of a function can be generated by adding an arbitrary constant to any one of its antiderivatives enables us to formulate the following definition:

DEFINITION 2.10 INDEFINITE INTEGRAL The collection of all antiderivatives of f is called the **indefinite integral** of f and is denoted by $\int f(x)dx$. In other words

$$\int f(x)dx = g(x) + C$$

where $g(x)$ is any antiderivative of $f(x)$. The number C in the above notation represents an arbitrary (real) number and is called the **constant of integration**.

The reason for the form $\int f(x)dx$ will surface in the next section.

For example:

Since x^3 is an antiderivative of $3x^2$: $\int 3x^2dx = x^3 + C$:

Since $(x^8)' = 8x^7$: $\int 8x^7dx = x^8 + C$

Answers: (a) $x^5 + C$
(b) $x^{-4} + C$

CHECK YOUR UNDERSTANDING 2.43

Determine:

(a) $\int 5x^4\,dx$ (b) $\int -4x^{-5}dx$

How can you justify the claim that $\sqrt{49} = 7$? Easy: $7^2 = 49$. By the same token:

Here is a special case:
$$\int 1\,dx = x + C$$
(Recall that $1 = x^0$)

THEOREM 2.14 For any number $n \neq -1$:
$$\int x^n dx = \frac{x^{n+1}}{n+1} + C$$

PROOF:

$$\left(\frac{x^{r+1}}{\boldsymbol{r+1}}\right)' = \frac{1}{\boldsymbol{r+1}}(x^{r+1})' = \frac{1}{r+1}\cdot(r+1)x^r = x^r$$

(first step: $[c\cdot f(x)]' = c\cdot f'(x)$; second step: $(x^n)' = nx^{n-1}$)

For example:

$$\int x^9 dx = \frac{x^{10}}{10} + C \qquad \int x^{-5}dx = \frac{x^{-4}}{-4} + C \qquad \int x^{\frac{2}{3}}dx = \frac{x^{\frac{5}{3}}}{\frac{5}{3}} + C$$

(exponent: up one; divided by the "upped one")

Turning around the differentiation theorems:
$$[f(x) \pm g(x)]' = f'(x) \pm g'(x)$$
$$[cf(x)]' = cf'(x)$$
brings us to the following result:.

THEOREM 2.15
$$\int [f(x) \pm g(x)]dx = \int f(x)dx \pm \int g(x)dx$$
$$\int cf(x)dx = c\int f(x)dx$$

For example:

$$\int (5x^3 + x^2 - 2x + 3)dx = 5\int x^3 dx + 1\int x^2 dx - 2\int x dx + 3\int 1 dx$$
$$= 5\cdot\frac{x^4}{4} + \frac{x^3}{3} - 2\cdot\frac{x^2}{2} + 3x + C$$
$$= \frac{5}{4}x^4 + \frac{x^3}{3} - x^2 + 3x + C$$

(Note that the constants associated with the four integrals in the first line of the solution are combined into a single constant C in the second row.)

One generally does not bother to insert the middle step, but simply writes:
$$\int (5x^3 + x^2 - 2x + 3)dx$$
$$\frac{5x^4}{4} + \frac{x^3}{3} - x^2 + 3x + C$$

Answer:
$\frac{x^6}{3}+x^4-\frac{x^3}{9}+2x+C$

CHECK YOUR UNDERSTANDING 2.44

Determine:

$$\int(2x^5+4x^3-\tfrac{1}{3}x^2+2)dx$$

Please note that **only constant factors** can be "extracted" from an integral. In particular, as you can easily verify:

$$\int(4+x^7)dx \neq 4+\int x^7dx \text{ and } \int x(x^2+x)dx \neq x\int(x^2+x)dx$$

Moreover, as it is with derivatives, it is important to remember that:

The integral of a product, (or quotient) is NOT the product (or quotient) of the integrals.

In particular, as you can easily verify:

$$\int[(2x-5)(x+4)]dx \neq \int(2x-5)dx\cdot\int(x+4)dx$$

and: $$\int\frac{2x^5-3x+1}{x^3}dx \neq \frac{\int(2x^5-3x+1)dx}{\int x^3dx}$$

But not all is lost:

EXAMPLE 2.36 Determine:

(a) $\int[(2x-5)(x+4)]dx$ (b) $\int\frac{2x^5-3x+1}{x^3}dx$

SOLUTION: The "trick" is to rewrite the given expression as powers of x, and then apply Theorems 2.14 and 2.15:

(a) $$\int[(2x-5)(x+4)]dx = \int(2x^2+3x-20)dx = \frac{2x^3}{3}+\frac{3x^2}{2}-20x+C$$

(b) $$\int\left(\frac{2x^5-3x+1}{x^3}\right)dx = \int\left(\frac{2x^5}{x^3}-\frac{3x}{x^3}+\frac{1}{x^3}\right)dx = \int(2x^2-3x^{-2}+x^{-3})dx$$

"up one divided by that up one:"

$$= \frac{2x^3}{3}-\frac{3x^{-1}}{-1}+\frac{x^{-2}}{-2}+C = \frac{2x^3}{3}+\frac{3}{x}-\frac{1}{2x^2}+C$$

Answers:
(a) $\frac{3}{4}x^4-\frac{17}{3}x^3+\frac{11}{2}x^2-5x+C$
(b) $x+\frac{1}{x^2}+\frac{2}{x^3}+C$

CHECK YOUR UNDERSTANDING 2.45

Determine:

(a) $\int[(3x^2-2x+1)(x-5)]dx$ (b) $\int\frac{x^4-2x-6}{x^4}dx$

DIFFERENTIAL EQUATIONS

An equation that involves the derivative of an unknown function f is said to be a **differential equation**. To solve such an equation is to find f, and here is the main ingredient of the solution process:

$$\text{If } f'(x) = g(x), \text{ then } \int g(x)dx = f(x) + C$$

Information that will enable you to determine the value of C may be provided. Consider the following example.

EXAMPLE 2.37 Solve the differential equation:

$$f'(x) = 2x^2 + 3x - 1, \text{ if } f(1) = 2$$

SOLUTION: $f(x) = \int (2x^2 + 3x - 1)dx = \frac{2x^3}{3} + \frac{3x^2}{2} - x + C$

To find C, we use the given information that $f(1) = 2$:

If $x = \mathbf{1}, f(x) = \mathbf{2}$:

$$\mathbf{2} = \frac{2 \cdot \mathbf{1}^3}{3} + \frac{3 \cdot \mathbf{1}^2}{2} - \mathbf{1} + C$$

$$C = 2 - \frac{2}{3} - \frac{3}{2} + 1 = \mathbf{\frac{5}{6}}$$

And so we have: $f(x) = \frac{2x^3}{3} + \frac{3x^2}{2} - x + \mathbf{\frac{5}{6}}$

CHECK YOUR UNDERSTANDING 2.46

Solve the differential equation:

$$f'(x) = 5x^4 - 2, \text{ if } f(0) = 1$$

Answer: $f(x) = x^5 - 2x + 1$

MARGINAL ANALYSIS

On page 164 we noted that:

To go **from cost** (or revenue, or profit) **to marginal cost** (or marginal revenue, or marginal profit) you **differentiate**.

Turning this around, we have:

To go **from marginal cost** (or marginal revenue, or marginal profit) **to cost** (or revenue, or profit) you **integrate**.

EXAMPLE 2.38
MONTHLY COST

Suppose that the marginal cost (in dollars) of producing x units is given by $\overline{MC}(x) = \frac{x}{100} + 35$, and that the monthly fixed cost for the company is $17,000. Find the cost of producing 300 units monthly.

SOLUTION: Integrating the marginal cost function, we have:

$$C(x) = \int\left(\frac{x}{100} + 35\right)dx = \frac{x^2}{200} + 35x + C$$

The condition "fixed cost for the company is \$17,000" translates into: $C(0) = 17{,}000$ (dollars). This information enables us to solve for the constant of integration, C:

It costs the company \$17,000 each month to operate its facility (insurance, taxes, etc.)

$$C(x) = \frac{x^2}{200} + 35x + C$$

when $x = \mathbf{0}$, cost $= \mathbf{17000}$:

$$\mathbf{17000} = \frac{\mathbf{0}^2}{200} + 35 \cdot \mathbf{0} + C$$

$$C = 17000$$

We now have the cost function:

$$C(x) = \frac{x^2}{200} + 35x + 17000$$

Evaluating at 300 gives the monthly cost for the production of 300 units:

$$C(300) = \frac{(300)^2}{200} + 35 \cdot 300 + 17000 = \$27{,}950$$

EXAMPLE 2.39
MONTHLY PROFIT

Assume that marginal cost and marginal revenue are given by:

$$\overline{MC}(x) = -\frac{x}{2000} + 40, \text{ and } \overline{MR}(x) = \frac{x}{300} + 75$$

Find the monthly profit from the sale of 500 units, if it costs the company \$15,000 to produce 200 units monthly.

SOLUTION: We begin by finding the cost function:

$$C(x) = \int\left(-\frac{x}{2000} + 40\right)dx = -\frac{x^2}{4000} + 40x + C$$

when $x = \mathbf{200}$, cost $= \mathbf{15000}$:

$$\mathbf{15000} = -\frac{(\mathbf{200})^2}{4000} + 40(\mathbf{200}) + C$$

$$C = 15000 + \frac{(200)^2}{4000} - 40(200) = 7010$$

Monthly cost function (in dollars):

$$C(x) = -\frac{x^2}{4000} + 40x + 7010$$

Turning to the revenue function, we have:

$$R(x) = \int\left(\frac{x}{300} + 75\right)dx = \frac{x^2}{600} + 75x + C$$

$$R(x) = \frac{x^2}{600} + 75x + C$$
$$0 = \frac{0^2}{600} + 75 \cdot 0 + C$$
$$0 = 0 + C$$
$$C = 0$$

At first glance, it appears that we are not given information which will enable us to find the above constant of integration. But we don't have to be told that $R(0) = 0$ (sell nothing, get nothing). Consequently, $C = 0$ (margin), and:

$$R(x) = \frac{x^2}{600} + 75x$$

This brings us to the profit function:

$$P(x) = R(x) - C(x) = \frac{x^2}{600} + 75x - \left(-\frac{x^2}{4000} + 40x + 7010\right)$$
$$= \frac{23x^2}{12000} + 35x - 7010$$

Evaluating the profit function at $x = 500$, we find that:

$$P(500) = \frac{23(500)^2}{12000} + 35(500) - 7010 = \frac{65815}{6} \approx \$10{,}969$$

CHECK YOUR UNDERSTANDING 2.47

Assume that marginal cost and marginal revenue are given by:

$$\overline{MC}(x) = -\frac{x}{500} + 20, \text{ and } \overline{MR}(x) = 45$$

Find the monthly profit from the sale of 600 units, if the monthly fixed cost for the company is \$7,000.

Answer: \$8,360

FREE FALLING OBJECTS

In the metric system: $a(t) = -9.8 \text{ m/sec}^2$

Due to the force of gravity, an object released near the surface of the earth will accelerate at a rate of (approximately) 32 feet per second per second: $a(t) = -32 \text{ ft/sec}^2$ (the negative sign indicates that the object is accelerating in a downward direction).

Based solely on the above (measured) force of gravity, and the force of mathematics, we can now express velocity and position of the object (while in flight) as a function of time:

THEOREM 2.16 If an object is thrown, in a vertical direction, with initial velocity v_0 (in feet per second), from a height of s_0 feet, then t seconds later the velocity (in feet per second) of the object is given by:

$$v(t) = -32t + v_0$$

and the position (in feet) of the object is given by:

$$s(t) = -16t^2 + v_0 t + s_0$$

As previously noted on page 162, a positive velocity indicates an upward movement, while a negative velocity denotes a downward movement.

PROOF: Since acceleration is the derivative of velocity with respect to time (see page 162), velocity is the integral of acceleration:

acceleration due to gravity

$$v(t) = \int a(t)dt = \int(-32)dt = -32t + C$$

When $t = 0$, $v = v_0$. So, $v_0 = -32 \cdot 0 + C$, or $C = v_0$, and this brings us to the velocity equation: $v(t) = -32t + v_0$.

Since velocity is the derivative of position with respect to time (see page 162), position is the integral of velocity:

$$s(t) = \int v(t)dt = \int(-32t + v_0)dt = -16t^2 + v_0 t + C$$

When $t = 0$, $s = s_0$. So, $s_0 = -16 \cdot 0^2 + v_0 \cdot 0 + C$; or: $C = s_0$, and this brings us to the position equation: $s(t) = -16t^2 + v_0 t + s_0$.

EXAMPLE 2.40 A stone is dropped from a height of 1600 feet. What is its speed on impact with the ground?

SOLUTION: Since the stone is dropped, $v_0 = 0$, and the velocity and position functions of Theorem 2.16 take the simple forms:

$$v(t) = -32t$$

and:

$$s(t) = -16t^2 + 1600$$

Since both the velocity and position functions are functions of time, the critical step in most gravity problems, is to find the particular t of interest.

Setting position to zero (ground level), we find the time it takes for the stone to hit the ground:

$$0 = -16t^2 + 1600$$

$$t^2 = 100, \text{ or } t = 10 \text{ (seconds)}$$

Evaluating the velocity function at $t = 10$, we find the impact velocity:

$$v(10) = -32 \cdot 10 = -320 \text{ (feet per second)}$$

By definition, speed is the magnitude of velocity. Thus, the impact speed is 320 feet per second.

CHECK YOUR UNDERSTANDING 2.48

A stone is thrown upward from an 80 foot building at a speed of 64 feet per second.

(a) Find the maximum height the stone reaches.

(b) At what speed will the stone hit the ground?

Answers: (a) 144 feet (b) 96 ft/sec.

EXERCISES

Exercises 1-15. Evaluate:

1. $\int 3dx$
2. $\int 3xdx$
3. $\int (3+3x)dx$
4. $\int 6x^5dx$
5. $\int (6x^5+5x^4)ds$
6. $\int (6x^5+5x^4+x-1)dx$
7. $\int \left(\frac{x^4}{5}-\frac{3}{x^5}\right)dx$
8. $\int (9x^9-x^{-9})dx$
9. $\int \left(3x^4-4x^{-4}+\frac{2}{x^5}\right)dx$
10. $\int x(3x-2)dx$
11. $\int x^2(2x-5)dx$
12. $\int (3x^2-2)(x^3+x)dx$
13. $\int \frac{3x^5+2x-1}{x^4}dx$
14. $\int \frac{(2x^3+1)(x^4+x^2)}{2x}dx$
15. $\int \frac{(x^4+x)(x+1)}{x^4}dx$

Exercises 16-19. Verify:

16. $\int -\frac{10}{(x-3)^2}dx = \frac{10}{x-3}+C$
17. $\int \frac{5}{(2x+5)^2}\,dx = \frac{x}{2x+5}+C$
18. $\int \frac{2x^4+6x^2+10x}{(x^2+1)^2}dx = \frac{2x^3-5}{x^2+1}+C$
19. $\int \frac{x}{(x^2+1)^3}\,dx = \frac{-1}{(2x^2+2)^2}+C$

Exercises 20-25. (Differential Equations) Solve:

20. $f'(x) = 3x+5,\ f(5) = 1$
21. $f'(x) = 3x+5,\ f(1) = 5$
22. $f'(x) = x^3+5x-2,\ f(0) = 1$
23. $f'(x) = x^3+5x-2,\ f(1) = 0$
24. $f'(x) = (2x+3)(x-1),\ f(1) = 0$
25. $f'(x) = \frac{3x^2+5x}{x^5},\ f(1) = 2$

26. **(Cost)** Suppose that the marginal cost for producing x units is given by $\overline{MC}(x) = -\frac{3x}{400}+50$, and that the monthly fixed cost for the company is \$25,000. Find the monthly cost function, and the cost of producing 500 units per month.

27. **(Cost)** Suppose that the marginal cost for producing x units is given by $\overline{MC}(x) = \frac{x}{250}+100$, and that the monthly cost for the production of 100 units is \$20,000. What is the monthly cost for the production of 125 units?

28. **(Revenue)** Suppose that the marginal revenue from the sale of x units is given by $\overline{MR}(x) = -\frac{x}{500}+50$. What revenue will be realized from the sale of 1000 units?

29. **(Revenue)** Suppose that the marginal revenue from the sale of x units is given by $\overline{MR}(x) = -\frac{x}{200} + 75$. What revenue will be realized from the sale of 100 units?

30. **(Profit)** Assume that marginal cost and marginal revenue are given by:
$$\overline{MC}(x) = -\frac{x}{750} + 75, \text{ and } \overline{MR}(x) = \frac{x}{500} + 150$$
Find the monthly profit from the sale of 200 units, if it costs the company $15,000 to produce 150 units monthly.

31. **(Profit)** Assume that marginal cost and marginal revenue are given by:
$$\overline{MC}(x) = -\frac{3x}{200} + 25, \text{ and } \overline{MR}(x) = -\frac{x}{500} + 75$$
Find the monthly profit from the sale of 500 units, if it costs the company $10,000 to produce 300 units monthly.

Exercises 32-34. (Demand Equation) If p denotes the price per unit, then the total revenue from the sale of x units is $R(x) = px$. Expressing p in terms of $R(x)$ and x, we come to what is called the **demand equation**: $p = \frac{R(x)}{x}$ (the average revenue per unit sold). Find the demand equation associated with the given marginal revenue function.

32. $\overline{MR(x)} = 75$

33. $\overline{MR(x)} = 75 - \frac{x}{1000}$

34. The demand equation of a company is $p = 50 - \frac{x}{10}$, and marginal cost is given by $\overline{MC(x)} = 25 - \frac{x}{300}$. Find the monthly profit function, if the monthly fixed cost for the company is $5000.

35. **(From Slope to Function)** The slope of the tangent line to the graph of a function f at $(x, f(x))$ is x^2. Find the function if its graph passes through the point (1,5).

36. **(From Slope to Function)** The slope of the tangent line to the graph of a function f at $(x, f(x))$ is $2x^3 + x - 1$. Find the function, if its graph passes through the point (0,1).

37. **(Impact Speed)** A stone is dropped from a height of 2304 feet. At what speed will it hit the ground?

38. **(Initial Speed)** At what speed should an object be tossed upwards, in order for it to reach a maximum height of 160 feet from the point of its release?

39. **(Bouncing Height)** An object is thrown downward from a 96 foot building at a speed of 16 feet per second. Upon hitting the ground, it bounces back up at three-quarters of its impact speed. How high will it bounce?

40. **(Collision Velocity)** An object is thrown downward from a 264 foot building at a speed of 24 feet per second, at the same time that an object is thrown up from the ground at 64 feet per second. Assuming that the two objects are in line with each other, determine the velocity of both objects when they collide.

41. **(Stopping Distance)** After its brakes are applied, a car decelerates at a constant rate of 30 feet per second per second. Compute the stopping distance, if the car was going 55 miles per hour when the brakes were applied (1 mile = 5280 feet).
Suggestion: Use the fact that the derivative of velocity is acceleration. Also: convert 55 miles per hour to feet per second.

42. **(Stopping Distance)** After its brakes are applied, a car decelerates at a constant rate of 30 feet per second per second. Compute the speed of the car at the point at which the brakes were applied, if the stopping distance turned out to be 120 feet. (Remember that the acceleration is the derivative of velocity.)

43. **(Solution Verification)** Verify that the function $y = -\frac{1}{2}x^2 - \frac{1}{2}x + \frac{1}{4}$ is a solution of the second order differential equation $y'' - y' - 2y = x^2 + 2x - 1$.

§9. The Definite Integral

A geometrical quest for slopes of tangent lines led us to the definition of the derivative. We now go on another geometrical quest, that of finding the area A in Figure 2.19(a) — an area which is bounded above by the graph of the function $y = f(x)$, below by the x-axis, and on the sides by the lines $x = a$ and $x = b$. As it was with the tangent line situation, we know what we are looking for, but still have to find it (**define** it). Here goes:

This "area quest" will lead us to the definition of another immensely useful object — the definite integral.
At first blush, the definite integral will not appear to have any connection whatsoever to the indefinite integral of the previous section. As you will see: there is a link

Loosely speaking, partition the interval $[a, b]$ into subintervals $[x_i, x_{i+1}]$ of length, $\Delta x_i = x_{i+1} - x_i$, as in Figure 2.19(b).

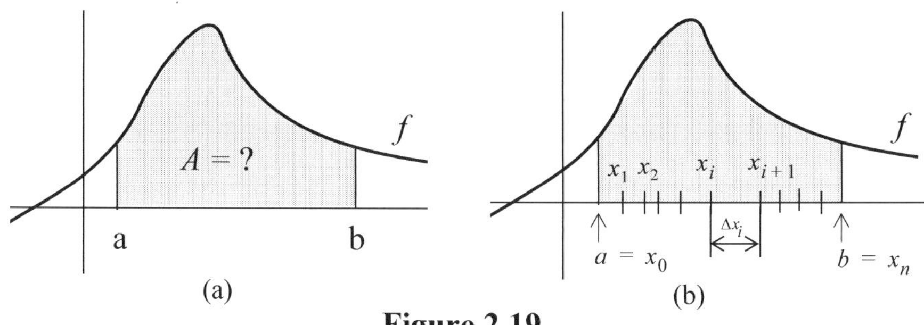

Figure 2.19

Pick an arbitrary point $\bar{x}_i$ in each subinterval $[x_i, x_{i+1}]$, and construct the rectangles of base Δx_i and height $f(\bar{x}_i)$, as is illustrated in Figure 2.20(a). We denote the sum of the areas of all those rectangles by the symbol $\sum_a^b f(x)\Delta x$ — a sum that yields an approximation for the area of the region which we are trying to define.

The Greek letter sigma "$\sum$" denotes "sum."

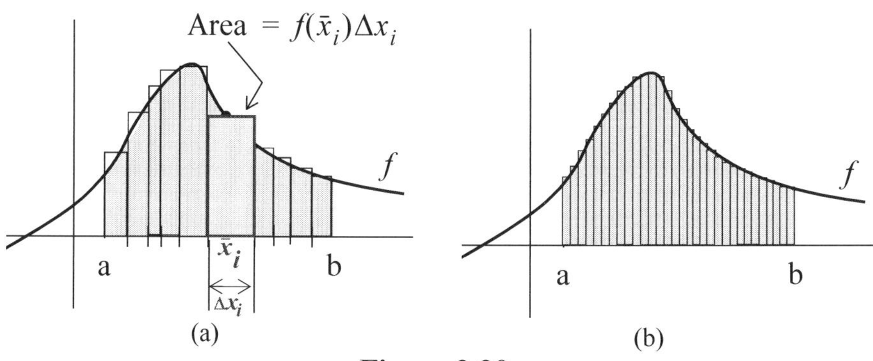

Figure 2.20

Clearly, the smaller we make those Δx_i's, the closer $\sum_a^b f(x)\Delta x$ will get to the area we seek [see Figure 2.20(b)]. And so we (naturally) define the area A to be:

The sums $\sum_a^b f(x)\Delta x$ are called **Riemann sums**, after the German mathematician Georg Riemann (1826-1866).

$$A = \lim_{\Delta x \to 0} \sum_a^b f(x)\Delta x$$

Limits of Riemann sums play many important roles throughout mathematics, bringing us to the following definition:

The symbol $\int_a^b f(x)dx$ is one "word." In particular, "dx" is just a "letter" in that word, that's all. The notation does, however, recall its origin: the sum symbol $\sum_a^b$ "evolving" into $\int_a^b$, and the Δx into "dx."

DEFINITION 2.11

DEFINITE INTEGRAL

A function f is said to be **integrable** over the interval $[a, b]$ if $\lim_{\Delta x \to 0} \sum_a^b f(x)\Delta x$ exists. In this case, we write:

$$\int_a^b f(x)dx = \lim_{\Delta x \to 0} \sum_a^b f(x)\Delta x$$

and call the number $\int_a^b f(x)dx$ the **integral** of f over $[a, b]$.

A "degenerate case:" $\int_a^a f(x)dx = 0$ for any function containing a in its domain.

We would not want this to be any other way, since there is "no area" between a and a.

In the above definition the function need not be positive throughout the interval $[a, b]$. Consequently, as is illustrated in the adjacent figure, $\int_a^b f(x)dx = \lim_{\Delta x \to 0} \sum_a^b f(x)\Delta x$ may end up being zero, or negative, and would not denote an area.

f, a, b: $\int_a^b f(x)dx = 0$ $\quad$ g, a, b: $\int_a^b g(x)dx < 0$

For additional insight on this topic, see Exercises 46-56.

In a way, it is "easier" for a function to be integrable than it is for it to be differentiable. In particular, though a continuous function need not be differentiable (see Figure 2.8, page 144), it can be shown that every continuous function is integrable:

THEOREM 2.17 If f is continuous on $[a, b]$, then it is integrable over that interval.

Both the definition of the derivative and that of the definite integral involve limits. The limit situation for the integral, however, is much more complicated than that of the derivative. For one thing, we have to worry about partitioning the given interval, and then we have to compute the Riemann sum for that partition, and then we have to see if all the Riemann sums approach something as the largest Δx of the partitions tends to zero. This gets way out of hand, even for relatively simple functions like $f(x) = x^2 + x$. Help is on the way; via the Fundamental Theorem of Calculus.

THE FUNDAMENTAL THEOREM OF CALCULUS

The derivative and the definite integral are really quite different: the derivative gives slopes of tangent lines to a curve, while the integral represents the area under a curve (at least for positive functions). At first glance, one would not assume that these two concepts are related to each other; but they are:

THEOREM 2.18
PRINCIPAL THEOREM OF CALCULUS

If f is integrable on $[a, b]$ then the function T given by $T(x) = \int_a^x f(t)dt$ is differentiable on (a, b) and has derivative $T'(x) = f(x)$ for all $x \in (a, b)$.

A formal proof of the above amazing result lies outside the scope of this text. We can, however, attempt to convince you of its validity. Our first order of business is to explain the nature of that strange looking function $T(x) = \int_a^x f(t)dt$. To keep our discussion on a geometric level, we assume that the graph of the function f lies above the t-axis over some interval $[a, b]$ [see Figure 2.21]. Note that the function $T(x)$ simply gives the indicated "**this Area**" over the interval $[a, x]$,.

Note that the horizontal axis is labeled t. We can't call it x, since we chose the variable x for our "main" function T.

We like to call T a *Trombone* function — as you slide the variable x back and forth, you get less or more area from the "integral instrument "

$$T(x) = \int_a^x f(t)dt$$

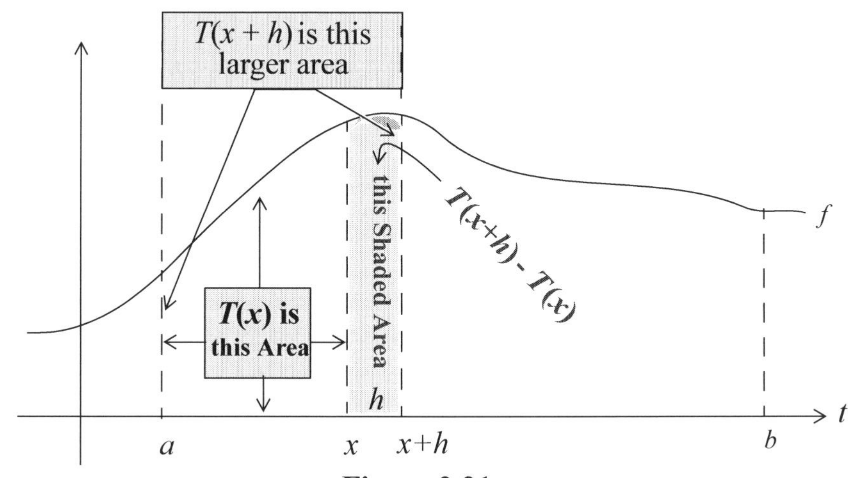

Figure 2.21

From Figure 2.21, we see that:

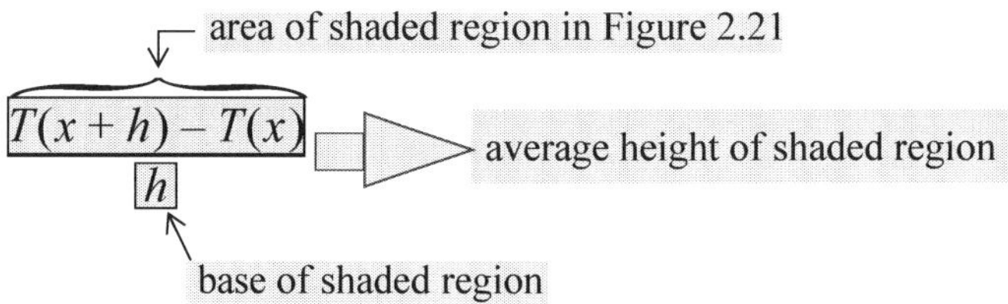

As h approaches 0, the average height of the shaded region must approach the height at x: $f(x)$. Hence, as advertised in Theorem 2.18:

$$T'(x) = \lim_{h \to 0} \frac{T(x+h) - T(x)}{h} = f(x)$$

EXAMPLE 2.41 Find the derivative of the function:

$$T(x) = \int_1^x \frac{(t+5)^5}{t^2+2t} dt$$

SOLUTION: Applying Theorem 2.18, with $f(t) = \frac{(t+5)^5}{t^2+2t}$, we have:

$$T'(x) = f(x) = \frac{(x+5)^5}{x^2+2x}$$

CHECK YOUR UNDERSTANDING 2.49

Find the derivative of the function:

$$T(x) = \int_3^x (3t^2+2)^7 dt$$

Answer: $(3x^2+2)^7$

What is so great about Theorem 2.18? For one thing it will enable us to establish the next theorem which says that:

IF you can find an antiderivative g of a function f, then you can determine the complicated limit $\lim_{\Delta x \to 0} \sum_a^b f(x)\Delta x$ by simply subtracting the number $g(a)$ from the number $g(b)$:

You can now see why similar notation and terminology are used for both the definite and indefinite integral. The connection is this theorem which links the definite integral $\int_a^b f(x)dx$ with one of the functions in the indefinite integral $\int f(x)dx$: the $g(x)$.

THEOREM 2.19 FUNDAMENTAL THEOREM OF CALCULUS If $g'(x) = f(x)$, then:

$$\int_a^b f(x)dx = g(b) - g(a)$$

PROOF: We are told that g is an antiderivative of f, and Theorem 2.18 gives us another. Theorem 2.13, page 192, tells us that these two antiderivatives can differ only by a constant $\boldsymbol{C}$, bringing us to:

$$g(x) = \int_a^x f(t)dt + \boldsymbol{C}$$

Evaluating both sides of the above equation at $x = a$, we have:

$$g(a) = \int_a^a f(t)dt + \boldsymbol{C} = 0 + \boldsymbol{C}, \text{ i.e: } \boldsymbol{C} = \boldsymbol{g(a)}$$

└ there is 0 area between a and a (Definition 2.11)

At this point we know that $g(x) = \int_a^x f(t)dt + \overset{C}{\mathbf{g(a)}}$. Evaluating both sides of this equation at $x = b$ brings us to:

$$g(b) = \int_{\boldsymbol{a}}^{\boldsymbol{b}} \boldsymbol{f(t)dt} + g(a)$$

$$\textbf{or: } \int_{\boldsymbol{a}}^{\boldsymbol{b}} \boldsymbol{f(t)dt} = g(b) - g(a) \quad (*)$$

Since the variable x is no longer in use, we can choose to substitute x for t in (*) to arrive at our desired result:

$$\int_a^b f(x)dx = g(b) - g(a)$$

NOTATION: The difference $g(b) - g(a)$ is represented by the symbol $g(x)\big|_a^b$, leading us to the form:

$$\int_a^b f(x)dx = g(x)\big|_a^b \quad \boxed{= g(b) - g(a)}$$

EXAMPLE 2.42 Evaluate:

$$\int_1^2 (3x^2 + 2)dx$$

SOLUTION: Since $g(x) = x^3 + 2x$ is an antiderivative of $f(x) = 3x^2 + 2$, we have:

$$\int_1^2 (3x^2 + 2)dx = (x^3 + 2x)\big|_1^2 = \overbrace{(\mathbf{2}^3 + 2 \cdot \mathbf{2})}^{g(\mathbf{2})} - \overbrace{(\mathbf{1}^3 + 2 \cdot \mathbf{1})}^{g(\mathbf{1})} = 9$$

GRAPHING CALCULATOR GLIMPSE 2.5

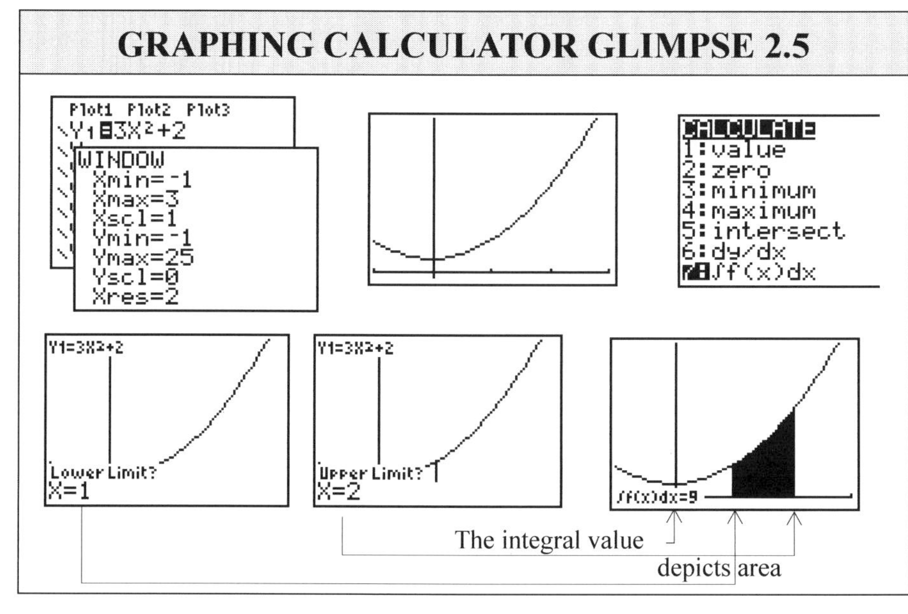

CHECK YOUR UNDERSTANDING 2.50

Referring to Example 2.42, see what happens if you use $x^3 + 2x + 100$, instead of $x^3 + 2x$, as the chosen antiderivative of $3x^2 + 2$.

Answer: Same result.

EXAMPLE 2.43 Determine the area of the region over the interval $[1, 2]$ that is bounded above by the graph of the function:

$$f(x) = \frac{x^3 + x^2 + 1}{x^2}$$

SOLUTION: Since the function f is positive over the indicated interval, the area in question is given by the integral:

Please remember that the integral of a quotient is **not** the quotient of the integrals.

$$\int_1^2 \frac{x^3 + x^2 + 1}{x^2}\,dx$$

Which we now evaluate:

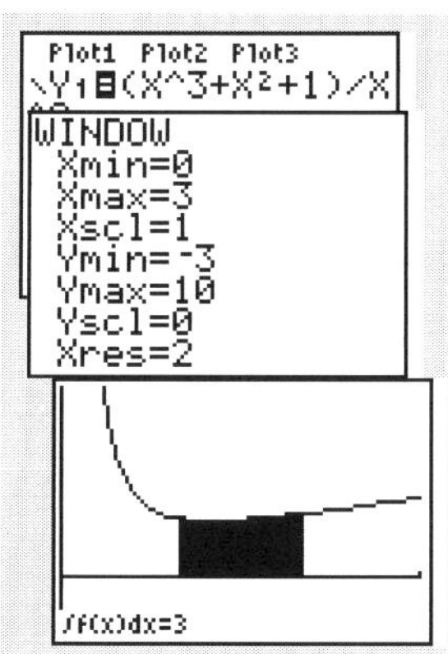

$$\int_1^2 \frac{x^3 + x^2 + 1}{x^2}\,dx = \int_1^2 \left(\frac{x^3}{x^2} + \frac{x^2}{x^2} + \frac{1}{x^2}\right)dx = \int_1^2 (x + 1 + x^{-2})\,dx$$

$$= \left(\frac{x^2}{2} + x + \frac{x^{-1}}{-1}\right)\Bigg|_1^2 = \left(\frac{x^2}{2} + x - \frac{1}{x}\right)\Bigg|_1^2$$

$$= \left(\frac{2^2}{2} + 2 - \frac{1}{2}\right) - \left(\frac{1}{2} + 1 - \frac{1}{1}\right) = 3 \text{ square units}$$

CHECK YOUR UNDERSTANDING 2.51

Determine the area of the region over the interval $[-1, 1]$ that is bounded above by the graph of the function:

$$f(x) = (x^2 + 1)(x^2 + 3)$$

Answer: $\frac{136}{15}$ square units

EXAMPLE 2.44 Evaluate:

$$\int_0^2 (x^3 - 4x^2 + 3x)dx$$

SOLUTION:

$$\int_0^2 (x^3 - 4x^2 + 3x)dx = \left(\frac{x^4}{4} - \frac{4x^3}{3} + \frac{3x^2}{2}\right)\Bigg|_0^2$$

$$= \left(\frac{2^4}{4} - \frac{4 \cdot 2^3}{3} + \frac{3 \cdot 2^2}{2}\right) - (0) = -\frac{2}{3}$$

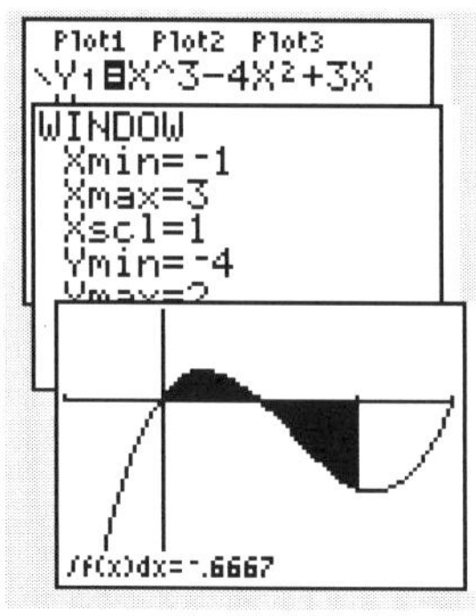

Of course, being negative, the above integral **cannot represent any area whatsoever**. Note that the function

$$f(x) = x^3 - 4x^2 + 3x = x(x-1)(x-3)$$

takes on negative values in the interval $[1, 2]$ (see margin).] As such, some of the terms in the Riemann sum, $\sum_a^b f(x)\Delta x$ will be negative.

CHECK YOUR UNDERSTANDING 2.52

Evaluate:

(a) $\int_{-2}^{1}(3x^2+2)dx$ (b) $\int_0^1 (x^3+x-1)dx$

Answers: (a) 15 (b) $-\frac{1}{4}$

NET-CHANGE DERIVED FROM RATE OF CHANGE

Claim: If f is differentiable on $[a, b]$, then $\int_a^b f'(x)dx = f(b) - f(a)$.

Why? Because $f(x)$ is an antiderivative of $f'(x)$, that's why.

Calling the difference $f(b) - f(a)$ the **net-change** of the function f over the interval $[a, b]$, we have just observed that:

THEOREM 2.20 The net-change of a differentiable function f from $x = a$ to $x = b$ is given by:

$$\text{Net- change} = \int_a^b f'(x)dx$$

EXAMPLE 2.45 Oil is leaking out of a ruptured tanker at a rate of $125 - \frac{t}{50}$ gallons per minute, where t is in minutes. How many gallons leak out during:

(a) the first hour? (b) the second hour?

Units can help point the way. We are given a rate in gallons per minute; and want to end up with total gallons over a specified period of time:

$$\text{gal} = \frac{\textbf{gal}}{\textbf{min}} \cdot \text{min}$$

$$\int_0^{60}\left(125 - \frac{t}{50}\right)dt$$

↑"add up those gallons"

SOLUTION: (a) Quantity of oil that leaks out in the first hour:

$$\int_0^{60}\left(125 - \frac{t}{50}\right)dt = 125t - \frac{t^2}{100}\Big|_0^{60} = 125(60) - \frac{60^2}{100} = 7464 \text{ gallons}$$

(b) Quantity of oil that leaks out in the second hour:

$$\int_{60}^{120}\left(125 - \frac{t}{50}\right)dt = 125t - \frac{t^2}{100}\Big|_{60}^{120}$$

$$= \left(125(120) - \frac{120^2}{100}\right) - \left(125(60) - \frac{60^2}{100}\right) = 7392 \text{ gallons}$$

Income accruing continuously is called a **continuous income stream**. The rate of change of a continuous income stream is then said to be its **rate of flow**. Using this terminology, Theorem 2.20 takes the following form:

THEOREM 2.21 Let $I(t)$ be an income stream, with rate of flow $I'(t)$. The income accrued during the time period from $t = a$ to $t = b$ is then given by:

$$\text{Income} = \int_a^b I'(t)dt$$

EXAMPLE 2.46 A printing company is considering purchasing a new hole-punching machine for \$2,000. It estimates that with the purchase of the machine, monthly income will increase at a rate of $190 + 2t$ dollars per month (t in months). How many months will it take for the machine to pay for itself?

SOLUTION:

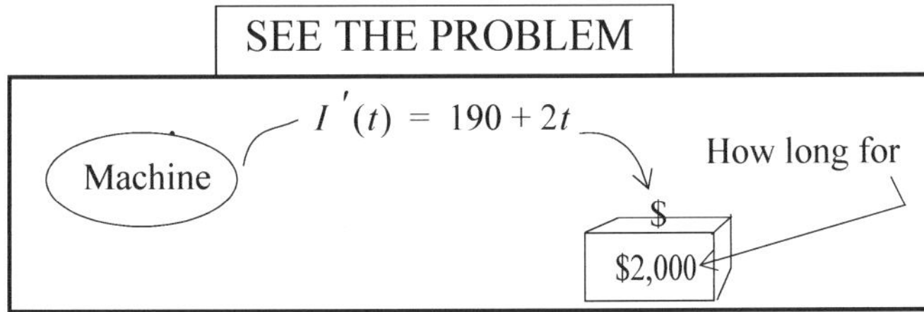

First find the increase of income in T months:

$$\text{Income increase} = \int_0^T (190 + 2t)dt = 190t + t^2\Big|_0^T = 190T + T^2$$

Then set that income to 2000, and solve for T:

$$\begin{aligned} 190T + T^2 &= 2000 \\ T^2 + 190T - 2000 &= 0 \\ (T - 10)(T + 200) &= 0 \\ T = 10 \text{ or } T &= -200 \end{aligned}$$

Ignoring the negative time period we conclude that the machine will pay for itself in 10 months.

CHECK YOUR UNDERSTANDING 2.53

The rate of production, in barrels per day, of an oil well is anticipated to be $75 - \frac{t}{2500}$ (t in days). Find the total income produced by the well in its first 30 days of operation, if crude sells at \$85 per barrel.

Answer: \$191,250

A Touch of Theory

In the definition of $\int_a^b f(x)dx$, the lower limit of integration, a, was less than its upper limit of integration, b. Is there a reasonable way of defining an integral such as $\int_4^2 (2x+1)dx$? Yes, for if we formally apply the Fundamental Theorem of Calculus to that expression, we obtain:

$$\int_4^2 (2x+1)dx = x^2 + x\Big|_4^2 = (2^2+2)-(4^2+4) = 6-20 = -14$$

On the other hand:

$$\int_2^4 (2x+1)dx = x^2 + x\Big|_2^4 = (4^2+4)-(2^2+2) = 20-6 = 14$$

The above observations leads us to:

Switching the limits of integration introduces a minus sign.

DEFINITION 2.12 For f integrable on $[a, b]$:

$$\int_b^a f(x)dx = -\int_a^b f(x)dx$$

As was the case with indefinite integrals, we have:

THEOREM 2.22 For f and g integrable over $[a, b]$, and any constant c:

(a) $\int_a^b [f(x)+g(x)]dx = \int_a^b f(x)dx + \int_a^b g(x)dx$

(b) $\int_a^b [f(x)-g(x)]dx = \int_a^b f(x)dx - \int_a^b g(x)dx$

(c) $\int_a^b cf(x)\,dx = c\int_a^b f(x)$

THEOREM 2.23 If f is integrable over $[a, b]$, and $a < c < b$, then f is integrable over $[a, c]$ and $[c, b]$, and:

$$\int_a^b f(x)dx = \int_a^c f(x)dx + \int_c^b f(x)dx$$

"Area-proof":

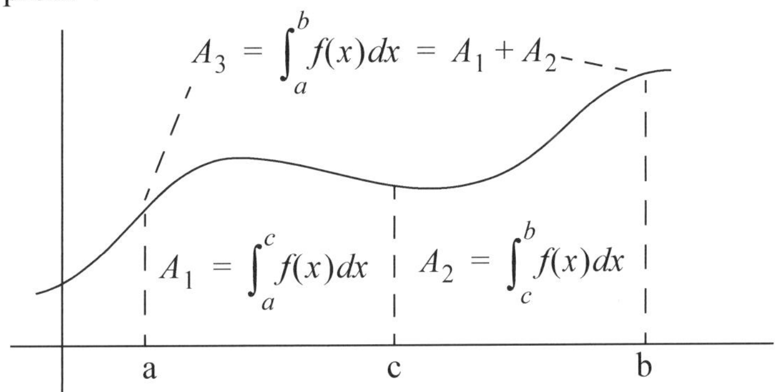

CHECK YOUR UNDERSTANDING 2.54

Let $\int_a^c f(x)dx = 5$, $\int_c^b f(x)dx = -3$ and $\int_a^c g(x)dx = 7$.

Evaluate:

(a) $\int_a^c -2f(x)dx + \int_c^a g(x)dx$ (b) $\int_a^b f(x)dx + \int_a^c 2g(x)dx$

Answers: (a) –17 (b) 16

EXERCISES

Exercises 1-18. Evaluate:

1. $\int_0^1 3dx$
2. $\int_1^2 3x\,dx$
3. $\int_{-1}^1 (3+3x)dx$
4. $\int_0^1 (x-5)dx$
5. $\int_1^2 (2x-5)dx$
6. $\int_{-1}^1 (x^2-5x)dx$
7. $\int_0^1 (x^2+3x-1)dx$
8. $\int_1^2 (x^2+3x-1)dx$
9. $\int_{-1}^1 x^3dx$
10. $\int_{-1}^1 x^4\,dx$
11. $\int_{-1}^2 3x^2dt$
12. $\int_{-1}^1 (x^5-2x+1)dx$
13. $\int_1^2 x(3x-2)dx$
14. $\int_{-1}^0 (3x-1)(x-1)dx$
15. $\int_0^{-1} (3x-1)(x-1)dx$
16. $\int_1^2 \frac{x^3-2}{x^2}\,dx$
17. $\int_1^2 \frac{(x^4+x)(x+1)}{x^4}dx$
18. $\int_1^2 (x+1)^2\,dx$

Exercises 19-22. **(Area)** Determine the area of the region bounded by the graph of the given function over the specified interval.

19. $f(x) = x^3, 0 \le x \le 1$
20. $f(x) = x^2, -1 \le x \le 1$
21. $f(x) = -x^3+x^2, -1 \le x \le 1$
22. $f(x) = \frac{1}{x^2},\ 1 \le x \le 4$

Exercises 23-25. **(Principal Theorem of Calculus)** Use the Principal Theorem of Calculus, Theorem 2.18, to find the derivative of the given function T.

23. $T(x) = \int_1^x \frac{1}{t^2+4}dt$
24. $T(x) = \int_5^x \sqrt{3t^4+1}\,dt$
25. $T(x) = \int_x^5 \sqrt{3t^4+1}\,dt$

Exercises 26-27. **(Principal and Fundamental Theorem of Calculus)**

(a) Use the Principal Theorem of Calculus, Theorem 2.18, to find the derivative of the given function $T(x)$.

(b) Use the Fundamental Theorem of Calculus, Theorem 2.19, to first express $T(x)$ in a form that does not involve an integral, and then differentiate that explicit function of F directly. Compare your answer with that of (a).

26. $T(x) = \int_1^x t^2dt$
27. $T(x) = \int_1^x (t^2+t)dt$

28. **(Tree Growth)** A transplanted tree grows at a rate of $g(t) = 1+\frac{t}{5}$ feet per year, t years after the transplant. How much does the tree grow:

(a) in the first two years? (b) during the third year?

29. **(Cost Increase)** In July, the price of gas increased at the rate of $0.06t + 0.001t^2$ cents per gallon, where t denotes the number of days from June 30. How much did the cost of a gallon increase through the month of July?

30. **(Depreciation)** The resale value of a car **decreases** at the rate of $1200 + 600t + 4t^3$ dollars per year, where $0 \le t \le 7$ denotes the number of years following the car's year of manufacture. How much did the car's value depreciate:

    (a) in the first three years? (d) during the third year?

31. **(Melting Ice)** A 360 cubic inch block of ice is melting at the rate of $\frac{t}{5}$ cubic inches per minute. How many minutes will it take for the block to totally melt?

32. **(Puppet Revenue)** A puppet manufacturer expects that n months from now she will be selling $S(n) = 50 + 2n$ puppets per month, at $p(n) = 25 + \frac{n}{2}$ dollars per puppet. What is her expected revenue for the year?

33. **(Advertising)** A store is launching an aggressive advertising campaign, and anticipates that the number of daily customers, N, will grow from its current value of 200, at a rate of $N'(t) = \frac{t}{100}$, where t is the number of days from the beginning of the campaign. How many days from the beginning of the campaign will it take before the number of daily customers doubles?

34. **(Declining Sales)** Because of fierce competition, the weekly sales at an appliance store are expected to decline at the rate of $S'(t) = -\frac{t^2}{100}$ units per week, where t is time in weeks. The store plans to go out of business when weekly sales drop below 500. Currently, the shop sells 900 units weekly. How many more weeks will the company remain in business?

35. **(Depreciation)** The resale value of a certain industrial machine decreases over a 10-year period at a rate that depends on the age of the machine. When the machine is x years old, the rate at which its value is dropping is $250(15 - x)$ dollars per year. If the machine was originally worth $28,000, how much will it be worth when it is 10 years old?

36. **(National Debt)** Assuming that the national debt is increasing at a rate of 1% each year. How many years will it take for the current debt to double?

37. **(Income Stream)** A printing company can rent a hole-punching machine for $1000/year that will increase monthly earnings at a rate of $190 + 2t$ dollars per month, or a $2,000/year machine that will increase monthly earnings at a rate of $200 + 4t$ dollars per month (t in months). Which should be rented, given that the company anticipates using the machine for exactly two years?

38. **(Net-Change in Cost)** Suppose that the marginal cost for producing x units is given by $\overline{MC}(x) = -\frac{3x}{400} + 50$. Determine the net-change in cost from a production of 100 units to a production of 200 units.

39. **(Net-Change in Revenue)** Suppose that the marginal revenue from the sale of x units is given by $\overline{MR}(x) = -\frac{x}{500} + 50$. Determine the net-change in revenue from the sale of 100 through 200 units.

40. **(Net-Change in Profit)** Assume that marginal cost and marginal revenue are given by:
$$\overline{MC}(x) = -\frac{3x}{200} + 25, \text{ and } \overline{MR}(x) = -\frac{x}{500} + 75$$
Determine the net-change in profit from the sale of 100 through 200 units.

41. **(Displacement vs. Distance)** A particle is moving along an East-West line with velocity $v(t) = t^2 - 3t + 2$ (in meters per second). Assume that a positive velocity indicates that the particle is moving in an easterly direction, and a negative velocity indicates a westerly direction. Assume also that a negative position indicates the particle is west of its starting point, while a positive position indicates the particle is east of its starting point.
   (a) Determine the net change in the particle's position during the first 4 seconds.
   (b) Determine the distance traveled during the first 4 seconds.

42. **(Theory)** Use the definition of the definite integral, Definition 2.11, to explain why $\int_0^2 f(x)dx = \int_0^2 g(x)dx = 2$ for each of the two functions depicted below.

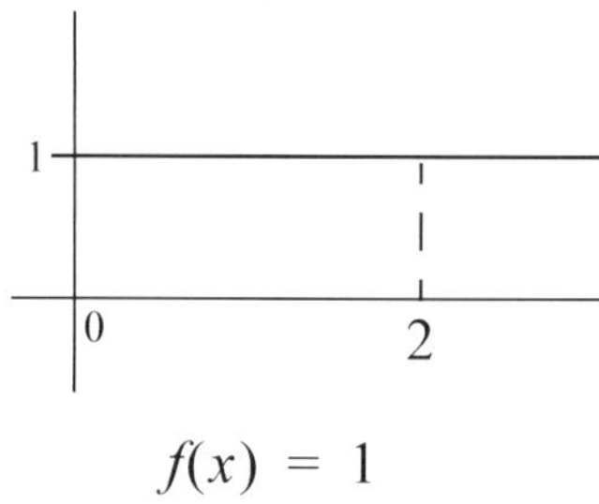

$f(x) = 1$

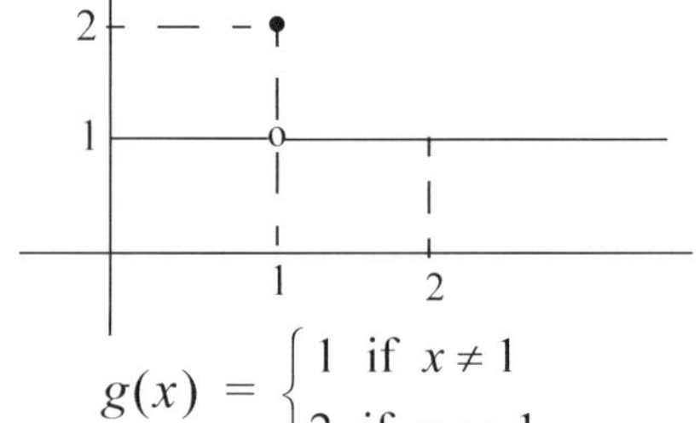

$$g(x) = \begin{cases} 1 & \text{if } x \neq 1 \\ 2 & \text{if } x = 1 \end{cases}$$

43. **(Theory)** Referring to Definition 2.11, explain why the function
$$f(x) = \begin{cases} 1 & \text{if } x \text{ is a rational number} \\ -1 & \text{if } x \text{ is not a rational number} \end{cases}$$

is **not** integrable over the interval $[0, 1]$ (or any other interval, for that matter).
Note: You need to know that any interval, no matter how small, contains both rational numbers (fractions: ratios of two integers), and irrational numbers (numbers that are not rational).

AREA BETWEEN CURVES

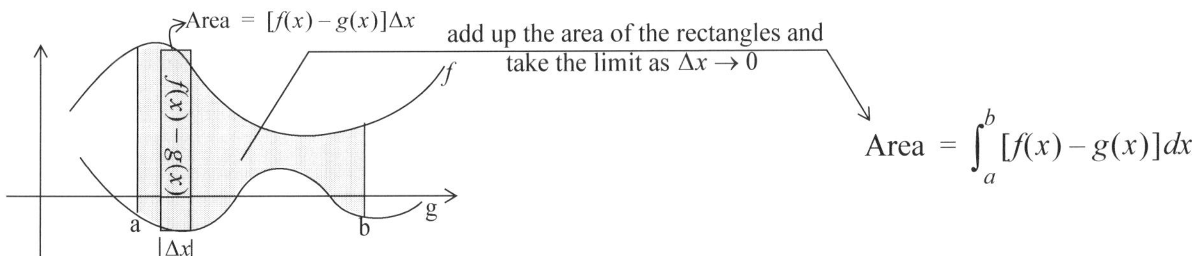

For example: The area bounded by the graphs of the functions $f(x) = x^3$ and $g(x) = x^2$ is given by: $A = \int_0^1 (x^2 - x^3)dx = \left(\frac{x^3}{3} - \frac{x^4}{4}\right)\Big|_0^1 = \frac{1}{3} - \frac{1}{4} = \frac{1}{12}$ (dominant: x^2; subordinate: x^3)

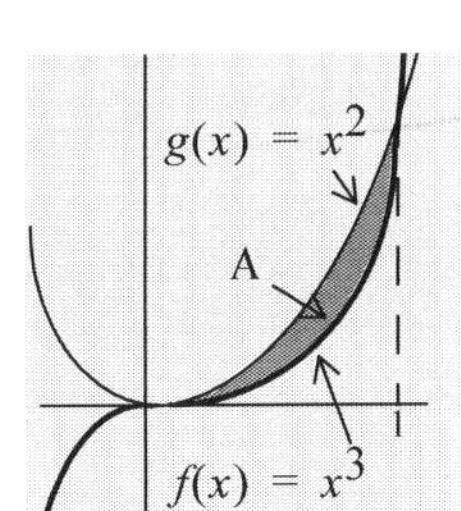

Exercises 44-49. Sketch the graphs of the given functions and lines, and determine the area of the finite region lying between them.

44. $f(x) = x^2, \ y = x$
45. $f(x) = x^2 - 2, \ y = x$
46. $f(x) = x^2 - x, \ y = x$
47. $f(x) = x^2 + x, \ g(x) = -x^2 + 1$
48. $f(x) = x^3, \ y = x$
49. $f(x) = x^3 + 1, \ g(x) = x^3 + x^2$

Since the graphs of f and g switch dominance about the point $x = b$, two integrals are needed to calculate the area of the region:

$$A = \int_a^b [f(x) - g(x)]dx + \int_b^c [g(x) - f(x)]dx$$

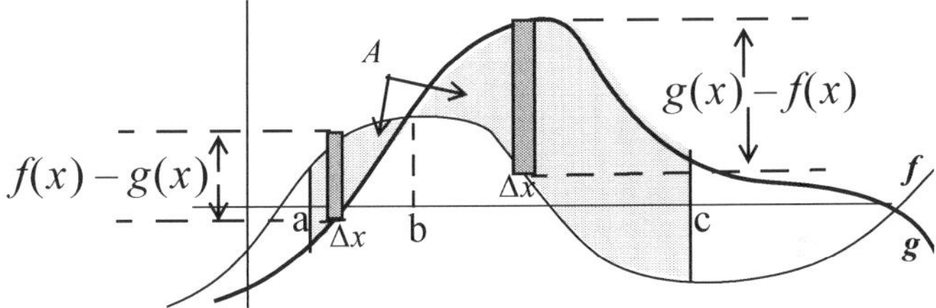

For example: The area bounded by the by the graphs of the functions $f(x) = x^3 - 3x + 1$ and $g(x) = x + 1$, between $x = -2$ and $x = 1$ is given by:

$$A = \int_{-2}^{0} [(x^3 - 3x + 1) - (x + 1)]dx + \int_0^1 [(x + 1) - (x^3 - 3x + 1)]dx$$

Exercises 50-54. Find the area of the finite region bounded by the given functions and lines.

50. $y = x, \ \ y = -x + 1, \ \ x = 0, x = 2$

51. $y = x, y = -x + 2, y = -\frac{x}{2}$

52. $f(x) = x^3, g(x) = -x^2 + 1, x = 0, x = 2$

53. $f(x) = x^3 - 1, y = -x - 1, y = 7$

54. $f(x) = x^3, g(x) = -x^2 + 1, x = -2, x = 2$

PART 2 SUMMARY

EXPONENTS

DEFINITION For any positive integer n and any number a:

$$a^n = \underbrace{a \cdot a \cdot \ldots \cdot a}_{n\text{-times}}$$

In addition, if $a \neq 0$: $a^0 = 1$ and $a^{-n} = \frac{1}{a^n}$

EXPONENT THEOREMS

(i) $a^m a^n = a^{m+n}$ — When multiplying add the exponents

(ii) $\frac{a^m}{a^n} = a^{m-n}$ — When dividing subtract the exponents

(iii) $(a^m)^n = (a^n)^m = a^{mn}$ — A power of a power: multiply the exponents

(iv) $(ab)^n = a^n b^n$ — A power of a product equals the product of the powers

(v) $\left(\frac{a}{b}\right)^n = \frac{a^n}{b^n}$ — A power of a quotient equals the quotient of the powers

LINES

SLOPE For any nonvertical line L and any two distinct points (x_1,y_1) and (x_2,y_2) on L, we define the **slope** of L to be the number m given by:

$$m = \frac{y_2 - y_1}{x_2 - x_1} \quad \left[\frac{\text{change in } y}{\text{change in } x}: \frac{\Delta y}{\Delta x}\right]$$

SLOPE-INTERCEPT EQUATION OF A LINE A point (x, y) is on the line of slope m and y-intercept b if and only if:

$$y = mx + b$$

LINEAR EQUATIONS AND INEQUALITIES

EQUATIONS Adding (or subtracting) the same quantity to (or from) both sides of an equation, or multiplying (or dividing) both sides of an equation by the same nonzero quantity will not alter the solution set of the equation.

INEQUALITIES When multiplying or dividing both sides of an inequality by a **negative** quantity, one must **reverse** the direction of the inequality sign.

Polynomial Inequalities

Step 1. Indicate the zeros of the polynomial on the number line. Place the letter ***c*** above the zeros of factors which are raised to an **o**dd power (the odd zeros) and the letter ***n*** above the zeros of factors raised to an **e**ven power (the even zeros).

Step 2. Determine the sign of the polynomial to the right of its last zero.

Step 3. Take the sign to the right of the last zero and "walk it to the left," changing the sign each time you cross over an odd zero and not changing it if you cross over an even zero.

Step 4. Read off the solution set of the inequality from the above sign information.

Limit

We write $\lim_{x \to c} f(x) = L$ to mean that the function values, $f(x)$, approach the number L, as the values of x **approach** the number c (from either side).

Limit Theorems

If $\lim_{x \to c} f(x)$ and $\lim_{x \to c} g(x)$ exist, then:

(i) $\lim_{x \to c} [f(x) \pm g(x)] = \lim_{x \to c} f(x) \pm \lim_{x \to c} g(x)$

(ii) $\lim_{x \to c} [f(x)g(x)] = \lim_{x \to c} f(x) \cdot \lim_{x \to c} g(x)$

(iii) $\lim_{x \to c} \frac{f(x)}{g(x)} = \frac{\lim_{x \to c} f(x)}{\lim_{x \to c} g(x)}$ if $\lim_{x \to c} g(x) \neq 0$

(iv) $\lim_{x \to c} [af(x)] = a \lim_{x \to c} [f(x)]$

Derivative

Derivative at a Point

The **derivative of a function *f* at *c*** is the number $f'(c)$ given by:

$$f'(c) = \lim_{h \to 0} \frac{f(c+h) - f(c)}{h}$$

Geometrical Interpretation

Slope of tangent line to the graph of f at the point $(c, f(c))$.

Derivative Function

The **derivative of a function *f*** is the function $f'(x)$ given by:

$$f'(x) = \lim_{h \to 0} \frac{f(x+h) - f(x)}{h}$$

(Providing that the above limits exist)

Derivative Formulas

$(x^n)' = nx^{n-1}$ $\qquad [f(x) \pm g(x)]' = f'(x) \pm g'(x)$

$[cf(x)]' = cf'(x)$ $\qquad [f(x)g(x)]' = f(x)g'(x) + g(x)f'(x)$

$$\left[\frac{f(x)}{g(x)}\right]' = \frac{g(x)f'(x) - f(x)g'(x)}{[g(x)]^2}$$

MAXIMA AND MINIMA

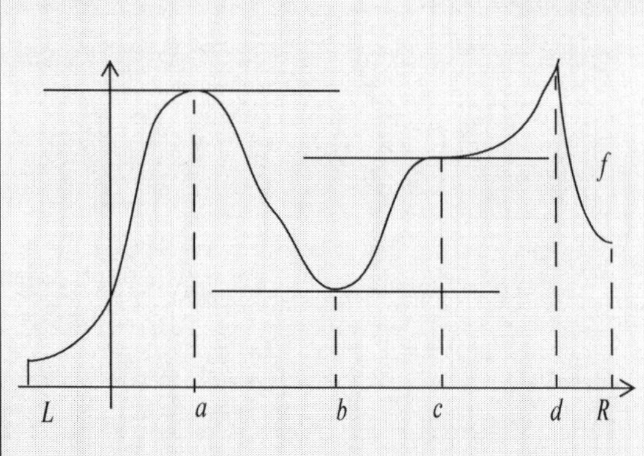

Where a (local) maximum or minimum occurs, if the function is differentiable, the derivative must be zero: see a and b.

The derivative of a function may be zero, without either a maximum or minimum occurring at that point: see c.

A maximum or minimum can occur without the derivative being zero at that point, but only because the derivative does not exist there: see d.

An endpoint maximum or minimum occurs at the end points.

FIRST DERIVATIVE TEST Let c be a critical point for f. If the derivative is positive immediately to the left of c, and negative immediately to the right of c, then the function has a maximum at c. If the derivative is negative immediately to the left of c, and positive immediately to the right of c, then the function has a minimum at c.

CONCAVITY AND POINTS OF INFLECTION

$$f''(x) > 0 : \text{Concave up}$$

$$f''(x) < 0 : \text{Concave down}$$

A point $(c, f(c))$ about which concavity changes (from up to down, or vice versa) is said to be a **point of inflection.**

OPTIMIZATION PROCEDURE

Step 1. See the problem.

Step 2. Express the quantity to be optimized in terms of any convenient number of variables.

Step 3. If necessary, use the given information to eliminate all but one variable in the expression of Step 2, to arrive at a function of one variable.

Step 4. Differentiate, set equal to zero, and solve (find the critical points). If necessary, analyze the nature of the critical point.

INDEFINITE INTEGRAL

ANTIDERIVATIVE An **antiderivative** of a function f is a function whose derivative is f.

INTEGRAL The indefinite **integral** of a function f, denoted by $\int f(x)dx$, is the collection of **all** antiderivatives of f:

$$\int f(x)dx = g(x) + C$$

where g is any antiderivative of f, and C, called the **constant of integration**, represents an arbitrary real number.

FORMULAS

$$\text{For any } n \neq -1 : \int x^n dx = \frac{x^{n+1}}{n+1} + C$$

$$\int [f(x) \pm g(x)]dx = \int f(x)dx \pm \int g(x)dx \qquad \int af(x)dx = a\int f(x)dx$$

DEFINITE INTEGRAL	
LIMIT OF RIEMANN SUMS	$\int_a^b f(x)dx = \lim_{\Delta x \to 0} \sum_a^b f(x)\Delta x$
PRINCIPAL THEOREM OF CALCULUS	If $F(x) = \int_a^x f(t)dt$, then $F'(x) = f(x)$.
FUNDAMENTAL THEOREM OF CALCULUS	If $g'(x) = f(x)$, then $\int_a^b f(x)dx = g(b) - g(a)$
NET-CHANGE FROM RATE OF CHANGE	The net-change of a differentiable function f from $x = a$ to $x = b$ is given by: $\text{Net-change} = \int_a^b f'(x)dx$

Review Exercises
Brief

1. Find the indicated limit.

   (a) $\lim_{x \to 0} \frac{3x^2 - x}{x + 7}$ (b) $\lim_{x \to 2} \frac{x^2 - 4}{2x^2 - x - 6}$

2. Use the definition of the derivative $f'(x) = \lim_{h \to 0} \frac{f(x+h) - f(x)}{h}$ to find the derivative of $f(x) = 2x^2 - 3x + 1$.

3. Find the equation of the tangent line to the graph of $f(x) = x^3 - 2x^2 + x - 1$ at $x = 2$.

4. Differentiate:

   (a) $f(x) = 5x^3 - 3x^2 + 4x - 2$ (b) $f(x) = 2x(x - x^2)$ (c) $f(x) = 2x^{-3} - \frac{3}{x} + \frac{1}{2}$

   (d) $f(x) = \frac{x^4 + 3x^2 + 1}{4x}$ (e) $f(x) = \frac{x^2 + 2}{x - 1}$ (need quotient rule)

5. Suppose that the total monthly cost (in dollars) for manufacturing x units is given by $C(x) = 12,000 + 50x - \frac{x^2}{100}$ and that the revenue realized from the sale of x units is given by $R(x) = 60x$. Determine the marginal profit at $x = 300$ and indicate what that value predicts.

6. Determine the values of x where the graph of the function $g(x) = x^3 - 3x^2$ has a horizontal tangent line.

7. Sketch the graph of the function $f(x) = x^3 - 6x^2 + 9x + 1$, labeling its maximum and minimum points, and points of inflection.

8. A rectangular field is to be enclosed with a fence. One side of the field is against an existing fence, so that no fence is needed on that side. The material for the fence costs \$2 per foot for the two ends and \$4 per foot for the side parallel to the existing fence. Find the dimensions of the field that will minimize fencing cost, if the field is to enclose an area of 3600 square feet.

9. A company charges \$200 for each box of tools on orders of 150 or fewer boxes. The cost per box is reduced by \$1 for each box ordered over 150. For what size order is revenue maximum?

10. Evaluate:

    (a) $\int (2x^3 + x^2 - 3x + 1)\,dx$ (b) $\int \frac{x^6 - 5x}{4x^3}\,dx$ (c) $\int_1^2 x(x + 4x^2)\,dx$

11. The marginal cost and marginal revenue (in dollars) are given by $\overline{MC}(x) = 20 - \frac{x}{50}$ and $\overline{MR}(x) = 30 - \frac{x}{25}$. Find the profit function if it costs the company \$3,000 to produce 100 units.

12. Find the area of the region bounded by the graph of the function $f(x) = 2x^2 + 3x + 1$ over the interval $[1, 2]$ and below by the x-axis, for $1 \leq x \leq 2$.

13. The current average price of grapes is \$2.55 per pound. Due to a water shortage, it is estimated that x weeks from now the price will be increasing at a rate of $20x$ cents per week. How much will grapes cost three weeks from now?

ANSWERS

1. (a) 0 (b) $\frac{4}{7}$ 2. $\lim_{h \to 0} \frac{[2(x+h)^2 - 3(x+h) + 1] - (2x^2 - 3x + 1)}{h} = \ldots = 4x - 3$ 3. $y = 5x - 9$

4. (a) $15x^2 - 6x + 4$ (b) $4x - 6x^2$ (c) $-\frac{6}{x^4} + \frac{3}{x^2}$ (d) $\frac{3}{4}x^2 + \frac{3}{4} - \frac{1}{4x^2}$ (e) $\frac{x^2 - 2x - 2}{(x-1)^2}$

5. $\overline{MP}(300) = \$16$. Prediction: a profit of approximately \$16 will be realized from the sale of the 301th unit.

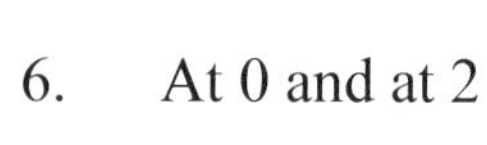
6. At 0 and at 2

7. 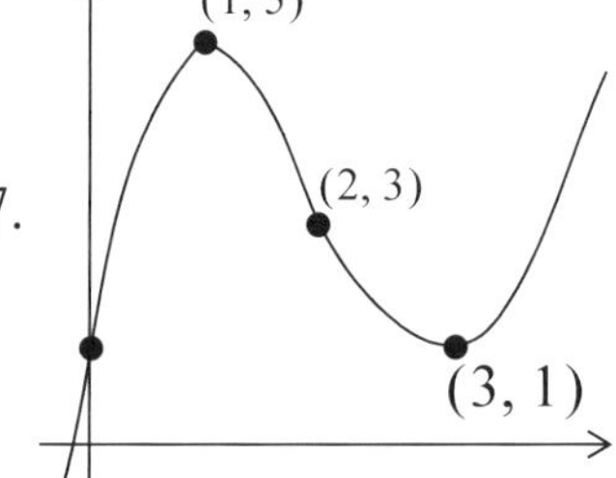

8. 60 feet by 60 feet 9. 175

10. (a) $\frac{1}{2}x^4 + \frac{1}{3}x^3 - \frac{3}{2}x^2 + x + C$ (b) $\frac{x^4}{16} + \frac{5}{4x} + C$ (c) $\frac{52}{3}$

11. $P(x) = 10x - \frac{x^2}{100} - 1100$ 12. $\frac{61}{6}$ square units 13. \$2.64

REVIEW EXERCISES

Exercises 1-4. Evaluate each of the following.

1. -5^2
2. $(-5)^{-2}$
3. $\left(\frac{2}{3}\right)^{-2}$
4. $\frac{2^{-1}+2}{9\cdot 3^{-1}}$

Exercises 5-8. Simplify.

5. $\frac{(ab)^2}{a^3(-b)^3}$
6. $\frac{(-ab)^2a^3}{(2ab)^3}$
7. $\frac{(ab)^2}{a^3-a^2b}$
8. $\frac{(ab)^2+a^2}{(-ab)^2}$

Exercises 9-14. Factor completely.

9. x^4-16
10. x^3-9x
11. $4x^4-9$
12. $6x^2-15x+6$
13. $4x^2-16x+16$
14. $4x^3-16x^2+16x$

Exercises 15-16. Find the equation of the line which:

15. Has slope 5 and y-intercept 4.
16. Contains the points (2,7) and (3,9).

Exercises 17-22. Solve the given equation.

17. $-2x+19 = 5x-3$
18. $5x+4 = -4x+1$
19. $x^2-2x-15 = 0$
20. $x^2+4x+3 = 0$
21. $7x^2-10x-8 = 0$
22. $2x^5-7x^4+3x^3 = 0$

Exercises 23-36. Solve the given inequality.

23. $x+4<3x+5$
24. $-2x+1\geq 5x+9$
25. $x(5x-3)(x+2)(x-7)<0$
26. $x^4-x^3-6x^2\geq 0$

Exercises 27-29. Evaluate.

27. $\lim_{x\to 2}\frac{x-2}{x+2}$
28. $\lim_{x\to 3}\frac{x^2-9}{2x^2-5x-3}$
29. $\lim_{x\to -4}\frac{2x^2+5x-12}{x^2+x-12}$

Exercises 30-32. Use Definition 2.5, page 140, to determine $f'(1)$ for the given function.

30. $f(x) = 3x+10$
31. $f(x) = 2x^2+x-1$
32. $f(x) = \frac{x}{2x+1}$

Exercises 33-35. Use Definition 2.6, page 141, to determine the derivative of the given function.

33. $f(x) = 2x + 1$

34. $g(x) = -x^2 + 3x + 2$

35. $h(x) = \frac{x+1}{2x}$

Exercises 36-41. Differentiate the given function.

36. $f(x) = 4x^3 + 2x^2 - x - 1$

37. $f(x) = -3x^2 + 2x - 5x^{-1} - 1$

38. $g(x) = (x^2 + 1)(x - 3)$

39. $g(x) = \frac{3x^5 - 2x + 1}{x^3}$

40. $g(x) = \frac{x^2}{x - 3}$

41. $g(x) = \frac{x^2 + x}{2x - 1}$

Exercises 42-43. Determine the tangent line to the graph of the given function at the indicated point.

42. $f(x) = 2x^3 - 3x$ at $x = 1$

43. $f(x) = 4x^3 + 2x^2 - x - 1$ at $x = 0$

Exercises 44-45. Determine where the graph of the given function has a horizontal tangent line.

44. $f(x) = 3x^4 - 2x^3 + 100$

45. $g(x) = x^3 - \frac{x^2}{2} - 2x + 1$

Exercises 46-47. Determine the (instantaneous) rate of change of the given function at the indicated point.

46. $f(x) = 3x^2 - 2x + 5$ at $x = 2$

47. $g(x) = \frac{x^2 + 1}{2x}$ at $x = -1$

Exercises 48-49. (Marginal Cost, Revenue, and Profit) From the given monthly cost and revenue functions (in dollars), determine the marginal profit for $x = 100$ units. Indicate what the values predict.

48. $C(x) = 550 + 26x - 0.05x^2$, $R(x) = 51x$

49. $C(x) = 1000 + 37x + \frac{x^2}{100}$, $R(x) = 50x - \frac{x^2}{200}$

50. **(Cost)** Suppose that the marginal cost of producing x units is given by $\overline{MC} = -\frac{x}{50} + 50$, and that the monthly cost for the production of 100 units is $10,000. What is the monthly cost for the production of 150 units?

51. **(Revenue)** Suppose that the marginal revenue from the sale of x units is given by $\overline{MR} = -\frac{x}{1000} + 75$. What revenue will be realized from the sale of 500 units?

52. **(Profit)** Assume that marginal cost and marginal revenue are given by:

$$\overline{MC} = -\frac{x}{100} + 45, \text{ and } \overline{MR} = -\frac{x}{500} + 80$$

Find the monthly profit from the sale of 500 units, if the monthly fixed cost of the company is $5,000.

Exercises 53-56. Sketch the graph of the function. Label maximum, minimum and inflection points.

53. $f(x) = x^4 + x^2 - 2$

54. $f(x) = x^4 - 4x^3$

55. $f(x) = x^3 + \frac{x^2}{2} - 2x + 1$

56. $f(x) = 3x^5 - 25x^3 + 60x$

Exercises 57-62. Evaluate:

57. $\int(x^5 + 3x^2 - x + 1)dx$

58. $\int(4x^3 - 2x + 3)dx$

59. $\int\left(x^4 - 2x^2 + \frac{1}{x^3}\right)dx$

60. $\int x^4(x^2 + x - 1)dx$

61. $\int(x^2 + 5)(x - 3)dx$

62. $\int\frac{4x^5 + x - 3}{x^3}dx$

Exercises 62-64. Solve the given differential equation for the unknown function *f*.

63. $f'(x) = x^2 + 5,\ f(3) = 1$

64. $f'(x) = 4x^3 + 2x,\ f(1) = 5$

Exercises 65-70. Evaluate.

65. $\int_0^1 (3x + 4)dx$

66. $\int_1^2 (4x^3 - 3x^2 + 1)dx$

67. $\int_1^2 \left(5x - \frac{1}{x^2}\right)dx$

68. $\int_0^1 (3x + 4)(2x + 1)dx$

69. $\int_1^2 \frac{3x + 4}{x^3}dx$

70. $\int_{-2}^{-1} \frac{x^5 + x^3 - 3}{x^2}dx$

Exercises 71-72. Find the area of the region bounded by the graph of the given function, over the given interval.

71. $f(x) = -x^2 + 4;\ [-2, 2]$

72. $f(x) = 2x^3 - 2x^2;\ [1, 2]$

73. Determine those values of x (if any) where the slope of the tangent line to the graph of the function $f(x) = x^3$ is twice the rate of change of the function $g(x) = 2x^3 - 3x$.

74. Determine those values of x (if any) where the slope of the tangent line to the graph of the function $f(x) = x^3$ is greater than the rate of change of the function $g(x) = 2x^3 - 3x$.

Exercises 75-76. (Maximum Profit) Determine the production level that will maximize profit, for the given cost and revenue function.

75. $C(x) = 800 + 95x + \frac{x^2}{20}$, $R(x) = 110x$

76. $C(x) = 100 + 50x$, $R(x) = 60x - \frac{x^2}{10}$

77. **(Maximum Profit)** Each week, x units of a certain commodity will be sold when the price is $p(x) = 100 - \frac{x}{10}$ dollars per unit. The weekly cost for producing x units is $C(x) = 1000 + 50x$ dollars. What weekly production level will yield maximum profit?

78. **(Maximum Profit)** Each week, x units of a certain commodity will be sold when the price is $p(x) = 100 - \frac{x}{10}$ dollars per unit. Determine the price per unit which maximizes (weekly) profit for the company, if it costs the company \$15 to produce each unit, and the fixed weekly cost for the company is \$7500.

79. **(Maximum Revenue)** An electronics store can sell 300 computers per month at \$1,200 per unit. For each \$20 decrease in price, an additional 15 units will be sold. What should the price per unit be, in order to maximize revenue?

80. **(Minimum Material Cost)** A closed crate with a square base is to have a volume of 270 cubic feet. The material for the bottom of the crate costs \$3 per square foot, and the material for the rest of the box costs \$2 per square foot. What is the minimum material cost for such a crate?

81. **(Largest Area)** Find the largest possible area of a rectangle with base on the x-axis and upper vertices on the curve $y = 3 - x^2$.

82. **(Maximum Area)** Show that among all rectangles of a given perimeter, the square has the largest area.

83. **(Gravity Problem)** Jared throws a stone upward from a 64 foot bridge at a speed of 48 feet per second. Find the speed of the stone when it hits the water. (See Theorem 2.16, page 197.)

84. **(Water Tank)** Water is leaking out of a full 10 gallon tank at a rate of $7 - 2t$ gallons per hour, where t is measured in hours. How long it take for the tank to empty?

Part 1

CYU Solutions

1.1 (a) Since E is a subset of S, $\#(E) \le \#(S)$. It follows that $0 \le Pr(E) = \frac{\#(E)}{\#(S)} \le 1$.

(b) (One possible answer) Roll a standard die. The probability that you roll an even or odd number is 1.

(c) (One possible answer) Roll a standard die. The probability that you roll a seven is 0.

1.2 Letting, for example, the symbol (2,4) represent the event of drawing a 2 and then a 4 from the hat; and the symbol (4,5) represent the drawing of a 4 followed by a 5 we come to the following equiprobable sample space, S:

$$S = \left\{ \begin{matrix} (1,2) & (1,3) & (1,4) & (1,5) \\ (2,1) & (2,3) & (2,4) & (2,5) \\ (3,1) & (3,2) & (3,4) & (3,5) \\ (4,1) & (4,2) & (4,3) & (4,5) \\ (5,1) & (5,2) & (5,3) & (5,4) \end{matrix} \right\}$$

drawing a 2 and then a 4 [pointing to (2,4)]

Note that there is no (4,4), for example, since the first number drawn is not replaced.

drawing a 4 and then a 5 [pointing to (4,5)]

With the above $4 \cdot 5 = 20$ element equiprobable sample space at hand, it is a trivial matter to find the specified probabilities:

no. of successes: in first row, in third, in fifth

(a) $Pr(\text{first is odd}) = \frac{4+4+4}{20} = \frac{12}{20}$

number of possibilities

in first row, in second row, 3rd row, 4th row

(b) $Pr(\text{second} > \text{first}) = \frac{4+3+2+1}{20} = \frac{10}{20}$

(no success in 5th row)

1.3 Out of 900 lady-fingers tested (the sample space), $900 - 32 = 868$ functioned properly (the successes). Based on this, we calculate the (empirical) probability of a success:

$$Pr(\text{Bang}) = \frac{868}{900} \approx 0.96$$

1.4 (a) $Pr(club) = \frac{13}{52}$ ← successes / ← possibilities

(b) Odds (in favor) $= \frac{13}{39}$ ← successes / ← failures. So: 13:39, or 13 to 39

(c) Odds (against) $= \frac{39}{13}$ ← failures / ← successes. So: 39:13, or 39 to 13

1.5 We are given: $Pr(Win) = \frac{35 \leftarrow \text{successes}}{100 \leftarrow \text{possibilities}}$. This tells us that there are 35 successes and $100 - 35 = 65$ failures. Thus: Odds(against winning) $= \frac{65 \leftarrow \text{failures}}{35 \leftarrow \text{successes}}$ (or 13:7).

1.6 There are 52 cards, 12 of which **are** face cards. Thus:

$$Pr(\textbf{not}\text{ a face card}) = \frac{52-12}{52} = \frac{40}{52}$$

1.7 You can count the successes directly in your head, without benefit of Theorem 1.2: There are 26 red cards (half of the deck), and 6 black face cards (half of the face cards), for a total of 32 successes. Thus:

$$Pr(\text{Red or Face}) = \frac{32}{52}$$

You can also use Theorem 1.2:

$$\underset{\text{Red or Face}}{Pr(R \cup F)} = P(R) + P(F) - P(R \cap F) = \overset{\text{Red}}{\frac{26}{52}} + \overset{\text{Face}}{\frac{12}{52}} - \overset{\text{both Red and Face}}{\frac{6}{52}} = \frac{32}{52}$$

1.8 To get the number of successes, just add the number of black face cards to the number of red aces (nothing is being counted twice):

$$Pr(\text{Black Face Card or Red Ace}) = \frac{6+2}{52} = \frac{8}{52}$$

Using Theorem 1.3, we have:

$$Pr(\text{BFC} \cup \text{RA}) = P(\text{BFC}) + P(\text{RA}) = \frac{6}{52} + \frac{2}{52} = \frac{8}{52}$$

Black Face Card or Red Ace

1.9 We know, from the table of Example 1.12 that the sample space consists of 7720 voters. Our success is that the voter is a male who is not a Democrat.
One way to count the successes is to add up all of the males, and then subtract those males that are Democrats:

$$Pr(\text{M and not D}) = \frac{\overbrace{(1538+1337+1106)}^{\text{all of the males}} - \overbrace{(593+515+249)}^{\text{the male democrats}}}{7720} = \frac{2624}{7720} \approx 0.34$$

Another way to count the successes is to add up the voters in the male rows that are either Republican or independent:

$$Pr(\text{M and not D}) = P([\text{M and R}] \text{ or } [\text{M and I}]) = \frac{(655+712+830)+(290+110+27)}{7720} = \frac{2624}{7720}$$

[Note: we have not, as yet, developed a theorem dealing with an "and" situation. But with a sample space at hand (the table of page 18), we were able to simply count the number of possibilities and the number of successes.]

1.10

(a)

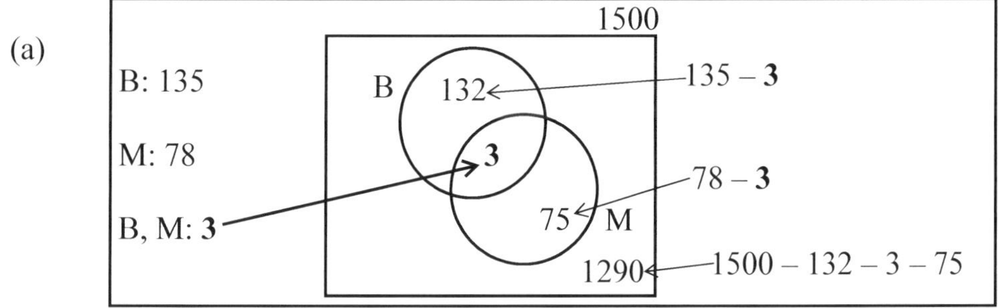

The rest is just counting:

(b) $Pr(\text{neither in M nor in B}) = \dfrac{1290}{1500} = 0.860$ (c) $Pr(\text{in B but out of M}) = \dfrac{132}{1500} = 0.088$

1.11

(a)

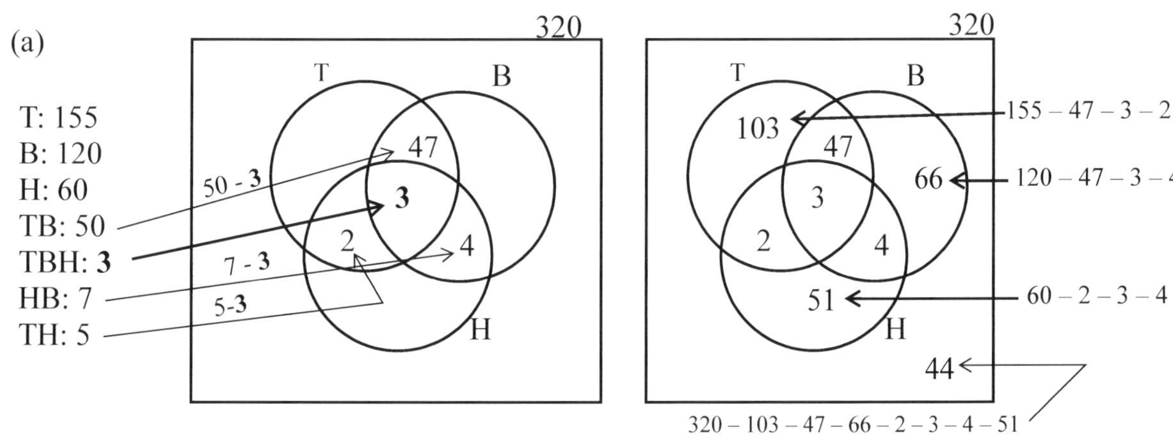

By looking at the above Venn diagram, you should have no difficulty in arriving at the following answers:

(b) There are $2 + 47 + 4 = 53$ students that play exactly two of the three sports.

(c) $Pr(\text{2 or 3 sports}) = \dfrac{47 + 4 + 2 + 3}{320} = \dfrac{56}{320} \approx 0.18$

1.12 The sample space for the rolling of a pair of dice appears below, two times. On the left, we shaded in the restricted sample space subject to the (a)-condition: at least one die is a 5. On the right, we shaded in the restricted sample space subject to the (b)-condition: the sum is 7. In both cases, we also boxed in the specified successes. The rest, is counting.

(a) 2 successes — 11 possibilities

(1, 1) (1, 2) (1, 3) (1, 4) (1, 5) (1, 6)
(2, 1) (2, 2) (2, 3) (2, 4) (2, 5) (2, 6)
(3, 1) (3, 2) (3, 3) (3, 4) (3, 5) (3, 6)
(4, 1) (4, 2) (4, 3) (4, 4) (4, 5) (4, 6)
(5, 1) (5, 2) (5, 3) (5, 4) (5, 5) (5, 6)
(6, 1) (6, 2) (6, 3) (6, 4) (6, 5) (6, 5)

$Pr(\text{sum is7}|\text{at least one 5}) = \dfrac{2}{11}$

(b) 2 successes — 6 possibilities

(1, 1) (1, 2) (1, 3) (1, 4) (1, 5) (1, 6)
(2, 1) (2, 2) (2, 3) (2, 4) (2, 5) (2, 6)
(3, 1) (3, 2) (3, 3) (3, 4) (3, 5) (3, 6)
(4, 1) (4, 2) (4, 3) (4, 4) (4, 5) (4, 6)
(5, 1) (5, 2) (5, 3) (5, 4) (5, 5) (5, 6)
(6, 1) (6, 2) (6, 3) (6, 4) (6, 5) (6, 5)

$Pr(\text{at least one 5}|\text{sum is 7}) = \dfrac{2}{6}$

1.13 (a) $Pr(\text{R}_{1st} \text{ and } \text{R}_{2nd}) = Pr(R_{1st}) \cdot Pr(R_{2nd}|R_{1st}) = \frac{5}{16} \cdot \frac{4}{15} = \frac{1}{12}$

(b) $Pr(\text{R and B}) = P(\text{R}_{1st} \text{ and } \text{B}_{2nd}) \text{ or } P(\text{B}_{1st} \text{ and } \text{R}_{2nd})$

$$\frac{5}{16} \cdot \frac{4}{15} + \frac{4}{16} \cdot \frac{5}{15} = 2\left(\frac{5 \cdot 4}{16 \cdot 15}\right) = \frac{1}{6}$$

one marble is gone

1.14 Using Theorem 1.6, and looking at the situation as 5 independent experiments (rolling one die five times), each with a probability of success $\frac{1}{6}$, we have:

$$Pr(\text{rolling a die five times and getting a 4 each time}) = \frac{1}{6} \cdot \frac{1}{6} \cdot \frac{1}{6} \cdot \frac{1}{6} \cdot \frac{1}{6} = \frac{1}{6^5} \approx 0.0001$$

1.15 Here is the probability that you make at least one basket out of 2 shots:

2 independent events

$$Pr(\text{at least one}) = 1 - Pr(\text{two misses}) = 1 - [Pr(\text{miss on first}) \cdot Pr(\text{miss on second})] = 1 - (0.6)^2$$

And here is the probability that you will get at least one basket in each of five games:

5 independent events

$$Pr(\text{get large cuddly bear}) = [1-(0.6)^2][1-(0.6)^2][1-(0.6)^2][1-(0.6)^2][1-(0.6)^2]$$
$$= [1-(0.6)^2]^5 \approx 0.11$$

1.16 We turn to a tree diagram:

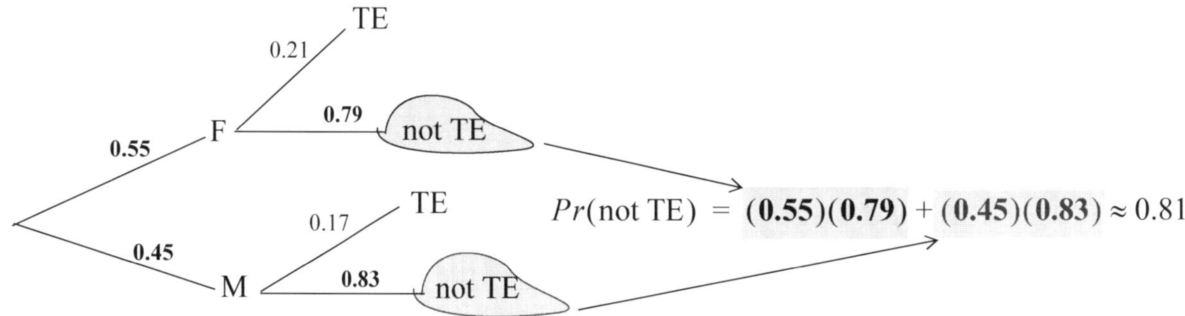

1.17 Referring to the tree diagram of Example 1.26, we focus our attention on the "black leaves:"

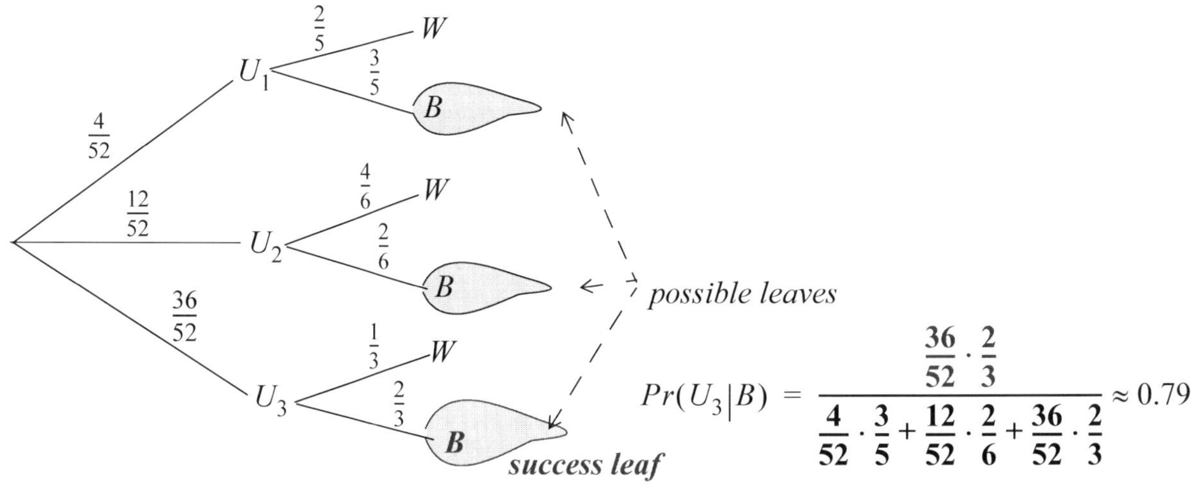

$$Pr(U_3|B) = \frac{\frac{36}{52} \cdot \frac{2}{3}}{\frac{4}{52} \cdot \frac{3}{5} + \frac{12}{52} \cdot \frac{2}{6} + \frac{36}{52} \cdot \frac{2}{3}} \approx 0.79$$

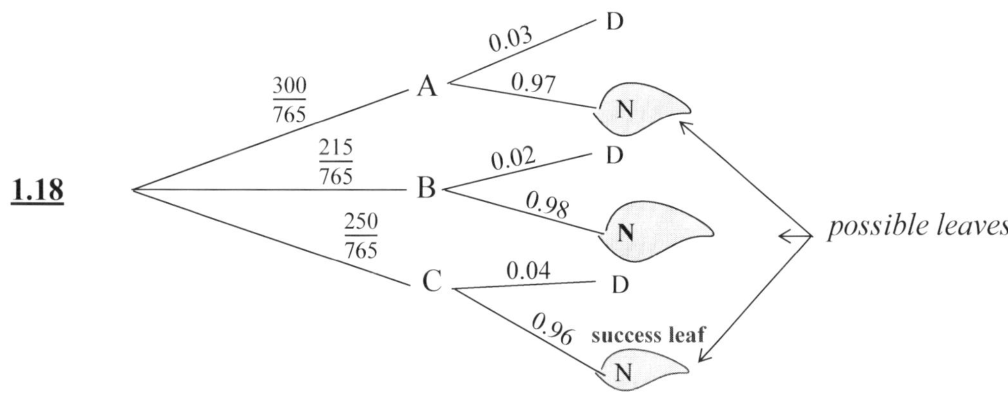

$$Pr(\mathrm{C} \mid \mathrm{N}) = \frac{\left(\frac{250}{765}\right)(0.96)}{\left(\frac{300}{765}\right)(0.97) + \left(\frac{215}{765}\right)(0.98) + \left(\frac{250}{765}\right)(0.96)} \approx 0.32$$

1.19 Here is one of the things you are trying to count: DD2431. In writing that sample down, we were aware of the fact that the second letter had to be the same as the first (choice of 1 for the second letter), and that every time we wrote down a digit, it was unavailable for future choices; in other words:

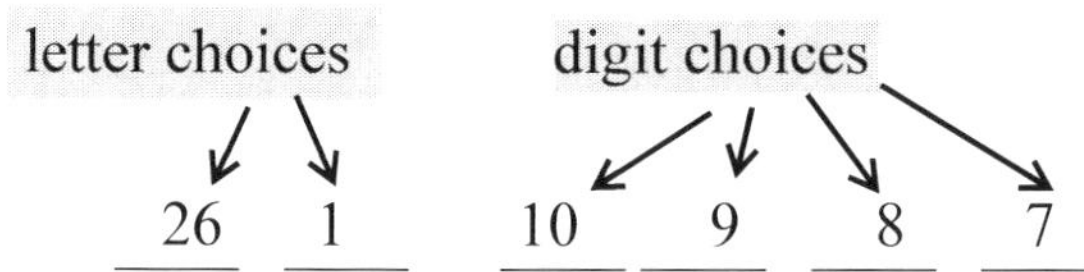

Turning to the Fundamental Counting Principle, we have:
Number of plates $= 26 \cdot 1 \cdot 10 \cdot 9 \cdot 8 \cdot 7 = 131{,}040$

1.20 One possible success: WMMMWW. The main focus was to make sure that the three M's were next to each other. We started them off in seat 2, but could have started that string of M's at seat 1, or 2, or 3, or 4; for **an initial choice of 4**. This done, let's focus on our possible success, and arrange the three men in the three M seats (shaded below), and the three women in the remaining 3 seats:

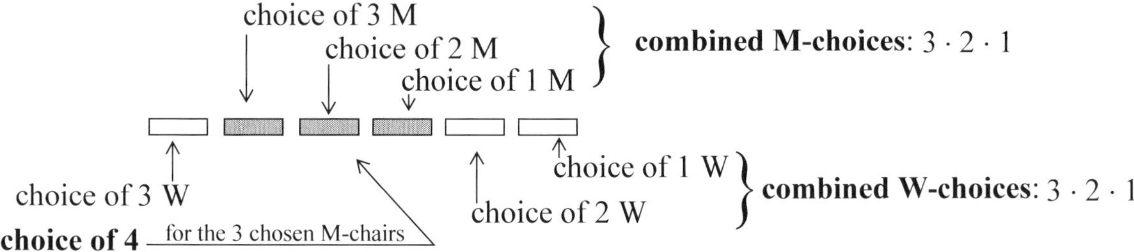

Turning to the Fundamental Counting Principle, we find the number of different ways of sitting the six people, given that the three men must be seated next to each other:

1.21 We've already counted the number of possibilities in Example 1.31:

$$\text{no. of possibilities: } 26^2 \cdot 10^4$$

Let's write down a success: CC2512. There is no problem with the first part of the "journey:" a choice of 26 for the first letter, followed by a choice of 1 for the second. The rest is number of ways the three men can be seated in the 3 adjacent chairs.

$$4(3 \cdot 2 \cdot 1)(3 \cdot 2 \cdot 1) = 144$$

The "first man" can be seated in any of 4 chairs

number of ways the three women can be seated in the remaining 3 chairs.

not so difficult either: any of the 10 digits will do for the first digit, or second, or third; but the last digit has to be the same as the first (a choice of 1):

same as　　same as

$$\text{no. of successes: } 26 \cdot 1 \cdot 10 \cdot 10 \cdot 10 \cdot 1$$

Knowing the number of possibilities (sample space), and the number of successes, we arrive at our answer:

$$Pr(\text{Letters same; first and last digit same}) = \frac{26 \cdot 1 \cdot 10 \cdot 10 \cdot 10 \cdot 1}{26^2 \cdot 10^4} = \frac{1}{260} \approx 0.0038$$

1.22 Let's lay three blocks down; say: ACD, and realize that we had a choice of 4 for the first block, a choice of 3 for the second, and a choice of 2 for the third—totaling: $4 \cdot 3 \cdot 2$ possible outcomes of the experiment. The number of successes are given to us explicitly in the problem, and there are 3 of them: CAD, BAD, and DAB. Thus:

$$Pr(\text{CAD or BAD or DAB}) = \frac{3}{4 \cdot 3 \cdot 2} = \frac{3}{24} = 0.125$$

1.23 Resorting to Theorem 1.9, we quickly arrive at the answer:

There are $7! = 5{,}040$ different ways of arranging 7 books on a shelf.

1.24 The number of possibilities is the number of ways the 15 children can be ordered:

$$\text{no. of possibilities} = 15!$$

Now we have to think a bit about the number of successes:

choice of 1: the tallest

choice of **13!**: the number of ways the remaining 13 children can be ordered.

choice of 1: the smallest

$$\underline{1}\ \_\ \_\ \_\ \_\ \_\ \_\ \_\ \_\ \_\ \_\ \_\ \_\ \_\ \underline{1}$$

Conclusion:

$$Pr(\text{tallest first and shortest last}) = \frac{1 \cdot 13! \cdot 1}{15!} = \frac{\mathbf{1}}{\mathbf{14 \cdot 15}} = \frac{1}{210} \approx 0.0048$$

$$\frac{\cancel{1} \cdot \cancel{2} \cdot \cancel{3} \ldots \cdot \cancel{12} \cdot \cancel{13}}{\cancel{1} \cdot \cancel{2} \cdot \cancel{3} \ldots \cdot \cancel{12} \cdot \cancel{13} \cdot 14 \cdot 15} = \frac{\mathbf{1}}{\mathbf{14 \cdot 15}}$$

1.25 There are as many ways of awarding a gold, silver, and bronze medal to a group of 15 individuals as there are of picking three objects from 15, when "order counts:"

$$P(15,3) = \frac{15!}{(15-3)!} = \frac{15!}{12!} = \frac{1\cdot 2\cdot 3\cdot \ldots \cdot 12\cdot 13\cdot 14\cdot 15}{1\cdot 2\cdot 3\cdot \ldots \cdot 12} = 13\cdot 14\cdot 15 = 2{,}730$$

1.26 Red can be hoisted first or second, but the two cases have to be treated separately, for if red is hoisted second, then blue must be hoisted last (there is got to be another flag between red and blue); whereas if red is hoisted first, then blue can be hoisted third or fourth. Here are the possible successes:

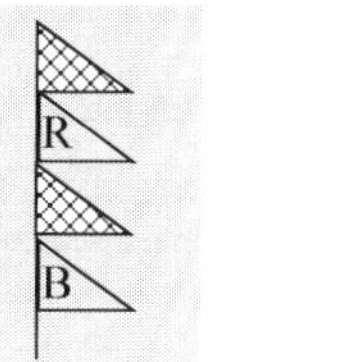

OR

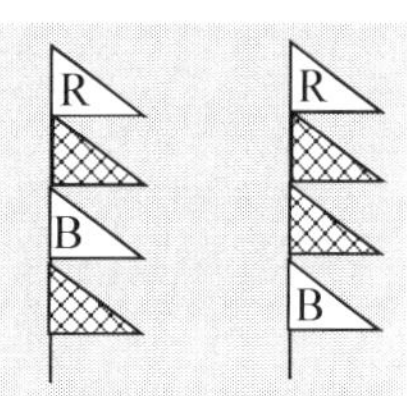

choice of 1 (R hoisted second) followed by a choice of $P(5,2)$ (pick two of the remaining 5 flags)

choice of 2 (R hoisted first and B third or B last) followed by a choice of $P(5,2)$ (pick two of the remaining 5 flags)

Having accounted for all of the success we have:

$$\text{Pr(R above B but not adjacent)} = \frac{1\cdot P(5,2)+2P(5,2)}{P(7,4)} = \frac{3P(5,2)}{P(7,4)} = \frac{3\cdot\frac{5!}{3!}}{\frac{7!}{3!}} = \frac{3\cdot 4\cdot 5}{4\cdot 5\cdot 6\cdot 7} = \frac{1}{14}$$

1.27 There are as many poker hands as there are ways of grabbing 5 objects (the 5 cards dealt) from 52 (the deck), when "order does not count:"

$$C(52,5) = \frac{52!}{(52-5)!5!} = \frac{52!}{47!5!} = \frac{47!\cdot 48\cdot 49\cdot 50\cdot 51\cdot 52}{47!\cdot 1\cdot 2\cdot 3\cdot 4\cdot 5} = 2\cdot 49\cdot 10\cdot 51\cdot 52 = 2{,}598{,}960$$

1.28

R_{25} W_9 P_4 B_7 → grab 8
success: 2 R, 2W, 2P, and 2B

The number of possibilities is the number of ways 8 objects can be grabbed from 45 when "order does not count:" $C(45,8)$. The number of successes is the number of ways 2 red jelly-beans can be grabbed from the 25, times the number of ways 2 white jelly-beans can be grabbed from the 9, times the number of ways 2 purple jelly-beans can be grabbed from the 4, times the number of ways 2 black jelly-beans can be grabbed from 7 (choices followed by choices—multiply): $C(25,2)\cdot C(9,2)\cdot C(4,2)\cdot C(7,2)$. Dividing the number of successes by the number of possibilities we have the answer (which you can check by pencil or calculator):

$$P(\text{two of each color}) = \frac{C(25,2)\cdot C(9,2)\cdot C(4,2)\cdot C(7,2)}{C(45,8)} = \frac{1{,}360{,}800}{215{,}553{,}195} \approx 0.0($$

1.29

n	p	r
20	0.84	17

$\rightarrow Pr(17 \text{ out of } 20) = C(20, 17) \cdot (0.84)^{17} \cdot (1 - 0.84)^3 \approx 0.24$

1.30 $Pr(\text{at least } 3) = 1 - [Pr(0) + Pr(1) + Pr(2)]$

n	p	r
10	$\frac{1}{5}$	0 or 1 or 2

$$= 1 - \left[C(10, 0)\left(\frac{1}{5}\right)^0\left(\frac{4}{5}\right)^{10} + C(10, 1)\left(\frac{1}{5}\right)^1\left(\frac{4}{5}\right)^9 + C(10, 2)\left(\frac{1}{5}\right)^2\left(\frac{4}{5}\right)^8\right] \approx 0.32$$

1.31 We first determine the probability that you win a penny when you play one game:

n	p	r
5	$\frac{1}{6}$	4 or 5

$$\rightarrow Pr(Win) = C(5, 4)\left(\frac{1}{6}\right)^4\left(\frac{5}{6}\right)^1 + C(5, 5)\left(\frac{1}{6}\right)^5\left(\frac{5}{6}\right)^0 \approx 0.0033$$

We then determine the probability of winning exactly two out of three games:

n	p	r
3	0.0033	2

$\rightarrow Pr(2 \text{ out of } 3) \approx C(3, 2)(0.0033)^2(1 - 0.0033)^1 \approx 0.000033$

1.32

$$E(\text{no. of Aces}) = 0 \cdot \frac{C(4, 0)C(48, 5)}{C(52, 5)} + 1 \cdot \frac{C(4, 1)C(48, 4)}{C(52, 5)} + 2 \cdot \frac{C(4, 2)C(48, 3)}{C(52, 5)}$$
$$+ 3 \cdot \frac{C(4, 3)C(48, 2)}{C(52, 5)} + 4 \cdot \frac{C(4, 4)C(48, 1)}{C(52, 5)} \approx 0.39$$

successes: grab any 3 of the 4 aces and 2 of the non-aces $\rightarrow \frac{C(4, 3)C(48, 2)}{C(52, 5)}$

possible outcomes: grab any 5 of 52 cards ↗

1.33

Time	1	2	3	4	5	6	7	8	9	Sum
Frequency	15	32	25	19	16	9	3	0	1	120
Probability	$\frac{15}{120}$	$\frac{32}{120}$	$\frac{25}{120}$	$\frac{19}{120}$	$\frac{16}{120}$	$\frac{9}{120}$	$\frac{3}{120}$	$\frac{0}{120}$	$\frac{1}{120}$	$\frac{120}{120} = 1$

$$E(\text{Time}) = 1\cdot\frac{15}{120}+2\cdot\frac{32}{120}+3\cdot\frac{25}{120}+4\cdot\frac{19}{120}+5\cdot\frac{16}{120}+6\cdot\frac{9}{120}+7\cdot\frac{3}{120}+8\cdot 0+9\cdot\frac{1}{120}\approx 3.28$$

In Example 1.45, the expected waiting time was 3.34 minutes, and here it is 3.28 minutes, so the situation did improve a bit.

1.34 $E(\text{Winnings}) = \$349\cdot\frac{1}{1000}+\$224\cdot\frac{2}{1000}+(-\$1)\cdot\frac{997}{1000} = -\$0.20$

1.35 Applying Theorem 1.13, with $n = 100$ and $p = \frac{1}{5}$, we find the expected number of matches to be $np = 100\cdot\frac{1}{5} = 20$. The experiment will be labeled successful if the number of matches equals or exceeds $20+0.25\cdot 20 = 25$.

1.36 We determine the expected winnings of Game 3:

$$\$0\left[C(3,1)\left(\frac{1}{6}\right)\left(\frac{5}{6}\right)^2\right]+\$8.50\left[C(3,2)\left(\frac{1}{6}\right)^2\left(\frac{5}{6}\right)\right]+\$23.50\left[C(3,3)\left(\frac{1}{6}\right)^3\left(\frac{5}{6}\right)^0\right]$$

$$-\$1.50\left(1-\left[C(3,1)\left(\frac{1}{6}\right)\left(\frac{5}{6}\right)^2+C(3,2)\left(\frac{1}{6}\right)^2\left(\frac{5}{6}\right)+C(3,3)\left(\frac{1}{6}\right)^3\left(\frac{5}{6}\right)^0\right]\right)\approx -\$0.17$$

Since Game 2's expected winnings turned out to be \$0.64, you should go with Game 2.

1.37 We determine the expected annual profit at both sites:

$$E(\text{profit at A}) = 45(0.25)-4.2 = 7.05 \text{ billion dollars}$$
$$E(\text{profit at B}) = 60(0.17)-5.5 = 4.70 \text{ billion dollars}$$

Conclusion: Site A should be chosen.

1.38 Beauford has three options:

(a) Price the shirts at \$21 (b) Price the shirts at \$15 (c) Price the shirts at \$7

We determine the expected profit resulting for each option.

Option (a):

cost per shirt ↓ (the 6 in $(21-6)$); Pr(in series and wins) ↓ (on $(0.35)\cdot(0.60)$)

$$E(\text{Profit at } \$21) = \$(21-6)(1000)[(0.35)\cdot(0.60)]$$
$$+\$[(21-6)(500)-6(500)][(0.35)\cdot(1-0.60)]-\$6(1000)(1-0.35) = -\$120$$

sell half of the shirts → $(21-6)(500)$; lose \$6 for each shirt not sold → $6(500)$; Pr(in series and loses) ↑ (on $(0.35)\cdot(1-0.60)$); Pr(not in series) ↑ (on $(1-0.35)$)

Option (b):

Pr(in series) ↓ (on (0.35)); Pr(not in series) ↓ (on $(1-0.35)$)

$$\text{E(Profit at } \$15) = \$(15-6)(1000)[(0.35)]-\$6(1000)(1-0.35) = -\$750$$

Option (c): Since all 1000 shirts will be sold at \$7, the profit realized at that price will be:

$$\$(7-6)(1000) = \$1000$$

Conclusion: Beauford should price the shirts at \$7.

Part 2
CYU Solutions

2.1 (a) $(-2ax^2)^4 = (-2)^4a^4(x^2)^4 = 16a^4x^8$

(b) $\left(\frac{-x^2}{2}\right)^3 = \frac{(-x^2)^3}{2^3} = \frac{-x^6}{8} = -\frac{x^6}{8}$ (note that, in general: $\frac{a}{-b} = \frac{-a}{b} = -\frac{a}{b}$)

(c) $(-x)^2\left(\frac{x}{2y}\right)^3 = x^2 \cdot \frac{x^3}{2^3y^3} = \frac{x^5}{8y^3}$

2.2 (a)
$$\begin{aligned} 3 - 2x + 5 - x &= -4x - 2 + 1 \\ -2x - x + 4x &= -2 + 1 - 3 - 5 \\ x &= -9 \end{aligned}$$

(b)
$$\begin{aligned} \frac{-3x}{5} - \frac{x}{2} + 1 &= \frac{2x+1}{10} \\ 10\left(\frac{-3x}{5} - \frac{x}{2} + 1\right) &= 10\left(\frac{2x+1}{10}\right) \\ -6x - 5x + 10 &= 2x + 1 \\ -6x - 5x - 2x &= 1 - 10 \\ -13x &= -9 \\ x &= \frac{9}{13} \end{aligned}$$

2.3
$$\begin{aligned} \frac{3x}{5} - \frac{2-x}{3} + 1 &< \frac{x-1}{15} \\ 15\left(\frac{3x}{5} - \frac{2-x}{3} + 1\right) &< 15\left(\frac{x-1}{15}\right) \\ 9x - 5(2-x) + 15 &< x - 1 \\ 9x - 10 + 5x + 15 &< x - 1 \\ 9x + 5x - x &< -1 + 10 - 15 \\ 13x &< -6 \\ x &< -\frac{6}{13} \end{aligned}$$

2.4 $m = \frac{5-(-2)}{-1-3} = \frac{7}{-4} = -\frac{7}{4}$

or: $m = \frac{-2-5}{3-(-1)} = \frac{-7}{4} = -\frac{7}{4}$

2.5 (a) $y = mx + b = 7x + 3$

(b) $y = 3x + b$. When $x = 2$, y must equal 5: $\begin{aligned} 5 &= 3 \cdot 2 + b \\ b &= 5 - 6 = -1 \end{aligned}$. Equation: $y = 3x - 1$

(c) $m = \frac{4-(-5)}{2-3} = -9$. So: $y = -9x + b$. Since $(2, 4)$ lies on the line: $\begin{aligned} 4 &= -9 \cdot 2 + b \\ b &= 4 + 18 = 22 \end{aligned}$. Equation: $y = -9x + 22$.

2.6 (a) $25x^2 - 1 = (5x)^2 - 1^2 = (5x+1)(5x-1)$

(b) $3x^2 - 2 = (\sqrt{3}x)^2 - (\sqrt{2})^2$
$= (\sqrt{3}x + \sqrt{2})(\sqrt{3}x - \sqrt{2})$

(c) $x^4 - 16 = (x^2+4)(x^2-4) = (x^2+4)(x+2)(x-2)$

2.7 (a) $2x^2 + 7x - 4 = (2x-1)(x+4)$

(b) $18x^4 - 6x^3 - 60x^2 = 6x^2(3x^2 - x - 10) = 6x^2(3x+5)(x-2)$

2.8 (a)

$$25x^2 - 1 = 0$$
$$(5x+1)(5x-1) = 0$$
$$5x + 1 = 0 \text{ or } 5x - 1 = 0$$
$$x = -\frac{1}{5} \text{ or } x = \frac{1}{5}$$

(b)

$$3x^2 - 2 = 0$$
$$(\sqrt{3}x + \sqrt{2})(\sqrt{3}x - \sqrt{2}) = 0$$
$$\sqrt{3}x + \sqrt{2} = 0 \text{ or } \sqrt{3}x - \sqrt{2} = 0$$
$$x = -\frac{\sqrt{2}}{\sqrt{3}} \text{ or } x = \frac{\sqrt{2}}{\sqrt{3}}$$
$$x = \pm\frac{\sqrt{2}}{\sqrt{3}} = \pm\frac{\sqrt{2}}{\sqrt{3}} \cdot \frac{\sqrt{3}}{\sqrt{3}} = \pm\frac{\sqrt{6}}{3}$$

(c)

$$2x^2 + 7x - 4 = 0$$
$$(2x-1)(x+4) = 0$$
$$x = \frac{1}{2} \text{ or } x = -4$$

2.9 (a) $(x-3)(x+2)(-x+5) < 0$

$(-2, 3) \cup (5, \infty)$

(b) $(x+1)^2(x+2)^3(x-4)^2 \le 0$

$(-\infty, -2] \cup \{-1, 4\}$

2.10 (a) $f(-2) = 3(-2)^2 - 5 = 3 \cdot 4 - 5 = 12 - 5 = 7$

(b) $f(x+h) = 3(x+h)^2 - 5 = 3(x^2 + 2xh + h^2) - 5 = 3x^2 + 6xh + 3h^2 - 5$

(c) $$\frac{f(x+h) - f(x)}{h} = \frac{\overbrace{\mathbf{3x^2 + 6xh + 3h^2 - 5}}^{\textbf{see (b)}} - (3x^2 - 5)}{h}$$

$$= \frac{3x^2 + 6xh + 3h^2 - 5 - 3x^2 + 5}{h} = \frac{6xh + 3h^2}{h} = \frac{h(6x+3h)}{h} = 6x + 3h$$

2.11 (a) As x approaches -1, $4x^2 + x$ approaches 3 ($4x^2$ tends to 4, and x to -1). Thus:

$$\lim_{x \to -1} (4x^2 + x) = 3$$

(b) As x approaches 2, $x + 3$ approaches 5 and $x + 2$ approaches 4. Thus:

$$\lim_{x \to 2} \frac{x+3}{x+2} = \frac{5}{4}$$

(c) As x approaches 3, $x(3x^2 + 1)$ approaches $3(3 \cdot 3^2 + 1) = 84$. Thus:

$$\lim_{x \to 3} [x(3x^2 + 1)] = 84$$

2.12 (a) $\lim_{x \to 1} \frac{x^2+3x-4}{x^2+1} = \frac{1^2+3\cdot1-4}{1^2+1} = \frac{0}{2} = 0$

not zero! Just plug in

(b) $\lim_{x \to 1} \frac{x^2+3x-4}{x^2-1} = \lim_{x \to 1} \frac{(x+4)(x-1)}{(x+1)(x-1)} = \lim_{x \to 1} \frac{(x+4)}{(x+1)} = \frac{5}{2}$

(c) $\lim_{x \to 1} \frac{x^2-1}{\frac{x}{2}-\frac{1}{2}} = \lim_{x \to 1} \frac{2(x^2-1)}{2\left(\frac{x}{2}-\frac{1}{2}\right)} = \lim_{x \to 1} \frac{2(x+1)(x-1)}{x-1} = \lim_{x \to 1} 2(x+1) = 4$

2.13 (a) $\lim_{x \to 3} \frac{x^2-9}{x-3} = \lim_{x \to 3} \frac{(x+3)(x-3)}{x-3} = \lim_{x \to 3}(x+3) = 6$

(b) $\lim_{x \to 3} \frac{x-3}{x^2-6x+9} = \lim_{x \to 3} \frac{x-3}{(x-3)(x-3)} = \lim_{x \to 3} \frac{1}{(x-3)}$ DNE (denominator goes to 0 while numerator does not)

(c) Does not exist: $\lim_{x \to 3^-} f(x) = 5$ while $\lim_{x \to 3^+} f(x) = 8$

(d) Since $\lim_{x \to 3^-} f(x) = 2$ and $\lim_{x \to 3^+} f(x) = 2$, $\lim_{x \to 3} f(x) = 2$

2.14

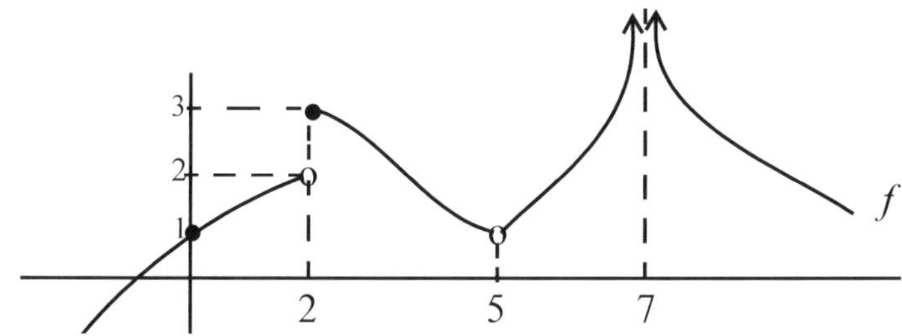

(a) As you approach 0 from either side, the function values approach 1. Thus: $\lim_{x \to 0} f(x) = 1$.

(b) As you approach 2 from the left, the function values approach 2, but as you approach from the right, the function values approach 3. Thus: $\lim_{x \to 2} f(x)$ does not exist.

(c) As you approach 5 from either side, the function values approach 1. Never mind that the function is not defined at 5; for the limit does not care what happens there—it is only concerned about what happens as you **approach** 5. Thus: $\lim_{x \to 5} f(x) = 1$.

(d) As you approach 7 from either side, the function values get larger and larger, and cannot tend to any number. Thus: $\lim_{x \to 7} f(x)$ does not exist. Given that the function values get larger and larger in the positive direction, it is also acceptable to write $\lim_{x \to \infty} f(x) = \infty$.

2.15 (a) $\lim_{x \to 3} \frac{x^2-9}{x-3} = \lim_{x \to 3} \frac{(x-3)(x+3)}{x-3} = \lim_{x \to 3}(x+3) = 6$. But function is not defined at 3, so: not continuous. (It has a removable discontinuity at that point.)

(b) $\lim_{x \to 3} \frac{x-3}{x^2-6x+9} = \lim_{x \to 3} \frac{x-3}{(x-3)(x-3)} = \lim_{x \to 3} \frac{1}{(x-3)}$ DNE

(c) $\lim_{x \to 3^-} f(x) = 2 \cdot 3 - 1 = 5$ and $\lim_{x \to 3^+} f(x) = 3 + 3 = 6$. Not continuous at 3. (It has a jump discontinuity at that point.)

(d) $\lim_{x \to 3^-} f(x) = 3 - 1 = 2$ and $\lim_{x \to 3^+} f(x) = 2$. So, the limit equals 2. Moreover, $f(3) = 2$. Conclusion the function is continuous at 3.

2.16 For $f(x) = -3x^2 + 6x - 1$:

$$f'(2) = \lim_{h \to 0} \frac{f(2+h) - f(2)}{h} = \lim_{h \to 0} \frac{[-3(2+h)^2 + 6(2+h) - 1] - (-1)}{h}$$

$$= \lim_{h \to 0} \frac{[-3(4 + 4h + h^2) + 12 + 6h - 1] + 1}{h}$$

$$= \lim_{h \to 0} \frac{-12 - 12h - 3h^2 + 12 + 6h - 1 + 1}{h} = \lim_{h \to 0} \frac{-3h^2 - 6h}{h} = \lim_{h \to 0} (-3h - 6) = -6$$

2.17 For $f(x) = x^2 + x + 1$:

$$f'(x) = \lim_{h \to 0} \frac{f(\boldsymbol{x} + \boldsymbol{h}) - f(x)}{h} = \lim_{h \to 0} \frac{(\boldsymbol{x} + \boldsymbol{h})^2 + (\boldsymbol{x} + \boldsymbol{h}) + \boldsymbol{1} - (x^2 + x + 1)}{h}$$

$$= \lim_{h \to 0} \frac{x^2 + 2xh + h^2 + x + h + 1 - x^2 - x - 1}{h}$$

$$= \lim_{h \to 0} \frac{2xh + h^2 + h}{h} = \lim_{h \to 0} \frac{h(2x + h + 1)}{h} = \lim_{h \to 0} (2x + h + 1) = 2x + 1$$

We then have:

$f'(0) = 2 \cdot 0 + 1 = 1$ $\quad f'(1) = 2 \cdot 1 + 1 = 3$ $\quad$ and $f'(-1) = 2(-1) + 1 = -1$

2.18 From CYU 2.17 we know that $f'(x) = 2x + 1$. So:

$$y = mx + b = 5x + b$$

$$f'(2) = 2(2) + 1 = 5$$

To find b, we need a point of the line. We know a point on the line: the point of tangency to the curve:

$$(2, f(2)) = (2, 7)$$

$$f(2) = 2^2 + 2 + 1 = 7$$

Substituting 2 for x and 7 for y in the equation $y = 5x + b$, we can solve for b:

$$7 = 5(2) + b$$

$$b = 7 - 10 = -3$$

Tangent line to the graph of $f(x) = x^2 + x + 1$ at $x = 2$: $y = 5x - 3$

2.19 (a) $(-4x^3+2x^2-3x+5)' = (-4x^3)'+(2x^2)'-(3x)'+(5)' = -12x^2+4x-3$

(b) $[x^2(3x^3+2x-5)]' = (3x^5+2x^3-5x^2)' = 15x^4+6x^2-10x$

(c) $\left(\frac{5x^3+2x-7}{2x^2}\right)' = \left(\frac{5}{2}x+x^{-1}-\frac{7}{2}x^{-2}\right)' = \frac{5}{2}-x^{-2}+7x^{-3} = \frac{5}{2}-\frac{1}{x^2}+\frac{7}{x^3}$

2.20 We know that the derivative gives slopes of tangent lines. Differentiating, we have:

$$f'(x) = \left(\frac{x^3-x}{x^2}\right)' = (x-x^{-1})' = 1-(-x^{-2}) = 1+\frac{1}{x^2}$$

The tangent line to the graph of f at $x = 2$ is of the form:

$$y = \underbrace{f'(2)}_{\text{slope}}x+b = \frac{5}{4}x+b$$

$$f'(2) = 1+\frac{1}{2^2} = 1+\frac{1}{4} = \frac{5}{4}$$

Knowing that the tangent line passes through the point

$$(2,f(2)) = \left(2,\frac{3}{2}\right)$$

$$f(2) = \frac{2^3-2}{2^2} = \frac{6}{4} = \frac{3}{2}$$

enables us to find b:

$$\frac{3}{2} = \frac{5}{4}\cdot 2+b \Rightarrow b = \frac{3}{2}-\frac{5}{2} \Rightarrow b = -1$$

Tangent line to the graph of f at $x = 2$: $y = \frac{5}{4}x-1$

2.21 By expanding:

$$f'(x) = [(x^3+x)(3x^2+x-1)]' = (3x^5+x^4+2x^3+x^2-x)' = 15x^4+4x^3+6x^2+2x-1$$

Using the product rule:

$$\begin{aligned} f'(x) = [(x^3+x)(3x^2+x-1)]' &= (x^3+x)(3x^2+x-1)'+(3x^2+x-1)(x^3+x)' \\ &= (x^3+x)(6x+1)+(3x^2+x-1)(3x^2+1) \\ &= (6x^4+x^3+6x^2+x)+(9x^4+3x^2+3x^3+x-3x^2-1) \\ &= 15x^4+4x^3+6x^2+2x-1 \end{aligned}$$

2.22

$$\begin{aligned} f'(x) = \left(\frac{2x^2+3x}{\mathbf{2x^2+3}}\right)' &= \frac{(\mathbf{2x^2+3})(2x^2+3x)'-(2x^2+3x)(\mathbf{2x^2+3})'}{(\mathbf{2x^2+3})^2} \\ &= \frac{(\mathbf{2x^2+3})(4x+3)-(2x^2+3x)(\mathbf{4x})}{(\mathbf{2x^2+3})^2} \\ &= \frac{(8x^3+6x^2+12x+9)-(8x^3+12x^2)}{(2x^2+3)^2} \\ &= \frac{-6x^2+12x+9}{(2x^2+3)^2} \end{aligned}$$

2.23 $f(x) = 5x^3 - x + \frac{1}{x} = 5x^3 - x + x^{-1} \Rightarrow$

$$f'(x) = 15x^2 - 1 - x^{-2}$$
$$f''(x) = 30x + 2x^{-3} = 30x + \frac{2}{x^3}$$
$$f'''(x) = 30 - 6x^{-4} = 30 - \frac{6}{x^4}$$

2.24 (a) Average temperature $= \dfrac{80° + 68°}{2} = 74°$ F

(b) $\dfrac{\text{change in temperature}}{\text{change in time}} = \dfrac{68° - 80°}{(6-0)\text{ hr}} = \dfrac{-12°}{6\text{ hr}} = -2°$ F per hour

2.25 The important thing to note here is that (a) and (b) are asking for the same thing: the derivative of the given function (just different notation), and that (c) and (d) are asking for the same thing: the derivative of the function evaluated at 2 (just different notation).

(a) and (b): For $y = f(x) = 3x^4 + 2x^3 - x + 1$: $f'(x) = \dfrac{dy}{dx} = 12x^3 + 6x^2 - 1$

(c) and (d): $f'(2) = \left.\dfrac{dy}{dx}\right|_2 = 12 \cdot 2^3 + 6 \cdot 2^2 - 1 = 119$

2.26 Setting position to zero, we can determine the time it takes the stone to hit the ground:

$$-16t^2 + 144 = 0$$
$$t^2 = \frac{144}{16} = 9$$
$$t = 3$$

Differentiating the position function, we obtain the velocity function: $v(t) = s'(t) = (-16t^2 + 144)' = -32t$. Evaluating the velocity function at $t = 3$ gives us the velocity of the stone on impact: $v(3) = -32(3) = -96$ ft/sec.

2.27 (a) The rate of change of the profit function is $P'(x) = 20 - \dfrac{x}{150}$. The rate of change when $x = 4500$ is $P'(4500) = 20 - \dfrac{4500}{150} = -10$.

(b) Profit will decrease by approximately \$10 from the sale of the 4501th unit.

2.28 (a) and (b): To get the marginal revenue function, we need but differentiate the revenue function:

$$\overline{MR}(x) = R'(x) = (5.9x - 0.02x^2)' = 5.9 - 0.04x$$

Evaluating at 25 we have: $R'(25) = 5.9 - 0.04(25) = 4.9$. This tells us that approximately \$4,900 will be realized from the sale of the 26th sailboat.

2.29 Profit function:

$$P(x) = R(x) - C(x) = 400x - (2500 + 325x - 0.005x^2) = 0.005x^2 + 75x - 2500$$

Marginal profit function: $P'(x) = (0.005x^2 + 75x - 2500)' = 0.01x + 75$

In particular: $P'(100) = 0.01(100) + 75 = 76$

Interpretation: A profit of approximately \$76 will be realized from the sale of the 101th unit.

2.30 For $f(x) = \frac{x^4}{4} + \frac{x^3}{3} - x^2 + 1$, $f'(x) = x^3 + x^2 - 2x = x(x^2 + x - 2) = x(x+2)(x-1)$.
Setting the derivative to 0 we find the critical points: $x = -2, x = 0, x = 1$, with corresponding critical values:

$$f(-2) = \frac{(-2)^4}{4} + \frac{(-2)^3}{3} - (-2)^2 + 1 = 4 - \frac{8}{3} - 4 + 1 = -\frac{5}{3},\ f(0) = 1, \text{ and } f(1) = \frac{1}{4} + \frac{1}{3} - 1 + 1 = \frac{7}{12}$$

2.31

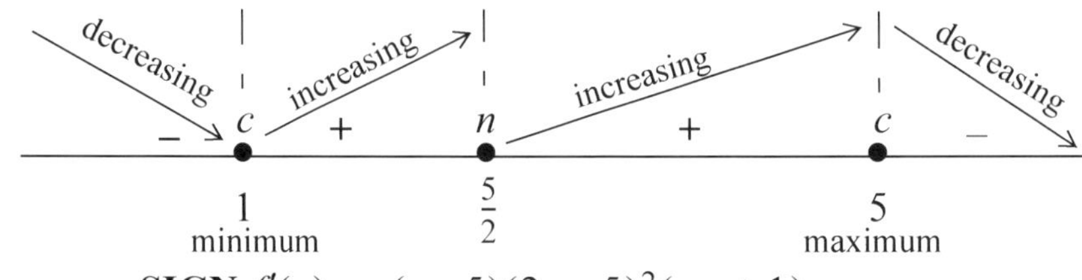

SIGN $f'(x) = (x-5)(2x-5)^2(-x+1)$

2.32 (a) Since the leading term of $f(x) = \boldsymbol{x}^3 - x$ is $\boldsymbol{x^3}$ the graph of $f(x) = x^3 - x$ resembles that of $g(x) = x^3$ as $x \to \pm\infty$.

(a) Since the leading term of $f(x) = (\boldsymbol{2x^3} + x)(\boldsymbol{x^2} - 5x + 1)(\boldsymbol{x} - 1)$ is $\boldsymbol{2x^3 \cdot x^2 \cdot x = 2x^6}$ the graph of f resembles that of $g(x) = 2x^6$ as $x \to \pm\infty$.

2.33 Differentiating and factoring $f(x) = \frac{x^5}{5} - \frac{x^3}{3} + 1$ we have:

$$f'(x) = x^4 - x^2 = x^2(x^2 - 1) = x^2(x+1)(x-1)$$

Then:

+	c	−	n	−	c	+
	−1 maximum		0		1 minimum	

SIGN $f'(x) = x^2(x+1)(x-1)$

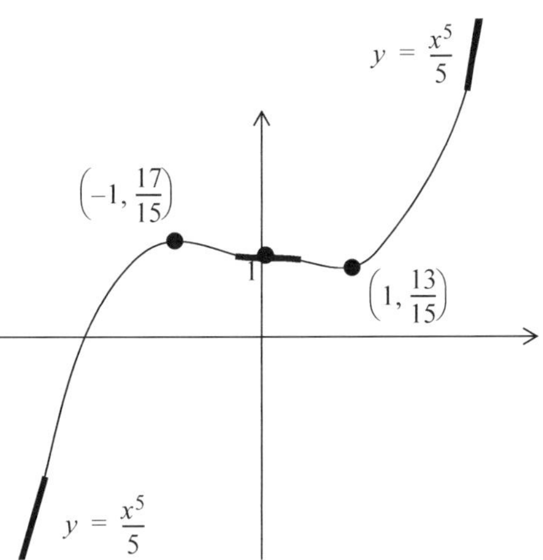

From the above we see that the graph achieves a maximum at $x = -1$, and evaluating the function at that point gives us the maximum value:

$$f(-1) = \frac{(-1)^5}{5} - \frac{(-1)^3}{3} + 1 = -\frac{1}{5} + \frac{1}{3} + 1 = \frac{17}{15}$$

Evaluating the function at $x = 1$ we find its minimum value: $f(1) = \frac{1}{5} - \frac{1}{3} + 1 = \frac{13}{15}$. Plotting the above maximum and minimum points, the y-intercept (note that a horizontal tangent line occurs at that intercept), and the nature of the graph as $x \to \pm\infty$ (leading term): $y = \frac{x^5}{5}$, lead us to the adjacent graph of the given function $f(x) = \frac{x^5}{5} - \frac{x^3}{3} + 1$.

2.34 $f(x) = 3x^5 - 5x^3$

$f'(x) = 15x^4 - 15x^2$

$f''(x) = 60x^3 - 30x = 30x(2x^2 - 1) = 30x(\sqrt{2}x + 1)(\sqrt{2}x - 1)$

SIGN $f''(x)$: $-$ c $+$ c $-$ c $+$ at $-\frac{1}{\sqrt{2}}$, 0, $\frac{1}{\sqrt{2}}$

Inflection points at: $-\frac{1}{\sqrt{2}}$, 0, and $\frac{1}{\sqrt{2}}$

2.35

Differentiating and factoring we have:

$$f'(x) = \left(\frac{x^4}{4} + \frac{x^3}{3} - x^2 + 1\right)'$$

$$= x^3 + x^2 - 2x = x(x+2)(x-1)$$

$-$ (shaded) | -2 min endpoint | $+$ | c 0 max | $-$ | c 1 min | $+$ | 2 max endpoint | (shaded)

SIGN $f'(x) = x(x+2)(x-1)$

We are only concerned with the non-shaded region at the left. In that region, there are two contenders for absolute minimum: at $x = -2$ and at $x = 1$. By evaluating the function at those two points, we find that -2 wins out: $f(-2) = -\frac{5}{3}$ while $f(1) = \frac{7}{12}$ (see 2.30 above).

There are also two contenders for absolute maximum: at $x = 0$ and at $x = 2$ (end-point maximum). Since $f(0) = 1$ and

$$f(2) = \frac{2^4}{4} + \frac{2^3}{3} - (2)^2 + 1 = 4 + \frac{8}{3} - 4 + 1 = \frac{11}{3}$$

the absolute maximum occurs at 2.

2.36 Differentiating the profit function, $P(x) = 5.7x - 0.025x^2$ and setting it to zero, we find that the only critical point occurs at $x = 114$: $P'(x) = 5.7 - 0.05x = 0$

$$x = \frac{5.7}{0.05} = 114$$

Knowing that the graph of the profit function is a parabola opening downward, we conclude that the maximum occurs at $x = 114$, with maximum value (maximum profit) of:

$$P(114) = 5.7(114) - 0.025(114)^2 = 324.9 \text{ thousand dollars: } \$324.900.$$

2.37

Step 1:

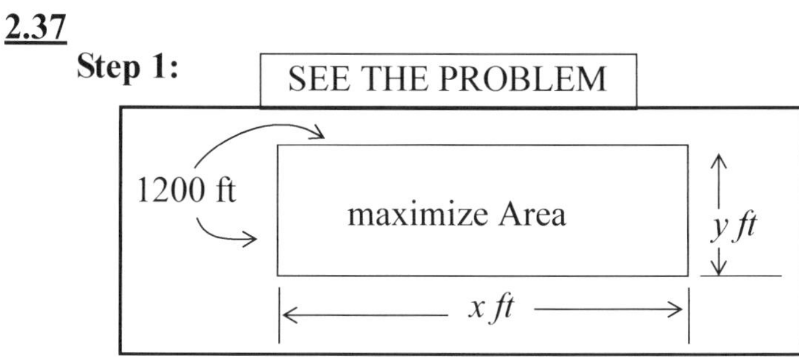

Step 2: $A = xy$

Step 3: $2x + 2y = 1200$

$2y = 1200 - 2x$

$y = \mathbf{600} - x$

$A(x) = x(\mathbf{600} - x) = 600x - x^2$ **(*)**

Step 4: Differentiate, set equal to zero, and solve: $A'(x) = 600 - 2x = 0$

$x = 300$

SIGN A': 0 $-$ c $+$; 300 min

At this point we know that maximum area occurs at $x = 300$. Substituting this value in (*) we obtain the maximum area: $600(300) - 300^2 = 90{,}000 \text{ ft}^2$.

2.38 **Step 1:**

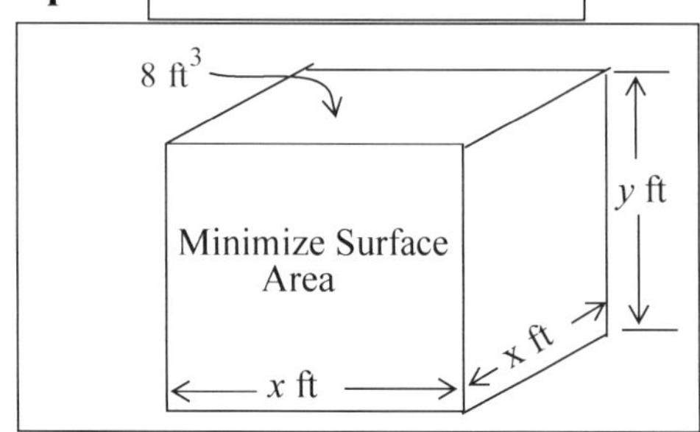

Step 2:

area of top and bottom ↓ ↓ area of the four sides

$$S = 2x^2 + 4xy$$

Step 3: $x^2y = 8$ (volume)

$$y = \frac{8}{x^2}$$

$$S(x) = 2x^2 + 4x\left(\frac{8}{x^2}\right) = 2x^2 + \frac{32}{x} \quad (*)$$

Step 4: Differentiate set equal to zero and solve:

$$S'(x) = (2x^2 + 32x^{-1})' = 4x - 32x^{-2} = 4x - \frac{32}{x^2} = 0$$

$$4x = \frac{32}{x^2}$$

$$4x^3 = 32$$

$$x^3 = 8$$

$$x = \mathbf{2}$$

0 — 2 + (min) SIGN S′

Returning to (*) we find the minimum amount of cardboard: $S(2) = 2 \cdot \mathbf{2}^2 + \frac{32}{\mathbf{2}} = 24 \text{ ft}^2$

2.39 **Step 1:** SEE THE PROBLEM: $24 ⟶ 150 $24+$x ⟶ 150 – 5x , where x denotes the number of dollars above $24 charged for rental.

Step 2: We express Revenue in terms of x:

cost of renting a car ↓ ↓ number of cars rented

$$R(x) = (24 + x)(150 - 5x) = -5x^2 + 30x + 3600$$

Step 3: Not Applicable.

Step 4: Differentiate, set equal to 0, and solve:

$$R'(x) = (-5x^2 + 30x + 3600)' = -10x + 30 = 0$$

$$x = 3$$

Conclusion: To maximize revenue, the agency should charge $27 per day.

2.40 **Step 1:** SEE THE PROBLEM

price per boat ↓
Boats per year: $n = 500 - \frac{p}{100}$ (*)
Cost for n boats: 500,000 + 10,000n
Maximize Profit

Step 2: Profit = Revenue - Cost:

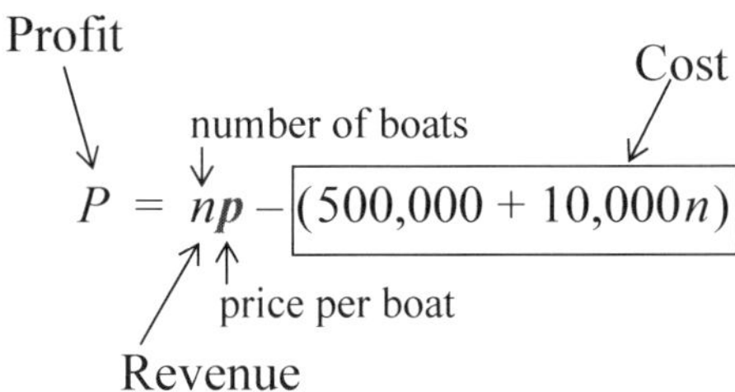

Step 3: We use (*) to express price in terms of n: $n = 500 - \frac{p}{100}$

$$p = 50{,}000 - 100n$$

Bringing us to:

$$\begin{aligned} P &= np - (500{,}000 + 10{,}000n) \\ &= n(50{,}000 - 100n) - (500{,}000 + 10{,}000n) \\ &= -500{,}000 + 40{,}000n - 100n^2 \end{aligned}$$

Step 4. Differentiating the profit function and setting the derivative equal to 0 reveals the one critical point:

$$\begin{aligned} P'(n) &= 40{,}000 - 200n = 0 \\ n &= 200 \end{aligned}$$

+ c –
0 200 max 500
SIGN P'

We conclude that a maximum profit will be realized with the manufacturing of 200 boats.

<u>**2.41**</u>

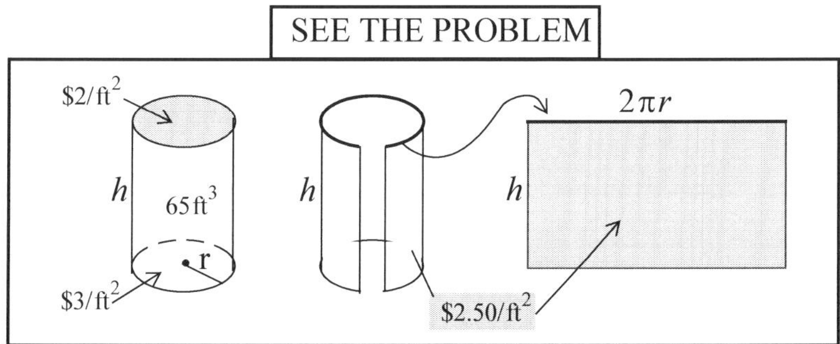

Total cost C is to be minimized, and from the above we see that:

$$\begin{aligned} C &= \text{cost of the top} + \text{cost of the bottom} + \text{cost of the side} \\ \$C &= \$2.00(\pi r^2) + \$3.00(\pi r^2) + \$2.50(2\pi rh) \\ C &= 5\pi r^2 + 5\pi rh \qquad (*) \end{aligned}$$

To express C as a function of only one variable, we need to eliminate either r or h. Since the volume of a cylinder is the area of its base times its height, we have:

$$\begin{aligned} \pi r^2 h &= 65 \\ h &= \frac{65}{\pi r^2} \qquad (**) \end{aligned}$$

Substituting in (*):

$$C = 5\pi r^2 + 5\pi r\left(\frac{65}{\pi r^2}\right) = 5\pi r^2 + \frac{325}{r}$$

Taking the easy way out:

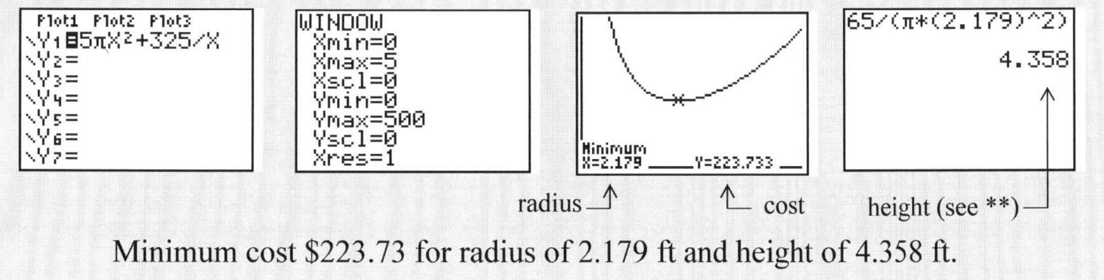

Minimum cost $223.73 for radius of 2.179 ft and height of 4.358 ft.

2.42 x^8+a and x^8+b, for any two different numbers a and b. One possible answer: x^8 and x^8+9.

2.43 (a) Since $(x^5)' = 5x^4$, $\int 5x^4 dx = x^5 + C$

(b) Since $(x^{-4})' = -4x^{-5}$, $\int -4x^{-5} dx = x^{-4} + C$

2.44 $\int\left(2x^5+4x^3-\frac{1}{3}x^2+2\right)dx = 2\int x^5 dx + 4\int x^3 dx - \frac{1}{3}\int x^2 dx + 2\int 1 dx$

$$= 2\cdot\frac{x^6}{6} + 4\cdot\frac{x^4}{4} - \frac{1}{3}\cdot\frac{x^3}{3} + 2x + C = \frac{x^6}{3} + x^4 - \frac{x^3}{9} + 2x + C$$

2.45 (a) $\int[(3x^2-2x+1)(x-5)]dx = \int(3x^3-17x^2+11x-5)dx = \frac{3}{4}x^4 - \frac{17}{3}x^3 + \frac{11}{2}x^2 - 5x + C$

(b) $\int\frac{x^4-2x-6}{x^4}dx = \int\left(\frac{x^4}{x^4} - 2\frac{x}{x^4} - \frac{6}{x^4}\right)dx = \int(1-2x^{-3}-6x^{-4})dx$

$$= x - 2\cdot\frac{x^{-2}}{-2} - 6\cdot\frac{x^{-3}}{-3} + C = x + \frac{1}{x^2} + \frac{2}{x^3} + C$$

2.46 $f(x) = \int(5x^4-2)dx = x^5 - 2x + C$. Since $f(0) = 1$: $1 = 0^5 - 2\cdot 0 + C$, or $C = 1$.

Thus: $f(x) = x^5 - 2x + 1$

2.47 Since $\overline{MC}(x) = -\frac{x}{500} + 20$: $C(x) = \int\left(-\frac{x}{500} + 20\right)dx = -\frac{x^2}{1000} + 20x + C$

Since $C(0) = 7000$: $7000 = \frac{-0^2}{1000} + 20\cdot 0 + C \Rightarrow C = 7000$, so: $C(x) = -\frac{x^2}{1000} + 20x + 7000$

Since $\overline{MR}(x) = 45$: $R(x) = \int 45 dx = 45x + C$

But $R(0) = 0$ (sell nothing, make nothing), so $C = 0$, and: $R(x) = 45x$

Thus: $P(x) = R(x) - C(x) = 45x - \left(-\frac{x^2}{1000} + 20x + 7000\right) = 25x + \frac{x^2}{1000} - 7000$

And: $P(600) = 25\cdot 600 + \frac{(600)^2}{1000} - 7000 = \$8,360$

2.48 (a) Differentiating the position function $s(t) = -16t^2 + 64t + 80$ gives us the velocity function $v(t) = -32t + 64$. Setting velocity to zero we determine the time it takes for the stone to reach its maximum height: $-32t + 64 = 0$, or $t = 2$. Evaluating the position function at $t = 2$ yields the maximum height: $s(2) = -16 \cdot 2^2 + 64 \cdot 2 + 80 = 144$ feet.

(b) Setting the position function to zero (ground level) we determine the time it takes for the stone to hit the ground:

$$\begin{aligned} -16t^2 + 64t + 80 &= 0 \\ -16(t^2 - 4t - 5) &= 0 \\ -16(t-5)(t+1) &= 0 \\ t = 5 \quad \text{or} \quad &\boxed{t = -1} \end{aligned}$$

Knowing it takes 5 seconds for the stone to hit the ground, we can determine its velocity at impact: $v(5) = -32 \cdot 5 + 64 = -96\frac{\text{ft}}{\text{sec}}$. Since speed is the magnitude of velocity, we conclude that the stone hits the ground at a speed of 96 feet per second.

2.49 Applying Theorem 2.18, with $f(t) = (3t^2 + 2)^7$, we arrive at the answer:

$$T'(x) = f(x) = (3x^2 + 2)^7$$

2.50 You will still get 9:

the +100 cancels out the -100

$$\int_1^2 (3x^2 + 2)dx = (x^3 + 2x + 100)\Big|_1^2 = (\mathbf{2}^3 + 2 \cdot \mathbf{2} + 100) - (\mathbf{1}^3 + 2 \cdot \mathbf{1} + 100) = 9$$

2.51 $$Area = \int_{-1}^{1} (x^2 + 1)(x^2 + 3)dx = \int_{-1}^{1} (x^4 + 4x^2 + 3)dx$$

$$= \frac{x^5}{5} + \frac{4x^3}{3} + 3x\Big|_{-1}^{1} = \left(\frac{1}{5} + \frac{4}{3} + 3\right) - \left(-\frac{1}{5} - \frac{4}{3} - 3\right) = \frac{136}{15}$$

2.52 (a) $\int_{-2}^{1} (3x^2 + 2)dx = x^3 + 2x\Big|_{-2}^{1} = (1^3 + 2 \cdot 1) - [(-2)^3 + 2(-2)] = 3 - (-12) = 15$

(b) $\int_0^1 (x^3 + x - 1)dx = \frac{x^4}{4} + \frac{x^2}{2} - x\Big|_0^1 = \left(\frac{1}{4} + \frac{1}{2} - 1\right) - (0) = -\frac{1}{4}$

2.53 Barrels produced: $\int_0^{30} \left(75 - \frac{t}{2500}\right)dt = \left(75t - \frac{t^2}{5000}\right)\Big|_0^{30} = 75 \cdot 30 - \frac{30^2}{5000} \approx 2250$

Income: $\$(85 \cdot 2250) = \$191,250$

2.54 (a) $\int_a^c -2f(x)dx + \int_c^a g(x)dx = -2\int_a^c f(x)dx - \int_a^c g(x)dx = -2 \cdot 5 - 7 = -17$

(b) $\int_a^b f(x)dx + \int_a^c 2g(x)dx = \left[\int_a^c f(x)dx + \int_c^b f(x)dx\right] + 2\int_a^c g(x)dx = 5 - 3 + 2 \cdot 7 = 16$

ANSWERS PART 1

EXERCISES 1.1, PAGE 9

1. {Jan, Feb, Mar, Apr, May, Jun, Jul, Aug, Sept, Oct, Nov, Dec} **3.** {CW, VF, VW, SF, SW}

5. {H1, H2, H3, H4, H5, H6, T1, T2, T3, T4, T5, T6}

7. {AA, AB, AC, BA, BB, BC, CA, CB, CC } **9.** {ABC, ACB, BAC, BCA, CAB, CBA}

11. $\frac{26}{52}$ **13.** $\frac{6}{52}$ **15.** $\frac{6}{36}$ **17.** $\frac{6}{36}$ **19.** $\frac{8}{20}$ **21.** $\frac{8}{20}$ **23.** $\frac{9}{25}$

25. $\frac{8}{25}$ **27.** $\frac{1}{6}$ **29.** $\frac{4}{6}$ **31.** $\frac{1}{6}$ **33.** $\frac{5}{6}$ **35.** $\frac{4}{10}$ **37.** $\frac{2}{10}$

39. $\frac{1}{10}$ **41.** $\frac{6}{10}$ **43.** $\frac{3}{10}$ **45.** $\frac{7}{10}$ **47.** $\frac{1}{6}$ **49.** $\frac{1}{6}$

51. $\frac{1}{5}$ **53.** $\frac{2}{6}$ **55.** $\frac{2}{6}$ **57.** $\frac{6}{20} = 0.3$ **59.** $\frac{6}{21} \approx 0.29$ **61.** $\frac{101}{301} \approx 0.34$

63. (a) $\frac{496}{2187} \approx 0.23$ (b) $\frac{263}{2187} \approx 0.12$ (c) $\frac{629}{2187} \approx 0.29$ (d) $\frac{50}{2187} \approx 0.023$

65. (a) 1:12 (one to twelve) (b) 10:3 (ten to three) **67.** $\frac{3}{8}$ **69.** 1:3 (one to three)

EXERCISES 1.2, PAGE 23

1. $\frac{12}{52}$ **3.** $\frac{36}{52}$ **5.** $\frac{22}{52}$ **7.** $\frac{32}{36}$ **9.** $\frac{24}{36}$ **11.** $\frac{12}{36}$ **13.** $\frac{2}{6}$

15. $\frac{4}{6}$ **17.** $\frac{4}{6}$ **19.** $\frac{7}{11}$ **21.** $\frac{3}{11}$ **23.** $\frac{1}{11}$ **25.** $\frac{2}{11}$ **27.** $\frac{9}{11}$

29. (a) $\frac{340{,}626}{634{,}410} \approx 0.54$ (b) $\frac{124{,}137}{634{,}410} \approx 0.20$ (c) $\frac{59{,}834}{634{,}410} \approx 0.094$ (d) $\frac{26{,}723}{634{,}410} \approx 0.042$
(e) $\frac{353{,}124}{634{,}410} \approx 0.557$ (f) $\frac{4718}{634{,}410} \approx 0.0074$ (g) $\frac{60{,}940}{634{,}410} \approx 0.096$

31. (a) $\frac{580}{585}$ (b) $\frac{530}{585}$ (c) $\frac{525}{585}$ (d) $\frac{5}{585}$ **33.** (a) 14 (b) 0 (c) $\frac{7}{21}$

35. (a) 400 (b) 365 (c) 460 (d) 290

37. (a) $\frac{9}{110}$ (b) $\frac{12}{110}$ (c) $\frac{77}{110}$ (d) $\frac{16}{110}$ (e) $\frac{93}{110}$ (f) $\frac{33}{110}$

39. (a) $\frac{206}{1000}$ (b) $\frac{794}{1000}$ (c) $\frac{41}{1000}$ (d) $\frac{15}{1000}$ (e) $\frac{80}{1000}$ (f) $\frac{191}{1000}$ (g) $\frac{375}{1000}$ (h) $\frac{419}{1000}$

EXERCISES 1.3, PAGE 36

1. (a) $\frac{13}{52}$ (b) $\frac{13}{26}$ (c) $\frac{13}{39}$ **3.** (a) $\frac{4}{11}$ (b) $\frac{4}{6}$ **5.** (a) $\frac{18}{36}$ (b) $\frac{18}{27}$ (c) $\frac{18}{18}$

7. (a) $\frac{1}{500}$ (b) $\frac{1}{450}$ (c) $\frac{1}{200}$ **9.** (a) $\frac{5120}{8994} \approx 0.57$ (b) $\frac{5120}{17{,}804} \approx 0.29$ (c) $\frac{8745}{17{,}795} \approx 0.50$ (d) $\frac{14{,}751}{28{,}621} \approx 0.52$ (e) $\frac{8044}{15{,}996} \approx 0.50$ (f) $\frac{23{,}686}{29{,}034} \approx 0.82$

11. (a) $\frac{4}{165} \approx 0.02$ (b) $\frac{7}{33} \approx 0.21$ (c) $\frac{28}{165} \approx 0.17$ (d) $\frac{28}{55} \approx 0.51$ **13.** $\frac{11}{850} \approx 0.013$

15. $\frac{22}{425} \approx 0.052$ **17.** $\frac{24}{5525} \approx 0.0043$ **19.** $\frac{1}{220} \approx 0.0045$ **21.** $\frac{1}{286} \approx 0.0035$

23. $\frac{7}{2162} \approx 0.0032$ **25.** $\frac{2}{3243} \approx 0.00062$ **27.** (a) $\frac{1}{6}$ (b) $\frac{1}{5}$ (c) $\frac{1}{5}$ (d) $\frac{1}{5}$

29. (a) $\left(\frac{4}{52}\right)^5 \approx 0.000003$ (b) $\left(\frac{8}{52}\right)^5 \approx 0.000086$ **31.** (a) $\left(\frac{1}{10}\right)^5 = 0.00001$ (b) $\left(\frac{1}{10}\right)^4 \approx 0.0001$

33. (a) $\left(\frac{280}{355}\right)^2 \approx 0.6221$ (b) $\left(\frac{75}{355}\right)^2 = 0.0446$ (c) 0.6226 for (a) and 0.0451 for (b).

35. (a) $\frac{5}{6} \cdot \frac{5}{6} \cdot \frac{1}{6} \approx 0.12$ (b) $\frac{1}{6} + \frac{5}{6} \cdot \frac{1}{6} + \frac{5}{6} \cdot \frac{5}{6} \cdot \frac{1}{6} \approx 0.42$ (c) $1 - \left(\frac{1}{6} + \frac{5}{6} \cdot \frac{1}{6}\right) \approx 0.69$

37. (a) $\frac{4}{52} \approx 0.077$ (b) $\frac{48}{52} \cdot \frac{4}{51} \approx 0.072$ (c) $\frac{48}{52} \cdot \frac{47}{51} \cdot \frac{4}{50} \approx 0.068$ (d) $\frac{48}{52} \cdot \frac{47}{51} \cdot \frac{46}{50} \cdot \frac{4}{49} \approx 0.064$ (e) 49

39. (a) $\frac{40}{52} \cdot \frac{39}{51} \cdot \frac{12}{50} \approx 0.141$ (b) $\frac{12}{52} + \frac{40}{52} \cdot \frac{12}{51} + \frac{40}{52} \cdot \frac{39}{51} \cdot \frac{12}{50} \approx 0.553$ (c) $1 - \left(\frac{12}{52} + \frac{40}{52} \cdot \frac{12}{51}\right) \approx 0.588$

41. (a) 1 (b) $\frac{14}{15} \cdot \frac{6}{8} \cdot \frac{2}{3} \approx 0.47$ **43.** (a) $\frac{131}{150} \approx 0.87$ (b) $\frac{19}{150} \approx 0.13$

EXERCISES 1.4, PAGE 50

1. (a) $\frac{10}{15}\cdot\frac{9}{14}\approx 0.043$ (b) $\frac{5}{15}\cdot\frac{4}{14}\approx 0.095$ (c) $\frac{6}{15}\cdot\frac{4}{14}\approx 0.11$ (d) $2\left(\frac{6}{15}\cdot\frac{4}{14}\right)\approx 0.23$

3. (a) $\frac{5}{100}\cdot\frac{4}{99}\cdot\frac{3}{98}\approx 0.00006$ (b) $\frac{5}{100}\cdot\frac{4}{99}\cdot\frac{95}{98}\approx 0.002$ (c) $\frac{95}{100}\cdot\frac{94}{99}\cdot\frac{5}{98}\approx 0.046$

(d) $\frac{95}{100}\cdot\frac{94}{99}\cdot\frac{93}{98}\approx 0.86$ (e) $3\left(\frac{5}{100}\cdot\frac{95}{99}\cdot\frac{94}{98}\right)\approx 0.046$

5. $\frac{3}{8}$ **7.** (a) $\frac{4}{45}$ (b) $\frac{4}{45}$ (c) $\frac{8}{45}$ (d) $\frac{4}{15}$ (e) $\frac{4}{15}$ **9.** (a) $\frac{2}{3}$ (b) 0 (c) $\frac{1}{2}$

11. (a) $\frac{1}{2}\cdot\frac{5}{15}+\frac{1}{2}\cdot\frac{12}{17}\approx 0.52$ (b) $\frac{12}{17}\approx 0.71$ (c) $\dfrac{\frac{1}{2}\cdot\frac{12}{17}}{\frac{1}{2}\cdot\frac{5}{15}+\frac{1}{2}\cdot\frac{12}{17}}\approx 0.68$ (d) $\dfrac{\frac{1}{2}\cdot\frac{4}{15}+\frac{1}{2}\cdot\frac{6}{15}}{1-\left(\frac{1}{2}\cdot\frac{5}{15}+\frac{1}{2}\cdot\frac{12}{17}\right)}\approx 0.70$

13. (a) $\frac{5}{15}\cdot\frac{4}{14}\approx 0.095$ (b) $\frac{1}{2}\cdot\frac{5}{15}\cdot\frac{4}{14}+\frac{1}{2}\cdot\frac{12}{17}\cdot\frac{11}{16}\approx 0.29$ (c) $\dfrac{\frac{1}{2}\cdot\frac{5}{15}\cdot\frac{4}{14}}{\frac{1}{2}\cdot\frac{5}{15}\cdot\frac{4}{14}+\frac{1}{2}\cdot\frac{12}{17}\cdot\frac{11}{16}}\approx 0.16$

(d) $\dfrac{\frac{1}{2}\left(1-\frac{12}{17}\cdot\frac{11}{16}\right)}{1-\left(\frac{1}{2}\cdot\frac{5}{15}\cdot\frac{4}{14}+\frac{1}{2}\cdot\frac{12}{17}\cdot\frac{11}{16}\right)}\approx 0.36$ **15.** (a) $\frac{6}{25}$ (b) $\frac{1}{5}$ (c) $\frac{1}{20}$ (d) $\frac{5}{6}$ (e) $\frac{1}{3}$ (f) 1

17. (a) $(0.9)(0.95)+(0.1)(0.85)\approx 0.94$ (b) $(0.9)(0.95)+(0.1)(0.15)\approx 0.87$

(c) $\dfrac{(0.9)(0.95)}{(0.9)(0.95)+(0.1)(0.15)}\approx 0.98$ (d) $\dfrac{(0.1)(0.15)}{(0.9)(0.95)+(0.1)(0.15)}\approx 0.02$

19. (a) $\frac{3}{80}\approx 0.038$ (b) $\frac{20}{80}(0.75)+\frac{10}{80}(0.9)+\frac{12}{80}(0.5)+\frac{3}{80}(0.75)+\frac{35}{80}(0.2)\approx 0.32$

(c) $\dfrac{\frac{12}{80}(0.5)}{\frac{20}{80}(0.75)+\frac{10}{80}(0.9)+\frac{12}{80}(0.5)+\frac{3}{80}(0.2)}\approx 0.23$ (d) $\dfrac{\frac{3}{80}(0.75)}{\frac{20}{80}(0.75)+\frac{12}{80}(0.5)+\frac{3}{80}(0.75)}\approx 0.097$

21. (a) $\frac{75}{100}$ (b) $\frac{75}{100}\cdot\frac{2}{7}+\frac{25}{100}\cdot\frac{1}{7}=\frac{1}{4}$ (c) $\dfrac{\frac{75}{100}\cdot\frac{2}{7}}{\frac{75}{100}\cdot\frac{2}{7}+\frac{25}{100}\cdot\frac{1}{7}}\approx 0.84$

23. $\frac{1}{3}\left[\frac{5}{12}\left(\frac{4}{11}+\frac{4}{11}+\frac{4}{13}\right)+\frac{4}{12}\left(\frac{5}{11}+\frac{4}{12}+\frac{3}{12}\right)+\frac{3}{12}\left(\frac{5}{12}+\frac{4}{11}+\frac{3}{12}\right)\right]\approx 0.34$

EXERCISES 1.5, PAGE 60

1. $26 \cdot 25 \cdot 10$ **3.** $26 \cdot 10^4$ **5.** $50 \cdot 10^4$ **7.** $26 \cdot 25 \cdot 4^4$ **9.** $26 \cdot 10^3$

11. $26^2 \cdot 4$ **13.** 20,000,000 **15.** 270,000,000 **17.** 273,375 **19.** 1080

21. (a) $\frac{1}{10^2}$ (b) $\frac{9^6}{10^9} \approx 0.00053$ (c) $\frac{9^6 \cdot 8 \cdot 7 \cdot 6}{10^9} \approx 0.18$ (d) $\frac{9 \cdot 8 \cdot 7 \cdot 6 \cdot 5 \cdot 4 \cdot 3 \cdot 2 \cdot 1}{10^9} \approx 0.00036$

23. (a) $\frac{1}{4^3} \approx 0.016$ (b) $\frac{24}{4^4} \approx 0.093$ (c) $\frac{13}{4^3} \approx 0.20$ (d) $\frac{9}{4^3} \approx 0.14$ (e) $\frac{10}{4^2} \approx 0.63$

25. (a) $\frac{1}{100}$ (b) $\frac{1}{2}$ (c) $\frac{81}{100}$ (d) $\frac{19}{100}$ (e) $\frac{18}{100}$ **27.** (a) $\frac{2}{15}$ (b) $\frac{3}{10}$ (c) $\frac{1}{10}$

EXERCISES 1.6, PAGE 72

1. 24 **3.** 604,800 **5.** 35 **7.** $7! = 5040$ **9.** 0 **11.** $5! = 120$

13. $10! = 3{,}628{,}800$ **15.** $P(11, 5) = 55{,}440$ **17.** $P(7, 5) = 2520$

19. $C(27, 4) = 17{,}550$ **21.** $C(52, 13) = 635{,}013{,}559{,}600 \approx 6.35 \times 10^{11}$

(**23** and **25:** no unique answer) **27.** (a) $P(52, 2) = 2652$ (b) $C(52, 2) = 1326$ (c) 2 times.

29. (a) $C(5, 3) = 10$ (b) $C(5, 1) + C(5, 2) = 15$

31. $C(10, 3)C(8, 2)C(4, 0) = 3360$ **33.** $C(10, 1)C(8, 1)C(4, 1) = 320$

35. $C(6, 3)C(5, 2) = 200$ **37.** $C(7, 3)C(9, 4) = 4410$

39. $C(5, 2)C(9, 2)C(12, 2) = 23{,}760$ **41.** $C(4, 2)C(4, 3) = 24$

43. $C(13, 2)C(13, 1)C(13, 2) = 79{,}092$ **45.** $P(5, 4) = 120$ **47.** $4! = 24$

49. $P(6, 5) + 6! = 1440$ **51.** $\frac{240}{6!} = \frac{1}{3}$ **53.** $\frac{3! \cdot 3!}{6!} = \frac{1}{20}$ **55.** $\frac{P(5, 4)}{P(6, 4)} = \frac{1}{3}$ [or $\frac{C(5, 4)}{C(6, 4)}$]

57. $\frac{P(5, 3)}{P(6, 4)} = \frac{1}{6}$ **59.** $\frac{3}{P(6, 4)} = \frac{1}{120}$ **61.** $\frac{P(4, 2)P(4, 2)}{P(6, 4)} = \frac{2}{5}$ [or $\frac{C(4, 2)C(4, 2)}{C(6, 4)}$]

63. $\frac{2[4P(4, 3)] + P(4, 2)P(4, 2)}{P(6, 4)} = \frac{14}{15}$ [or $1 - \frac{P(4, 4)}{P(6, 4)}$] **65.** $\frac{C(12, 4)C(9, 0)}{C(21, 4)} = \frac{11}{133} \approx 0.083$

67. $\frac{C(9, 4) + C(7, 4) + C(5, 4)}{C(21, 4)} = \frac{166}{5985} \approx 0.028$

69. $\frac{C(9, 2)C(7, 1)C(5, 1) + C(9, 1)C(7, 2)C(5, 1) + C(9, 1)C(7, 1)C(5, 2)}{C(21, 4)} = \frac{9}{19} \approx 0.47$

71. $\dfrac{C(9,3)C(6,2)}{C(15,5)} = \dfrac{60}{143} \approx 0.42$ **73.** $1 - \dfrac{C(9,4)C(6,1) + C(9,5)C(6,0)}{C(15,5)} = \dfrac{101}{143} \approx 0.71$

75. $\dfrac{C(9,3)C(6,2)}{C(15,5)} = \dfrac{60}{143} \approx 0.42$ [not the same as $\dfrac{P(9,3)P(6,2)}{P(15,5)} = \dfrac{6}{143} \approx 0.042$]

77. $\dfrac{P(9,3)P(6,2)}{P(15,5)} = \dfrac{6}{143} \approx 0.042$ **79.** $\dfrac{C(48,2)}{C(52,2)} = \dfrac{188}{221} \approx 0.85$

81. $\dfrac{C(13,5)}{C(52,5)} = \dfrac{33}{66{,}640} \approx 0.0005$ **83.** $\dfrac{C(4,3)C(4,2)}{C(52,5)} = \dfrac{1}{108{,}290} \approx 0.000009$

85. $\dfrac{C(13,2)C(4,2)C(4,2)C(44,1)}{C(52,5)} = \dfrac{198}{4165} \approx 0.048$ (Not: $\dfrac{C(13,1)C(4,2)C(12,1)C(4,2)C(44,1)}{C(52,5)} \approx 0.096$)

87. $\dfrac{4}{C(52,5)} = \dfrac{1}{649{,}740} \approx 0.0000015$ **89.** $\dfrac{10 \cdot 4^5 - 10}{C(52,5)} = \dfrac{5}{1274} \approx 0.0039$

91. $\dfrac{4}{C(52,13)} \approx 6.3 \times 10^{-12}$ **93.** $\dfrac{C(16,13)}{C(52,13)} \approx 8.8 \times 10^{-10}$

95. (a) 1 (b) $\dfrac{C(19,5)C(5,0) + C(19,4)C(5,1)}{C(24,5)} = \dfrac{1292}{1771} \approx 0.73$
(c) $\dfrac{C(19,0)C(5,5) + C(19,1)C(5,4)}{C(24,5)} = \dfrac{4}{1771} \approx 0.0023$

97. (a) $P(7,3) = 210$ (b) $\dfrac{1}{7}$ (c) $\dfrac{3}{7}$

99. (a) $\dfrac{3}{7}$ (b) $\dfrac{C(3,2)C(4,1)}{C(7,3)} = \dfrac{12}{35}$ (c) $\dfrac{C(4,2)C(3,1)}{C(7,3)} = \dfrac{18}{35}$
(d) $1 - \dfrac{C(3,0)C(4,3)}{C(7,3)} = \dfrac{31}{35}$ (e) $\dfrac{C(3,3) + C(1,4)C(3,2)}{C(7,3)} = \dfrac{13}{35}$

101. (a) $\dfrac{1}{C(50,5)} = \dfrac{1}{2{,}118{,}760} \approx 0.00000047$ (b) $\dfrac{5 \cdot 45}{C(50,5)} = \dfrac{45}{423{,}752} \approx 0.00011$
(c) $1 - \left[\dfrac{1}{C(50,5)} + \dfrac{5 \cdot 45}{C(50,5)}\right] \approx 0.99989$

103. (a) $\dfrac{1}{10!} = \dfrac{1}{3{,}628{,}800} \approx 0.00000027$ (b) 0 (c) $\dfrac{C(10,2)}{10!} = \dfrac{1}{80640} \approx 0.000012$

105. $\dfrac{5}{12} \cdot \dfrac{C(3,2)}{C(14,2)} + \dfrac{4}{12} \cdot \dfrac{C(3,2)}{C(14,2)} + \dfrac{3}{12} \cdot \dfrac{C(5,2)}{C(14,2)} = \dfrac{19}{364} \approx 0.052$

107. (a) $\frac{4!}{2} = 12$ (b) $\frac{10!}{4} = 907{,}200$ (c) $\frac{11!}{4!4!2!} = 34{,}650$

109. The number of ways of grabbing 52 objects from 100 equals the number of ways of leaving 48 of the 100 in place. In general: $C(n, r) = C(n, n-r)$.

EXERCISES 1.7, PAGE 84

1. (a) $\left(\frac{1}{6}\right)^3\left(\frac{5}{6}\right)^3 \approx 0.0027$ (b) $C(6, 3)\left(\frac{1}{6}\right)^3\left(\frac{5}{6}\right)^3 \approx 0.054$

(c) $C(6, 5)\left(\frac{1}{6}\right)^5\left(\frac{5}{6}\right)^1 + C(6, 6)\left(\frac{1}{6}\right)^6\left(\frac{5}{6}\right)^0 \approx 0.00066$ (d) $1 - C(6, 6)\left(\frac{1}{6}\right)^6\left(\frac{5}{6}\right)^0 \approx 0.9998$

(e) $C(6, 4)\left(\frac{2}{6}\right)^4\left(\frac{4}{6}\right)^2 \approx 0.082$ (f) $\left(\frac{2}{6}\right)^2\left(\frac{4}{6}\right)^4 \approx 0.022$ (g) $\frac{5 \cdot 5^4}{6^6} \approx 0.067$ (h) $\left(\frac{1}{6}\right)^5 \approx 0.00012$

3. (a) $C(9, 2)\left(\frac{1}{9}\right)^2\left(\frac{8}{9}\right)^7 \approx 0.19$ (b) 0

5. (a) $(0.67)^7 \approx 0.061$ (b) $(0.33)^7 \approx 0.00043$
(c) $C(7, 1)(0.67)^1(0.33)^6 \approx 0.0061$ (d) $1 - (0.33)^7 \approx 0.99957$

7. (a) $(0.95)^4 \approx 0.81$ (b) $(0.05)^4 \approx 0.0000063$ (c) $C(4, 2)(0.95)^2(0.05)^2 \approx 0.014$
(d) $C(4, 2)(0.95)^2(0.05)^2 + C(4, 3)(0.95)^3(0.05)^1 + C(4, 4)(0.95)^4(0.05)^0 \approx 0.9992$

9. (a) $C(5, 4)(0.57)^4(0.43)^1 \approx 0.23$ (b) $C(5, 4)(0.57)^4(0.43)^1 + C(5, 5)(0.57)^5(0.43)^0 \approx 0.29$
(c) $1 - C(5, 5)(0.57)^5(0.43)^0 \approx 0.94$

11. (a) $C(100, 1)\left[\frac{C(4, 4) \cdot C(48, 9)}{C(52, 13)}\right]\left[1 - \frac{C(4, 4) \cdot C(48, 9)}{C(52, 13)}\right]^{99} \approx 0.20$

(b) $\left[1 - \frac{C(4, 4) \cdot C(48, 9)}{C(52, 13)}\right]^{100} \approx 0.77$ (c) $1 - \underbrace{\left[1 - \frac{C(4, 4) \cdot C(48, 9)}{C(52, 13)}\right]^{100}}_{\text{answer to (b)}} \approx 0.23$

13. (a) $C(12, 3)\left(\frac{1}{9}\right)^3\left(\frac{8}{9}\right)^9 \approx 0.10$

(b) $C(12, 0)\left(\frac{1}{9}\right)^0\left(\frac{8}{9}\right)^{12} + C(12, 1)\left(\frac{1}{9}\right)^1\left(\frac{8}{9}\right)^{11} + C(12, 2)\left(\frac{1}{9}\right)^2\left(\frac{8}{9}\right)^{10} \approx 0.86$

15. (a) Same as 13(a) (b) Same as 13(b)

17. (a) $C(365, 0)(0.0001)^0(0.9999)^{365} + C(365, 1)(0.0001)^1(0.9999)^{364} \approx 0.9994$
(b) $C(93, 92)(A)^{92}(1 - A)^1 + C(93, 93)(A)^{93}(1 - A)^0 \approx 0.9982$, where A is the answer (a).

EXERCISES 1.8, PAGE 97

1. 1.63 **3.** $\frac{3}{2}$ **5.** 1.75 **7.** 0.75 **9.** \$0.42 **11.** \$0.00

13. \$0.12 **15.** –\$0.70 **17.** –\$0.47 **19.** 37 **21.** 10

23. 3.5 **25.** (a) ≈ 3.583 (b) 25 **27.** (a) $\frac{125}{3}$ (b) $\frac{10}{3}$

29. For each $1 \le i \le 100$, let X_i be the random variable that assigns the value 1 if person i draws his or her key from the hat, and the value 0 if not. Here is the expected value of each X_i: $E(X_i) = 0 \cdot \frac{99}{100} + 1 \cdot \frac{1}{100} = \frac{1}{100}$. Let S denote the random variable which assigns the total number of people drawing their own key: $S = \sum_{i=1}^{100} X_i$. Applying the sum theorem, we find that: $E(S) = \sum_{i=1}^{100} E(X_i) = \sum_{i=1}^{100} \frac{1}{100} = 1$. Conclusion: On the average, exactly one person will draw his or her key from the hat (independently of the number of people involved!).

31. For each $1 \le i \le n$, let X_i be the random variable that assigns the value 1 if trial i results in a success, and 0 if not. Then: $E(X_i) = 0 \cdot (1-p) + 1 \cdot p = p$. Let S denote the random variable which assigns the total number of successes: $S = \sum_{i=1}^{n} X_i$. Applying the sum theorem, we find that: $E(S) = \sum_{i=1}^{n} E(X_i) = \sum_{i=1}^{n} p = np$.

33. Assume that A will receive \$2 if a head comes up on the first toss (with probability 1/2); \$4 if on the second (with probability 1/4); \$8 if on the third (with probability 1/8); and so on. Suppose that A pays nothing to play the game. The expected winnings of A is then easily seen to be infinite:

$$\$2 \cdot \frac{1}{2} + \$4 \cdot \frac{1}{4} + \$8 \cdot \frac{1}{8} + \$16 \cdot \frac{1}{16} + \ldots = \$(1+1+1+1+\ldots)$$

It follows that no matter how much you have to pay to play the game, you will still come out ahead, on the average. Yes, but would you pay, say, a million dollars to play?

35. TV (Reaches 80,500 potential customers, versus 63,250 with radio, and 41,600 with magazine).

37. Door-To-Door (\$26,175, versus \$14,010 by mailing, and \$19,000 by phone).

39. Aircraft A (Profit of \$5952.50, versus \$4995.00 for aircraft B).

41. Do not carry collision insurance (final cost with collision: \$257; without: \$84).

43. Have ten cars at hand (6 cars: \$72; 7 cars: \$82.40; 8 cars: \$91.48; 9 cars: \$97.44; 10 cars: **\$100.16**; 11 cars: \$99.36; 12 cars: 96.80) **45.** Switch

REVIEW EXERCISES, PAGE 108

1. (a) $\frac{8}{52}$ (b) $\frac{1}{52}$ (c) $\frac{16}{52}$ (d) $\frac{15}{52}$ **2.** (a) $\frac{1}{365}$ (b) $\frac{11}{365}$ (c) $\frac{31}{365}$ (d) $\frac{334}{365}$

3. (a) $\frac{8}{15}$ (b) $\frac{4}{15}$ (c) $\frac{4}{15}$ (d) 0 (e) $\frac{7}{15}$ (f) $\frac{8}{15}$ **4.** $\frac{5}{11}$ **5.** 1 to 5 **6.** 12 to 1 **7.** $\frac{1}{101}$

8. 35 to 65 **9.** (a)

100
T S
16 5 9
2 3 17
42
W 6

(b) $\frac{21}{100}$ (c) $\frac{85}{100}$ (d) $\frac{22}{100}$

10. (a)

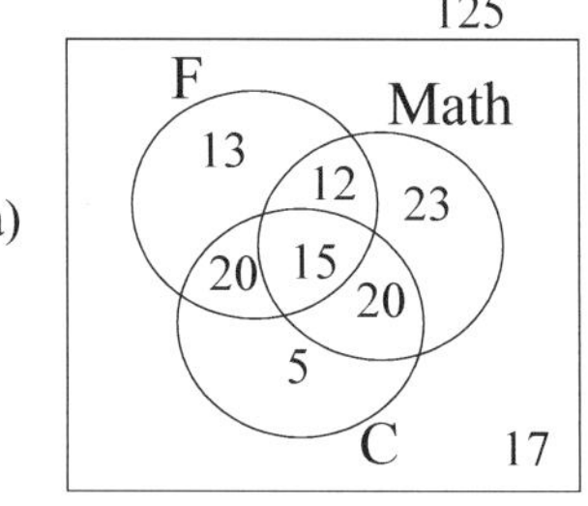

(b) $\frac{65}{125}$ (c) $\frac{33}{125}$ (d) $\frac{17}{125}$

11. $(0.95)^2$ **12.** $0.95 + (0.05)(0.95)^2$ **13.** (a) $\frac{3}{10}$ (b) $\frac{2}{5}$ (c) $\frac{1}{2}$

14. (a) $(0.4)(0.01) + (0.35)(0.02) + (0,25)(0.03) \approx 0.02$

(b) $\frac{(0.4)(0.99)}{(0.4)(0.99) + (0.35)(0.98) + (0.25)(0.97)} \approx 0.40$ (c) $\frac{(0.35)(0.98) + (0.25)(0.97)}{(0.4)(0.99) + (0.35)(0.98) + (0.25)(0.97)} \approx 0.78$

15. (a) $(0.7)(0.1) + (0.3)(0.9) = 0.34$ (b) $\frac{(0.7)(0.1)}{(0.7)(0.1) + (0.3)(0.9)} \approx 0.21$ **16.** 560

17. (a) $C(48, 5) = 1{,}712{,}304$ (b) $C(44, 5) = 1{,}086{,}008$ (c) $C(4, 2)C(4, 3) = 24$
(d) $C(4, 2)C(4, 2)C(44, 1) + C(4, 2)C(4, 3) = 1608$

18. $13!(4!)^{13} = 5{,}457{,}011{,}114{,}581{,}223{,}249{,}490{,}739{,}20$ **19.** (a) 384 (b) 2280 (c) 576

20. (a) 155 (b) 124 (c) 28 (d) 9 **21.** (a) $\frac{1}{9}$ (b) $\frac{1}{3}$ (c) 1

22. (a) $\frac{1}{10}$ (b) 9 to 1 (c) $\frac{6}{10}$ (d) 6 to 4

23. (a) $\frac{1}{C(12, 3)} = \frac{1}{220}$ (b) $\frac{5 \cdot 4 \cdot 3}{C(12, 3)} = \frac{3}{11}$ (c) $1 - \frac{C(9, 3)}{C(12, 3)} = \frac{34}{55}$

24. (a) $\frac{4! \cdot 8!}{12!} = \frac{1}{495}$ (b) $\frac{9 \cdot 4! \cdot 8!}{12!} = \frac{1}{55}$ (c) $\frac{3! \cdot 4! \cdot 5! \cdot 3!}{12!} = \frac{1}{4620}$

25. (a) $\frac{C(10, 4)}{C(14, 4)} \approx 0.21$ (b) $\frac{1}{C(14, 4)} \approx 0.001$ (c) $1 - \frac{C(4, 0)C(10, 4)}{C(14, 4)} \approx 0.79$

(d) $\frac{C(8, 4)}{C(12, 4)} \approx 0.14$ (e) $\frac{C(8, 4) + C(4, 1)C(2, 1)C(8, 2) + C(4, 2)C(2, 2)}{C(14, 4)} \approx 0.30$

(f) $\frac{C(5, 1)C(5, 3)}{C(5, 1)C(9, 3)} \approx 0.12$ (g) $\frac{C(4, 4)}{C(4, 3)C(10, 1) + C(4, 4)} \approx 0.025$

26. (a) $\dfrac{C(2,2)}{C(10,2)} = \dfrac{1}{45}$ (b) $\dfrac{C(2,1)C(3,1)}{C(10,2)} = \dfrac{2}{15}$

(c) $\dfrac{C(2,2)+C(2,1)C(3,1)+C(2,1)C(5,1)}{C(10,2)} = \dfrac{17}{45}$ (d) $1-\dfrac{C(3,2)}{C(10,2)} = \dfrac{14}{15}$ (e) $\dfrac{2}{5}$ (f) $\dfrac{3}{5}$

27. (a) $\dfrac{C(2,1)C(5,2)C(3,1)}{C(10,4)} = \dfrac{2}{7}$ (b) $\dfrac{C(2,2)C(3,2)}{C(10,2)} = \dfrac{1}{70}$ (c) $\dfrac{C(2,1)C(5,3)}{C(7,4)} = \dfrac{4}{7}$

28. (a) $\dfrac{C(8,1)}{C(10,3)} = \dfrac{1}{15}$ (b) $\dfrac{2\cdot 8}{C(10,3)} = \dfrac{2}{15}$ (c) $\dfrac{2\cdot 8+1\cdot 8}{C(10,3)} = \dfrac{17}{120}$ (d) $\dfrac{1}{C(10,3)} = \dfrac{1}{120}$

(e) $\dfrac{C(3,2)C(7,1)}{C(10,3)} = \dfrac{7}{40}$

29. (a) $\dfrac{1}{10}$ (b) $\dfrac{81}{100}$ (c) $\dfrac{1}{2}$ (d) $\dfrac{18}{25}$

30. (a) $\dfrac{13}{52}\cdot\dfrac{5}{12}+\dfrac{39}{52}\cdot\dfrac{C(5,1)C(7,1)+C(5,2)}{C(12,2)} \approx 0.62$ (b) $\dfrac{39}{52}\cdot\dfrac{C(5,2)}{C(12,2)} \approx 0.11$

(c) $\dfrac{\dfrac{13}{52}\cdot\dfrac{7}{12}}{\dfrac{13}{52}\cdot\dfrac{7}{12}+\dfrac{39}{52}\cdot\dfrac{C(7,2)}{C(12,2)}} \approx 0.38$

31. (a) $\dfrac{3}{7}\cdot\dfrac{5}{10}+\dfrac{4}{7}\cdot\dfrac{4}{10} \approx 0.44$ (b) $\dfrac{4}{7}\cdot\dfrac{1}{10} \approx 0.06$ (c) 0 (d) $\dfrac{\dfrac{3}{7}\cdot\dfrac{5}{10}}{\dfrac{3}{7}\cdot\dfrac{5}{10}+\dfrac{4}{7}\cdot\dfrac{5}{10}} \approx 0.43$

32. (a) $\dfrac{1}{4!} = \dfrac{1}{24}$ (b) 0 (c) $\dfrac{C(4,2)}{4!} = \dfrac{1}{4}$ (d) $\dfrac{4\cdot 2}{4!} = \dfrac{1}{3}$ (e) $\dfrac{3}{8}$

33. (a) $C(5,3)\left(\dfrac{2}{3}\right)^3\left(\dfrac{1}{3}\right)^2 \approx 0.33$ (b) $C(5,4)\left(\dfrac{2}{3}\right)^4\left(\dfrac{1}{3}\right)^1 + C(5,5)\left(\dfrac{2}{3}\right)^5\left(\dfrac{1}{3}\right)^0 \approx 0.46$

(c) $1 - C(5,0)\left(\dfrac{2}{3}\right)^0\left(\dfrac{1}{3}\right)^5 \approx 0.996$

(d) $C(5,3)\left(\dfrac{2}{3}\right)^3\left(\dfrac{1}{3}\right)^2 + C(5,4)\left(\dfrac{2}{3}\right)^4\left(\dfrac{1}{3}\right)^1 + C(5,5)\left(\dfrac{2}{3}\right)^5\left(\dfrac{1}{3}\right)^0 \approx 0.79$

34. (a) $C(10,4)\left(\dfrac{4}{52}\right)^4\left(\dfrac{48}{52}\right)^6 \approx 0.0045$ (b) $1-\left(\dfrac{48}{52}\right)^{10} \approx 0.55$

(c) $C(10,0)\left(\dfrac{4}{52}\right)^0\left(\dfrac{48}{52}\right)^{10} + C(10,1)\left(\dfrac{4}{52}\right)^1\left(\dfrac{48}{52}\right)^9 + C(10,2)\left(\dfrac{4}{52}\right)^2\left(\dfrac{48}{52}\right)^8 \approx 0.93$

35. (a) $C(100,10)(0.1)^{10}(0.9)^{90} \approx 0.013$ (b) $C(100,10)(0.02)^{10}(0.98)^{99} \approx 0.000029$

(c) $1-(0.98)^{100} \approx 0.87$ (d) $C(100,0)(0.02)^0(0.98)^{100} + C(100,1)(0.02)^1(0.98)^{99} \approx 0.40$

36. (a) $\dfrac{C(5,3)C(95,4)}{C(100,7)} \approx 0.002$ (b) $1 - \dfrac{C(95,7)}{C(100,7)} \approx 0.310$

(c) $1 - \dfrac{1}{C(100,7)} \approx 0.99999999994$

37. (a) $C(5,4)\left(\frac{1}{17}\right)^4\left(\frac{16}{17}\right)^1 \approx 0.000056$ (b) $1 - C(5,0)\left(\frac{1}{17}\right)^0\left(\frac{16}{17}\right)^5 \approx 0.26$

(c) $C(5,0)\left(\frac{1}{17}\right)^0\left(\frac{16}{17}\right)^5 + C(5,1)\left(\frac{1}{17}\right)^1\left(\frac{16}{17}\right)^4 \approx 0.97$

38. (a) $C(10,4)\left(\frac{1}{22}\right)^4\left(\frac{21}{22}\right)^6 \approx 0.00068$ (b) $1 - C(10,0)\left(\frac{1}{22}\right)^0\left(\frac{21}{22}\right)^{10} \approx 0.372$

(c) $C(10,2)\left(\frac{3}{11}\right)^2\left(\frac{8}{11}\right)^8 \approx 0.026$ (d) $C(10,9)\left(\frac{13}{55}\right)^9\left(\frac{42}{15}\right)^1 + \left(\frac{13}{55}\right)^{10} \approx 0.000065$

39. (a) $\dfrac{C(3,3)C(9,7)}{C(12,10)} = \dfrac{1}{22}$ (b) $\dfrac{C(3,2)C(9,8) + C(3,3)C(9,7)}{C(12,10)} = \dfrac{4}{11}$

(c) $1 - \dfrac{C(3,3)C(9,7)}{C(12,10)} = \dfrac{21}{22}$ (d) $\dfrac{C(4,1)C(3,3)C(5,1)}{C(4,1)C(11,4)} = \dfrac{1}{66}$

40. (a) $\left(\frac{893}{990}\right)^6 = 0.54$ (b) $\left(\frac{494}{495}\right)^6 \approx 0.988$ (c) $C(6,2)(\frac{1}{495})^2(1 - \frac{1}{495})^4 = 0.000061$

41. (a) $(0.6)^4 + C(4,3)(0.6)^3 \cdot \frac{1}{2} + C(4,2)(0.6)^2(0.4)^2 \cdot \frac{1}{6} \approx 0.62$ (b) $\dfrac{(0.6)^4}{\text{answer of (a)}} \approx 0.21$

42. \$0.50 **43.** \$8.75 **44.** 0.23 **45.** Play, for you can expect to win \$0.25

46. Company B with expected profit of \$2,250. (The expected profit of company A is \$1,250, and company C has a guaranteed profit of \$500).

ANSWERS PART 2

EXERCISES 2.1, PAGE 127

1. -100 **3.** $\frac{25}{9}$ **5.** $\frac{9}{128}$ **7.** $\frac{5}{6}$ **9.** $\frac{1}{8}$ **11.** $-\frac{1}{ab}$ **13.** $\frac{13}{5}$ **15.** $\frac{8}{5}$

17. $\frac{8}{7}$ **19.** $x > 9$, or $(9, \infty)$ **21.** $x \le -\frac{10}{13}$, or $\left(-\infty, -\frac{10}{13}\right]$ **23.** $x < 10$, or $(-\infty, 10)$

25. 2 **27.** 12 **29.** 0 **31.** $m_1 = 1$; $m_2 = -1$; $m_3 = 2$; $m_4 = \frac{2}{5}$

33. $(1, 7)$ and $(-2, -5)$ **35.** No unique answer. **37.** $y = 2$ **39.** $y = -x + 3$ **41.** $y = 3$

43. $y = 3x$ **45.** $y = 3x - 4$ **47.** $y = -\frac{9}{5}x - \frac{2}{5}$ **49.** $y = 3x - 13$ **51.** $y = \frac{1}{2}x$

53. (a) $y - 1 = 3(x + 2)$ (b) $y = 3x + 7$ **55.** $(3x + 2)(3x - 2)$ **57.** $(2x + 5)(2x - 5)$

59. $(-2x + 1)(2x + 1)$ **61.** $(\sqrt{2}x + \sqrt{5})(\sqrt{2}x - \sqrt{5})$ **63.** $(x + 3)(x + 4)$

65. $(x - 4)(x - 3)$ **67.** $(2x + 1)(3x + 2)$ **69.** $(2x + 1)(3x - 5)$ **71.** $x(2x + 5)(3x + 8)$

73. $x = \frac{5}{2}, x = -7$ **75.** $x = \pm\frac{4}{5}$ **77.** $x = \pm\frac{\sqrt{30}}{5}$ **79.** $x = -5$

81. $x = -3, x = 7$ **83.** $x = -\frac{1}{3}, x = 0, x = 5$ **85.** $x = -\frac{1}{2}, x = 0, x = 3$

87. $x = -5, x = -2, x = \frac{1}{5}, x = \frac{5}{2}$ **89.** $x = \pm\frac{5}{2}, x = \pm\sqrt{2}$ **91.** $(-\infty, -1) \cup \left(\frac{1}{3}, \infty\right)$

93. $\left(-\infty, \frac{5}{3}\right]$ **95.** $(-\infty, -3] \cup \{4\}$ **97.** $\left(-\frac{4}{3}, 0\right) \cup (2, 3)$ **99.** $\left[-\frac{1}{3}, 0\right] \cup [5, \infty)$

101. 8 **103.** $x^2 + x + 2$ **105.** $2xh + h^2 - h$ **107.** -1 **109.** $-2x^2 - 7x - 4$

111. $-4hx - 2h^2 - 3h$ **113.** $f(-5) = -8$, $f(-1) = -4$, $f(0) = 0$, $f(2) = 4$, $f(10) = 20$

115. $f(-5) = -10$, $f(-1) = -2$, $f(0) = 0$, $f(2) = -2$, $f(10) = 100$

Exercises 2.2, page 137

1. 4 **3.** $\frac{2}{3}$ **5.** 0 **7.** 50 **9.** $\frac{5}{3}$ **11.** –4 **13.** $\frac{4}{5}$ **15.** $\frac{2}{3}$ **17.** 0 **19.** $\frac{1}{5}$ **21.** $\frac{3}{32}$

23. $\frac{1}{5}$ **25.** $\frac{1}{2}$ **27.** 0 **29.** $\lim\limits_{x \to 3} f(x) = 2$; not continuous (removable discontinuity)

31. Limit does not exist; not continuous (Jump discontinuity) **33.** $\lim\limits_{x \to 2} f(x) = 4$; continuous

35. $\lim\limits_{x \to 2} f(x) = 4$; not continuous (removable singularity)

37. No unique answer. **39.** No unique answer.

Exercises 2.3, PAGE 147

1. $f'(2) \approx 1, f'(4) \approx 0, f'(7) \approx 0$ **3.** 5 **5.** 16 **7.** –1 **9.** 0 **11.** 13 **13.** –5

15. $6x$ **17.** $-4x + 1$ **19.** 0 **21.** $y = -3x - 30$ **23.** $y = 2x$ **25.** $y = 6x - 4$ **27.** $y = 11$

29. The graph of the function $f(x) = x$ is the line $y = x$ which has slope 1. Since the tangent line at every point on the graph of the function $f(x) = x$ is the line $y = x$, $f'(x) = 1$. More formally, for $f(x) = x$: $\lim\limits_{h \to 0} \frac{f(x+h) - f(x)}{h} = \lim\limits_{h \to 0} \frac{(x+h) - x}{h} = \lim\limits_{h \to 0} \frac{h}{h} = \lim\limits_{h \to 0} 1 = 1$.

31. The graph of the function $f(x) = mx + b$ is a line of slope m. It follows that the tangent line at every point on the graph of the function has slope m. Therefore $f'(x) = m$. More formally:
$\lim\limits_{h \to 0} \frac{f(x+h) - f(x)}{h} = \lim\limits_{h \to 0} \frac{m(x+h) + b - (mx+b)}{h} = \lim\limits_{h \to 0} \frac{mh}{h} = \lim\limits_{h \to 0} m = m$.

33. No limit at 3 and at 4. Neither continuous nor differentiable at 1,2,3, and 4.

35. (slope 2, slope 3; 2, 2) $\lim\limits_{h \to 2^-} \frac{f(x+h) - f(x)}{h} = 2$ while $\lim\limits_{h \to 2^+} \frac{f(x+h) - f(x)}{h} = 3$.

37.

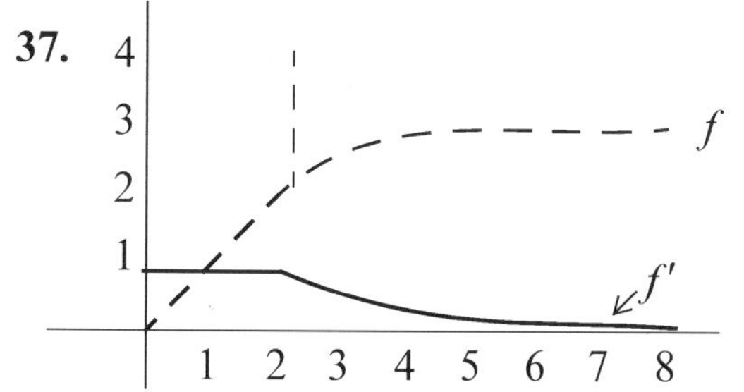

39. Let $h = x - c$. As $h \to 0: x \to c$ and $c + h \to x$. Thus:

$$f'(c) = \lim_{h \to 0} \frac{f(c+h) - f(c)}{h} = \lim_{x \to c} \frac{f(x) - f(c)}{x - c}$$

Exercises 2.4, PAGE 157

1. 2 **3.** $15x^4 + 12x^2$ **5.** $-\frac{6}{x^4}$ **7.** $21x^2 + 10x - 4 - \frac{4}{x^5}$ **9.** $1 - \frac{2}{x^2} - \frac{6}{x^3}$

11. $2x - \frac{6}{x^3} + \frac{15}{x^4}$ **13.** $18x^5 + 32x^3 + 9x^2 + 8x + 6$ **15.** $3 + \frac{5}{x^2}$ **17.** $\frac{3x^2 + 24x + 13}{(x+4)^2}$

19. $-\frac{30x}{(3x^2+1)^2}$ **21.** $\frac{5x^2 - 10x}{(2x+1)^2(3x-1)^2}$ **23.** $\frac{6x^3 + 9x^2 + 4x}{(3x+1)^2}$ **25.** $6x - 12$

27. $6x + 4 - \frac{2}{x^3}$ **29.** 8 **31.** $\frac{3}{2}$ **33.** 3 **35.** 13 **37.** $\frac{17}{6}$ **39.** $\frac{7}{4}$

41. 49 **43.** $y = 5x - 4$ **45.** $y = -5x - 8$ **47.** $y = \frac{1}{4}x + \frac{3}{4}$ **49.** 2 **51.** $-\frac{1}{2}, 1$

53.
$$[f(x) + g(x)]' = \lim_{h \to 0} \frac{[f(x+h) + g(x+h)] - [f(x) + g(x)]}{h}$$
$$= \lim_{h \to 0} \frac{f(x+h) - f(x)}{h} + \lim_{h \to 0} \frac{g(x+h) - g(x)}{h} = f'(x) + g'(x)$$

a similar argument can be given for part (b)

55.
$$(x^2)' = \lim_{h \to 0} \frac{(x+h)^2 - x^2}{h} = \lim_{h \to 0} \frac{x^2 + 2xh + h^2 - x^2}{h}$$
$$= \lim_{h \to 0} \frac{2xh + h^2}{h} = \lim_{h \to 0} \frac{h(2x+h)}{h} = \lim_{h \to 0} (2x + h) = 2x$$

Then: $(x^4)' = (x^2 \cdot x^2)' = x^2(x^2)' + x^2(x^2)' = x^2(2x) + x^2(2x) = 4x^3$.

Exercises 2.5, PAGE 167

1. (a) 0.08000 billion per year (b) 0.07800 billion per year (c) 0.07875 billion per year

3. (a) $\frac{9}{4}$ (b) 2 (c) $\frac{9}{4}$ (d) 2 **5.** (a) $\frac{2}{3}$ (b) $\frac{2}{3}$ (c) 0 (d) 0 **7.** 4π

9. (a) 64 ft/sec (b) 256 ft (c) 128 ft/sec

11. (a) Fixed monthly cost. (b) $23 - \frac{2x}{25}$ (c) \$15
(d) The approximate cost to produce the 101th unit. (e) \$14.96

13. (a) Fixed weekly cost. (b) $22.5 - 0.02x$ (c) \$21.50
(d) The approximate cost to produce the 51th unit (e) \$21.49

15. (a) $50 + 0.04x$ (b) \$54.00 (c) Approximate revenue from the sale of the 101th unit (d) \$54.02.

17. (a) $P(x) = 15x + 0.12x^2 - 2500$, $\overline{MP}(x) = 15 + 0.24x$ (b) \$39.00 (c) The approximate profit from the sale of the 101th unit (d) \$39.12

19. At $x = 1$, marginal cost is 1000; at $x = 7$ marginal cost is 0.

Exercises 2.6, PAGE 180

1. $x = -1, f(-1) = -\frac{7}{2}$; $x = 0, f(0) = 0$; $x = 3, f(3) = -\frac{135}{2}$

3. $x = 0,\ f(0) = 1$; $x = -1,\ f(-1) = 3$; $x = 1,\ f(1) = -1$

5. $x = 2, f(2) = \frac{1}{4}$

7. $x = 0, f(0) = 0$; $x = 2, f(2) = 4$

9. Maximum: $(-3, 54)$
Minimum: $(3, -54)$
Inflection point: $(0, 0)$
As $x \to \pm\infty$ the graph resembles that of $y = x^3$

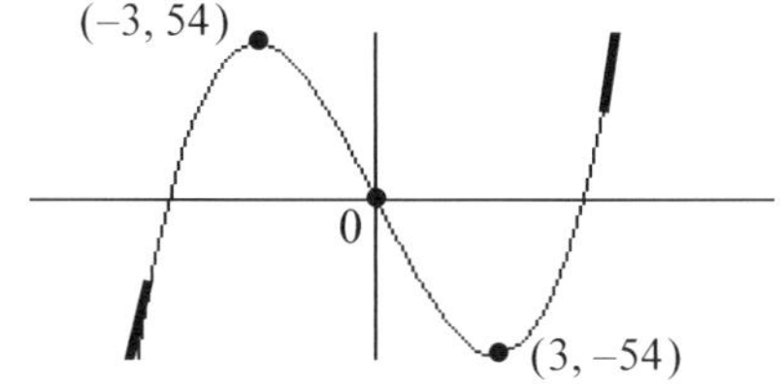

11. Minimum: $(-1, 2)$, $(1, 2)$
Maximum: $(0, 3)$
Inflection points: $\left(-\frac{1}{\sqrt{3}}, \frac{22}{9}\right), \left(\frac{1}{\sqrt{3}}, \frac{22}{9}\right)$
As $x \to \pm\infty$ the graph resembles that of $y = x^4$

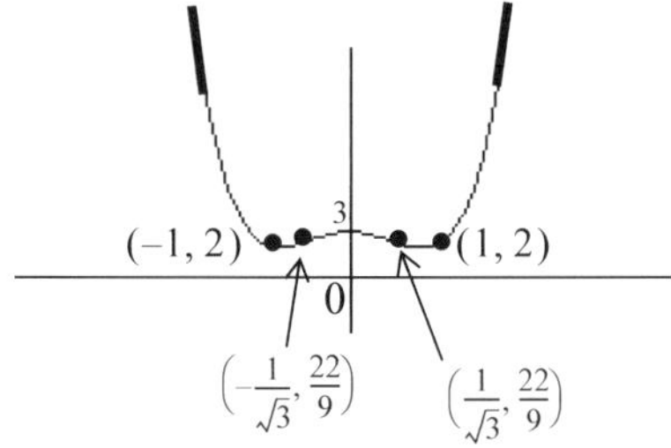

13. Maximum: $(-1, 12)$
Minimum: $(1, 8)$
As $x \to \pm\infty$
Inflection points: $(0, 10), \left(-\frac{1}{\sqrt{2}}, 10 + \frac{7\sqrt{2}}{8}\right), \left(\frac{1}{\sqrt{2}}, 10 - \frac{7\sqrt{2}}{8}\right)$
As $x \to \pm\infty$
the graph resembles that of $y = 3x^5$

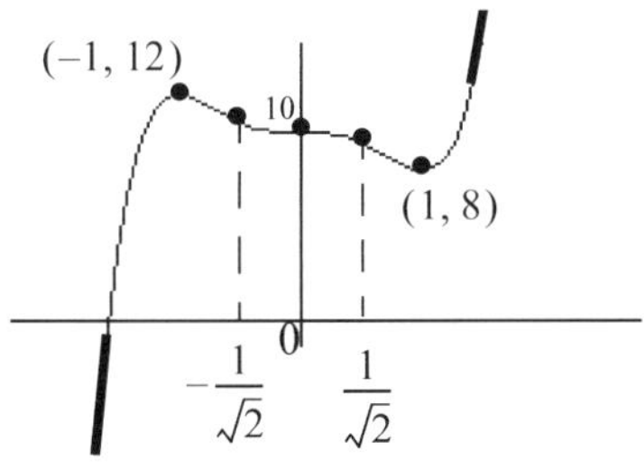

15. Minimum: $\left(1, \frac{4}{3}\right)$

Inflection points: $(0, 1)$, $\left(\frac{2}{3}, \frac{65}{81}\right)$

As $x \to \pm\infty$
the graph resembles that of $y = x^4$

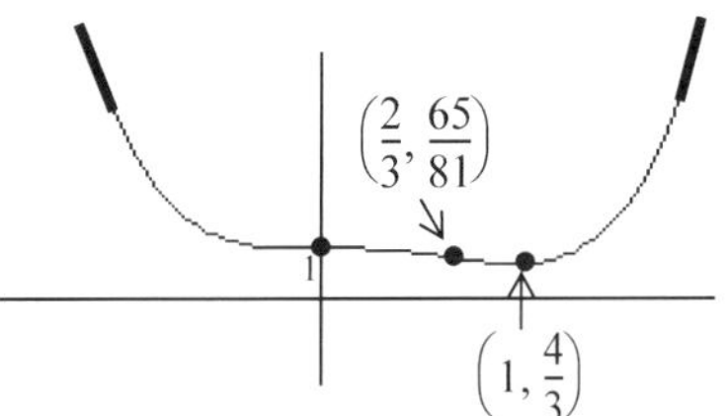

17. Since $f''(x) = (x^2)'' = (2x)' = 2$ is always positive, the graph is concave up everywhere.

Since $f''(x^{2n}) = (x^{2n})'' = (2nx^{2n-1})' = 2n(2n-1)x^{2n-2} = 2n(2n-1)(x^{n-1})^2$ is always positive, the graph is concave up everywhere. (positive, positive)

19. (a) Absolute maximum at -2, Absolute minimum at ± 1.
(b) Absolute maximum at 2, Absolute minimum at ± 1.
(c) Absolute maximum at 2, Absolute minimum at 1.
(d) Absolute maximum at 3, Absolute minimum at 2.

21. For $f(x) = ax^2 + bx + c$: $f'(x) = 2ax + b$ and $f''(x) = 2a$. It follows that the graph will be concave up if $a > 0$ and concave down if $a < 0$. Moreover, since $f'(x) = 0$ where $2ax + b = 0$, or at $x = -\frac{b}{2a}$, the graph will have a minimum at that point if $a > 0$ and a maximum if $a < 0$.

23. Maximum at $x = 0$. Minimum at $x = \frac{2}{3}$.

25. Maximum at $x = -1$. Minimum at $x = 1$. Second Derivative test is inconclusive at $x = 0$.

27. (i) $f'(x) = 3x^2$. SIGN f': + • + (at 0: no max nor min). $f''(0) = 0$ since $f''(x) = 6x$.

(ii) $f'(x) = 4x^3$. SIGN f': − • + (at 0: min). $f''(0) = 0$, since $f''(x) = 12x^2$.

(iii) $f'(x) = -4x^3$. SIGN f': + • − (at 0: max). $f''(0) = 0$, since $f''(x) = 12x^2$.

Exercises 2.7, PAGE 189

1. 500 units **3.** $2\,\frac{\text{quarts}}{\text{acre}}$ **5.** $\frac{2r_0}{3}$ **7.** 36 by 48 in.

9. 100 by 150 ft, with the 100 ft dimension parallel to the dividing fence.

11. $12,000 **13.** $27 **15.** 18 by 18 by 36 in. **17.** 10 by 10 by 10 ft

19. $r = h = \frac{15}{4+\pi}$ ft **21.** $r = \left(\frac{65}{2\pi}\right)^{1/3} \approx 2.2$ ft, $h = \left(\frac{260}{\pi}\right)^{1/3} \approx 4.4$ ft

23. 563 units **25.** ≈ 536.7 feet downstream from the power plant.

27. The cable from B should connect with a point on the power line that is 5.63 miles south of *b*.

Exercises 2.8, PAGE 199

1. $3x + C$ **3.** $3x + \frac{3x^2}{2} + C$ **5.** $x^6 + x^5 + C$ **7.** $\frac{x^5}{25} + \frac{3}{4x^4} + C$

9. $\frac{3x^5}{5} + \frac{4}{3x^3} - \frac{1}{2x^4} + C$ **11.** $\frac{x^4}{2} - \frac{5x^3}{3} + C$ **13.** $\frac{3x^2}{2} - \frac{1}{x^2} + \frac{1}{3x^3} + C$

15. $\frac{x^2}{2} + x - \frac{1}{x} - \frac{1}{2x^2} + C$ **21.** $f(x) = \frac{3x^2}{2} + 5x - \frac{3}{2}$ **23.** $f(x) = \frac{x^4}{4} + \frac{5x^2}{2} - 2x - \frac{3}{4}$

25. $f(x) = -\frac{3}{2x^2} - \frac{5}{3x^3} + \frac{31}{6}$ **27.** $22,511.25 **29.** $7475 **31.** $23,450

33. $75 - \frac{x}{2000}$ **35.** $f(x) = \frac{x^3}{3} + \frac{14}{3}$ **37.** 384 ft/sec **39.** 56.25 ft **41.** ≈ 108 ft

Exercises 2.9, PAGE 212

1. 3 **3.** 6 **5.** -2 **7.** $\frac{5}{6}$ **9.** 0 **11.** $\frac{3}{20}$ **13.** 4 **15.** -4

17. $\frac{27}{8}$ **19.** $\frac{1}{4}$ **21.** $\frac{2}{3}$ **23.** $\frac{1}{x^2+4}$ **25.** $-\sqrt{3x^4+1}$ **27.** $x^2 + x$ **29.** $0.39

31. One hour **33.** 200 days **35.** $3,000 **37.** The $2000 machine **39.** $4,970

41. (a) $\frac{16}{3}$ meters (b) $\frac{17}{3}$ meters

43. If you choose a rational number in each Δx of any partition of $[0, 1]$, then $\sum_0^1 f(x)\Delta x = \sum_0^1 1\Delta x = 1$. On the other hand, if you choose an irrational number in each Δx, then $\sum_0^1 f(x)\Delta x = \sum_0^1 0\Delta x = 0$. It follows that $\lim_{\Delta x \to 0} \sum_a^b f(x)\Delta x$ does not exist.

45.

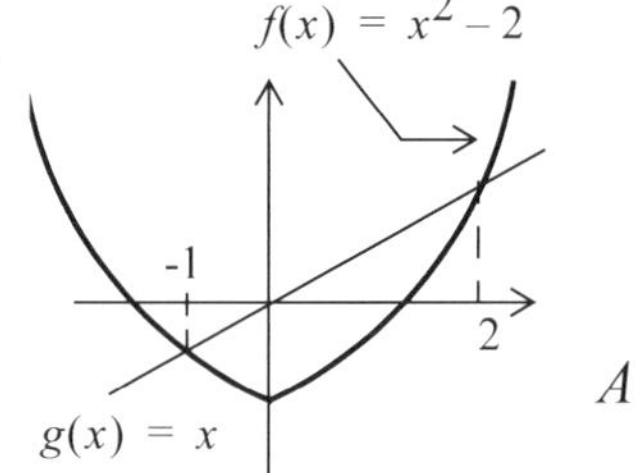

$A = \frac{9}{2}$

47.

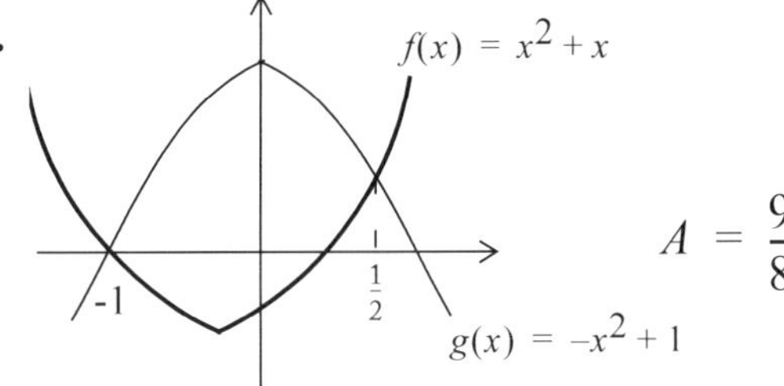

49.

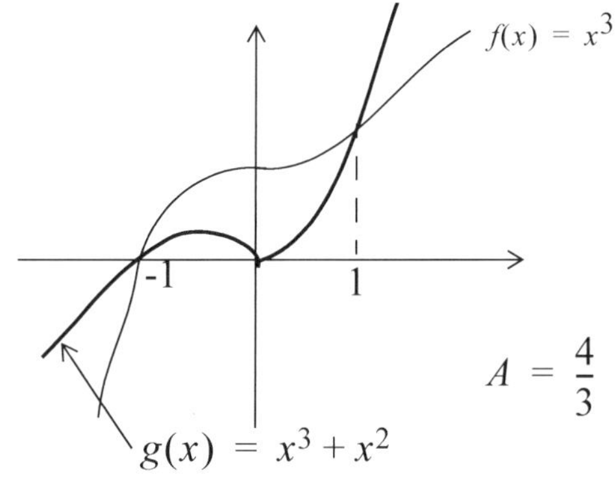

51.

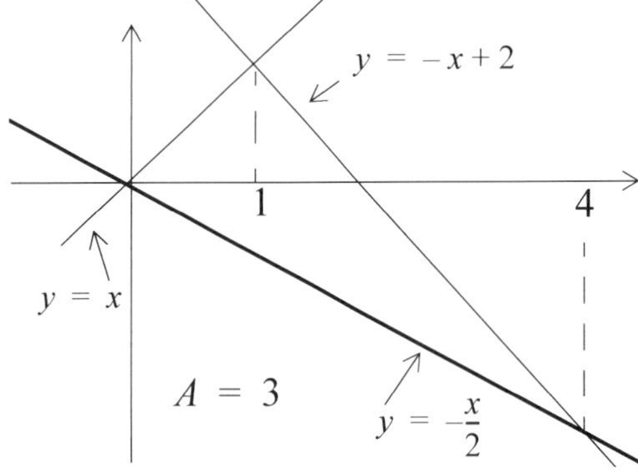

53.

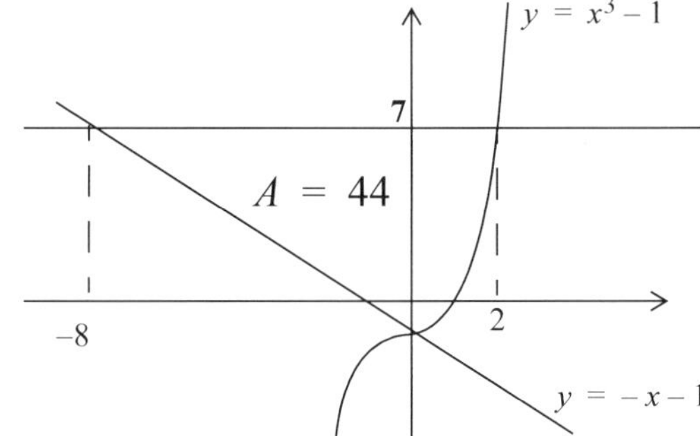

Review Exercises (Part 2), PAGE 222

1. -25 **2.** $\frac{1}{25}$ **3.** $\frac{9}{4}$ **4.** $\frac{5}{6}$ **5.** $-\frac{1}{ab}$ **6.** $\frac{a^2}{8b}$ **7.** $\frac{b^2}{a-b}$ **8.** $\frac{b^2+1}{b^2}$

9. $(x^2+4)(x+2)(x-2)$ **10.** $x(x+3)(x-3)$ **11.** $(2x^2+3)(\sqrt{2}x+\sqrt{3})(\sqrt{2}x-\sqrt{3})$

12. $3(x-2)(2x-1)$ **13.** $4(x-2)^2$ **14.** $4x(x-2)^2$ **15.** $y = 5x+4$

16. $y = 2x+3$ **17.** $x = 3$ **18.** $x = -\frac{1}{3}$ **19.** $x = -3, x = 5$ **20.** $x = -3, x = 1$

21. $x = -\frac{4}{7}, x = 2$ **22.** $x = 0, x = \frac{1}{2}, x = 3$ **23.** $\left(-\frac{1}{2}, \infty\right)$ **24.** $\left(-\infty, -\frac{8}{7}\right]$

25. $(-2, 0) \cup \left(\frac{3}{5}, 7\right)$ **26.** $[-2, 0] \cup [3, \infty)$ **27.** 0 **28.** $\frac{6}{7}$ **29.** $\frac{11}{7}$

30. 3 **31.** 5 **32.** $\frac{1}{9}$ **33.** 2 **34.** $-2x+3$ **35.** $-\frac{1}{2x^2}$ **36.** $12x^2+4x-1$

37. $-6x+2+\frac{5}{x^2}$ **38.** $3x^2-6x+1$ **39.** $6x+\frac{4}{x^3}-\frac{3}{x^4}$ **40.** $\frac{x^2-6x}{(x-3)^2}$ **41.** $\frac{2x^2-2x-1}{(2x-1)^2}$

42. $y = 3x-4$ **43.** $y = -x-1$ **44.** $x = 0, x = \frac{1}{2}$ **45.** $x = -\frac{3}{2}, x = 1$ **46.** 10 **47.** 0

48. $\overline{MP}(100) = \$35$. The company will realize a profit of approximately \$35 from the sale of the 101th unit.

49. $\overline{MP}(100) = \$11$. The company will realize a profit of approximately \$11 from the sale of the 101th unit.

50. \$12,375 **51.** \$37,375 **52.** \$13,500

53. Minimum: $(0, -2)$
No inflection points
As $x \to \pm\infty$ the graph resembles that of $y = x^4$

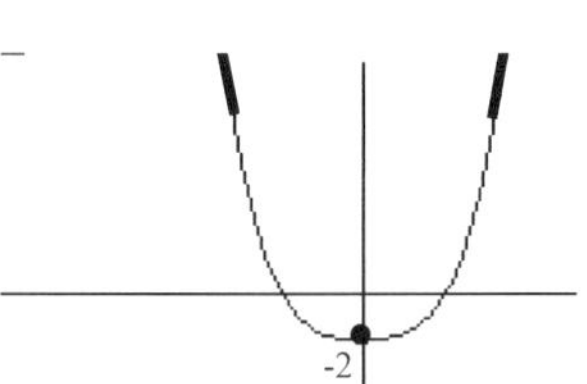

54. Minimum: $(3, -27)$
Inflection points: $(0, 0)$, $(2, -16)$
As $x \to \pm\infty$ the graph resembles that of $y = x^4$

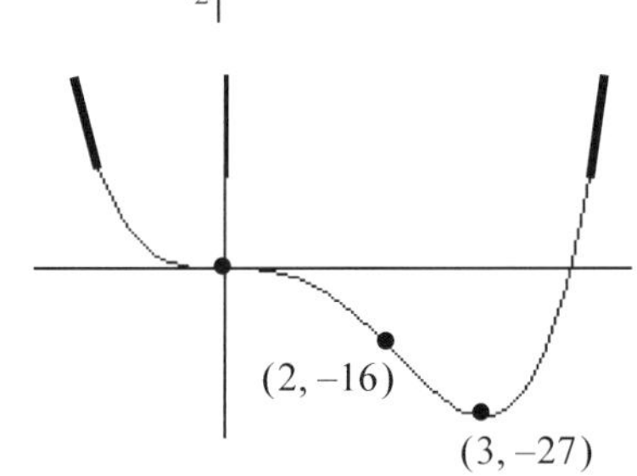

55. Minimum: $\left(\frac{3}{2}, \frac{5}{27}\right)$ Maximum: $\left(-1, \frac{5}{2}\right)$

Inflection point: $\left(-\frac{1}{6}, \frac{145}{108}\right)$

As $x \to \pm\infty$ the graph resembles that of $y = x^3$

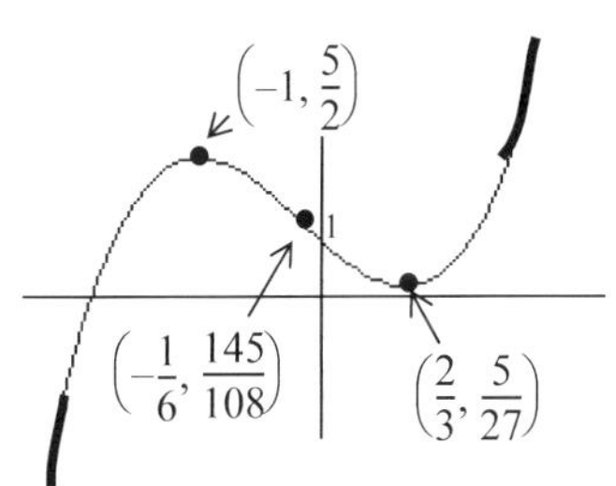

56. Maxima: (–2, –16), (2, 16) Minima: (–1, –38), (1, 38)

Inflection points at $x = 0$ and at $x = \pm\sqrt{\frac{15}{6}}$

As $x \to \pm\infty$ the graph resembles that of $y = 3x^5$

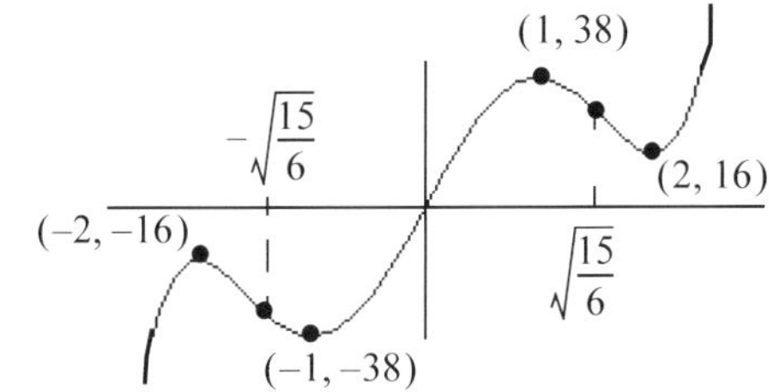

57. $\frac{x^6}{6} + x^3 - \frac{x^2}{2} + x + C$ **58.** $x^4 - x^2 + 3x + C$ **59.** $\frac{x^5}{5} - \frac{2x^3}{3} - \frac{1}{2x^2} + C$

60. $\frac{x^7}{7} + \frac{x^6}{6} - \frac{x^5}{5} + C$ **61.** $\frac{x^4}{4} - x^3 + \frac{5x^2}{2} - 15x + C$ **62.** $\frac{4x^3}{3} - \frac{1}{x} + \frac{3}{2x^2} + C$

63. $f(x) = \frac{x^3}{3} + 5x - 23$ **64.** $f(x) = x^4 + x^2 + 3$ **65.** $\frac{19}{4}$ **66.** 9 **67.** 7

68. $\frac{23}{2}$ **69.** 3 **70.** $-\frac{27}{4}$ **71.** $\frac{32}{3}$ **72.** $\frac{17}{6}$ **73.** $x = \pm\sqrt{\frac{2}{3}}$ **74.** $-1 < x < 1$

75. 150 units. **76.** 50 units. **77.** 250 units. **78.** \$57.50 **79.** \$800

80. \$540 **81.** 4. **83.** 80 feet per second **84.** 2 hours

INDEX

I

L

M

N

O

P

R

S

Made in the USA
San Bernardino, CA
19 July 2014